Sustainable Automotive Energy System in China

China Automotive Energy Research Center
of Tsinghua University

Sustainable Automotive Energy System in China

Springer

China Automotive Energy Research Center
of Tsinghua University
Beijing, People's Republic of China

ISBN 978-3-642-43362-7 ISBN 978-3-642-36847-9 (eBook)
DOI 10.1007/978-3-642-36847-9
Springer Heidelberg New York Dordrecht London

Printed on acid-free paper

Springer is part of Springer Science+Business Media (www.springer.com)

Editorial Committee

Editor-in-Chief

Zhang Xiliang (zhang_xl@mail.tsinghua.edu.cn)
China Automotive Energy Research Center,
Tsinghua University, Beijing 100084
P.R. China

Vice editors-in-Chief

Wang Hewu (wanghw@mail.tsinghua.edu.cn)
China Automotive Energy Research Center,
Tsinghua University, Beijing 100084
P.R. China

Ou Xunmin (ouxm@mail.tsinghua.edu.cn)
China Automotive Energy Research Center,
Tsinghua University, Beijing 100084
P.R. China

Members (Alphabetical Order)

Scientific Committee

Chairman

He Jiankun
Professor of Tsinghua University, Deputy Director of the National Panel on Climate Change of China, Vice-Chairman of the China Energy Research Society, Dean of the Low-Carbon Economy Institute of Tsinghua University, Chairman of the Steering Committee of the China Automotive Energy Research Center of Tsinghua University

Vice-Chairman

Ouyang Minggao
Professor of Tsinghua University, Deputy Director of the Academic Committee of Tsinghua University, Director of the State Key Laboratory of Automotive Safety and Energy, Chairman of the Expert Panel on Energy-Saving and New-Energy Vehicles Key Projects of the National 863 Program of China, Associate Director of the China Society of Automotive Engineers

Members

Ni Weidou
Professor of Tsinghua University, Academician of the Chinese Academy of Engineering

Chen Qingtai
Professor and Former Deputy Director of the Development Research Center of the State Council of China

Liu Yanhua
Counselor of the State Council of China, Former Deputy Minister of the Ministry of Science and Technology of China

Author Contributions by Chapter

Chapter	Title	Author(s)
1	Introduction	Zhang Xiliang
2	Motor Vehicle Development and Air Pollution Control	Huo Hong, Yao Zhiliang, He Kebin
3	Scenario Analyses of China's Vehicle Ownership and Vehicle Traffic Services	Wang Hewu, Hao Han, Ouyang Minggao
4	Vehicle Powertrain Technology	Wang Hewu, Du Jiuyu, Ouyang Minggao
5	Petroleum-Derived Liquid Fuels	Li Zheng, Fu Feng, Ma Linwei, Liu Pei, Zhou Zhai, Zhang Jianbing, Jiang Xiaolong
6	Natural Gas	Ma Linwei, Gao Dan, Li Weiqi, Li Zheng
7	Coal-Derived Liquid Fuels	Liu Pei, Ma Linwei, Liu Guangjian, Pan Lingyin, Li Zheng
8	Liquid Biofuels	Chang Shiyan, Zhao Lili, Zhang Ting, Zhang Xiliang
9	Energy for Electric Vehicles	Zhu Guiping, Lu Zongxiang, Wang Zanji
10	Hydrogen and Fuel-Cell Vehicle Technology	Wang Hewu, Huang Haiyan, Deng Xue, Ouyang Minggao
11	Life-Cycle Energy Consumption and Greenhouse Gas Emissions for Automotive Energy Pathways	Ou Xunmin, Zhang Xiliang
12	Scenario Analyses of Automotive Energy	Zhang Xiliang, Ou Xunmin, Zhang Jihong, Chai Qimin, Hao Han, Huo Hong, He Jiankun
13	Policy Recommendations Regarding Sustainable Development of China's Automotive Energy	Zhang Xiliang, Ou Xunmin

Preface

Energy safety and climate change have become topics of prime interest in current debates about international politics, economy, and the environment. In the ongoing process of modernization, China will continue to face challenges in providing a secure energy supply and in mitigating climate change over the long term. The development of future energy supply and usage in China will have a substantial impact on global energy markets and the local environment as well as implications for global climate change. China is currently one of the fastest growing regions in the global automotive market, and it has become one of the world's largest nations for automobile consumption and production. Automotive energy has therefore become a core energy and environmental issue in the country.

Commissioned by the National Energy Administration, the Ministry of Industry and Information Technology, the Ministry of Science and Technology, and other government departments, Tsinghua University's China Automotive Energy Research Center (CAERC) has studied the sustainable development of the country's automotive energy system. After examining such influences as energy, economy, environment, technology, society, industry leadership, and policy in addition to the life cycle of automotive energy in several technology strategies, CAERC has presented five scenarios for automotive energy development in China. These are the Reference Scenario (RS), the Electric Vehicles Development Scenario (EVDS), the Fuel Cell Vehicles Development Scenario (FCVDS), the Biofuels Development Scenario (BDS), and the Integrated Policy Scenario (IPS). In scenario analyses, CAERC assessed automotive energy demand management, improvement of vehicle fuel economy, and technical specifications of electric vehicles, fuel-cell vehicles, and vehicles powered by second-generation biofuels. CAERC has made policy recommendations and institutional arrangements for promoting sustainable transformation of China's automotive energy system. The aforementioned research efforts and CAERC's results are summarized in this present 13-chapter book.

This research was supported by the National Energy Administration, the Ministry of Industry and Information Technology, the Ministry of Science and Technology, General Motors (China) Company, and Shanghai Automotive Industry Company. Related supporting funds include National Key Basic Research and Development

Program (973) project 2010CB955600, National Science and Technology Support project 2007BAC03A04, a major project of the National Social Science Foundation of China (09&ZD029), projects of the National Natural Science Foundation of China (71041003, 71041028, and 71103109), a major project of the Humanities and Social Science Research Base of the Ministry of Education (2009JJD790029), the International Science and Technology Cooperation Program of the Ministry of Science and Technology (2010DFA72760), and Beijing Excellent Doctoral Dissertation Instructor project (20111000301). Thanks are also given to the Advisory Board of CAERC, especially to Dr. Michael Wang and Prof. Feng Fei.

Steering Committee Chairman
China Automotive Energy Research Center
Tsinghua University

Prof. He Jiankun

Contents

Chapter 1
Introduction

Zhang Xiliang

Abstract In the process of modernization, China will have to face the challenges of ensuring energy security and of mitigating climate change over the long term. China is currently one of the fastest growing regions in the global automotive market and has become the world's largest nation of automobile consumers and producers. Automotive energy has therefore become a core energy and environmental issue for the country. China intends to establish a sustainable automotive energy system. However, there is as of yet no currently recognized standard definition of such a system anywhere in the world. The resource endowment, population and geography, economic and social development levels, infrastructure characteristics of energy and transportation, and technological innovation capability of energy and transportation vary by country. We believe there are six basic standards for estimating the sustainability of China's automotive energy system: the transportation economy, energy efficiency, greenhouse gas emissions, security of energy supply, supply and demand matching of fuel types, and automotive industry leadership.

Keywords Automotive energy • System analysis • China

Energy security and climate change have become major issues for current international economy and environment policy. In the process of modernization, China will have to face the challenges of ensuring energy security and of mitigating climate change over the long term. Future energy development will have a significant effect on global energy markets and on addressing climate change. China is currently one of the fastest growing regions in the global automotive market and

Zhang Xiliang (✉)
China Automotive Energy Research Center, Tsinghua University,
Beijing 100084, P.R. China
e-mail: zhang_xl@mail.tsinghua.edu.cn

CAERC, Tsinghua University, *Sustainable Automotive Energy System in China*,
DOI 10.1007/978-3-642-36847-9_1, © Springer-Verlag Berlin Heidelberg 2013

has become the world's largest nation of automobile consumers and producers. Automotive energy has therefore become a core energy and environmental issue for the country.

1.1 Influences of Automotive Energy

There are four categories affecting development of the automotive industry: economic growth, population and geography, technological progress, and public policy.

1.1.1 Economic Growth

The demand for automotive transportation services determines car ownership and consumption of automotive energy in a country. Demand is also determined by macroeconomic characteristics such as total economic scale, economic structure, and industrial structure of the country. Over a certain period, economic scale, automotive transportation service demand, and consumption of automotive energy generally have strong positive correlations. One important example of this strong positive correlation is related to China's reform and opening up. By the middle of this century, China will achieve basic modernization and become a middle-income, developed country. In the future, the economy will remain in a growth stage, and the GDP growth rate will be a major driver of the country's automobile transportation services demand.

1.1.2 Population and Geography

In addition to economic scale and developmental level, the population and geography of a country are key influences on automobile transportation service demand and consumption of automotive energy. Characteristics of said population and geography include its size, demographic composition, land area, urbanization rate, urbanization patterns, and transportation infrastructure. In the United States, there are vast land areas, decentralized urban agglomerations, a highly developed highway network, and a high rate of car ownership at 785 cars per 1,000 people. In Japan, there is a small land area, high population density, dense urbanization, and a car ownership rate of 450 cars per 1,000 people less than America. The rate of car ownership in other countries such as Canada, the United Kingdom, Germany, and France lies between those of the aforementioned two countries. Therefore, population growth rate, population density, urbanization rate, urbanization patterns, and transportation infrastructure will be major drivers of China's demand for automobile transportation services and consumption of automotive energy.

1.1.3 Technological Progress

Future progress of automotive and energy technologies will significantly affect the development of automotive energy. Innovation and promotion of energy-saving technologies will continually improve automotive fuel economy and reduce emission levels of pollutants, including carbon dioxide. Advancement of new vehicle energy technology such as pure battery electric vehicles (BEVs) and fuel cell vehicles (FCVs) will significantly enhance automotive fuel economy and lower pollutant emissions. This will also bring new innovations such as automotive energy conversion and storage technologies, automotive energy infrastructures, and profound changes in the automotive industry itself. In addition, advances in bio-based alternative energy technologies will modify China's automotive energy supply structure. However, future automotive energy technology progress and market penetration rates are very uncertain.

1.1.4 Public Policy

The goal of China's automotive energy development is to establish a sustainable automotive energy system. Because of certain inconsistencies between the country's overall interests and individual interests, the market mechanism fails to internalize externalities and information supply. Establishment of a national sustainable automotive energy system should be supported by public policy. Policy orientation largely determines technical automotive progress. Under the same circumstances and with consideration for macroeconomic, population, and geography factors, different policy orientations will play distinct roles in technological innovation and selection of the technological direction for future automotive energy.

1.2 Sustainable Automotive Energy System

As stated above, China intends to establish a sustainable automotive energy system. However, there is as of yet no currently recognized standard definition of such a system anywhere in the world. The resource endowment, population and geography, economic and social development levels, infrastructure characteristics of energy and transportation, and technological innovation capability of energy and transportation vary by country. Therefore, the explanation, definition, and requirements of a sustainable automotive energy system also vary. China is a vast, developing country with a large population, relative lack of resources, and rapid economic development. Thus, the country's sustainable automotive energy system will have certain characteristics as follows:

1. Automotive traffic economy: social and economic development and the individual demands for automotive transport services should be satisfied by affordable (or as low as possible) cost. The sustainable automotive energy system will help reduce oil imports, which expend substantial foreign exchange reserves and economic resources. Consequently, the transition to a sustainable automotive energy system will significantly benefit the automotive traffic economy.
2. Efficiency of energy system: the future automotive energy system will expend minimum energy resources to satisfy social and economic development and individual demand for automotive transport services. The ideal sustainable automotive energy system should be one that achieves minimum or reduced energy consumption in the well-to-wheel (WTW) life cycle, under the same conditions as the automotive transport services demand.
3. Greenhouse gas emissions: carbon dioxide is the main greenhouse gas (GHG). According to the global CO_2 emission control target scenario, global average CO_2 emissions per capita in 2050 will be just greater than 1 t as presented by the fourth-version IPCC report. Controlling CO_2 emission has been a major impetus within automotive energy technology innovation. Automotive vehicles have become some of the most serious pollution contributors to air deterioration in cities. The ideal sustainable automotive energy system should be one in which CO_2 emissions of the WTW life cycle and motor vehicle pollutants are minimized or reduced, under similar conditions of automotive transport services demand satisfaction.
4. Security of energy supply: China's oil resources are relatively scarce. Its dependence on oil imports reached nearly 55 % in 2010 and will increase in the future. Because more than 95 % of automotive energy comes from petroleum-based fuels, an increase in the need for automotive energy is the main reason for the recent rise in dependence on oil imports. The sustainable automotive energy system should have the ability to respond to the risks of the international energy markets. From this standpoint, the automotive energy supply should be diversified and should reduce dependence on energy imports under feasible economic conditions.
5. Supply and demand matching of fuel type: the sustainable automotive energy system should not only ensure a supply and demand balance of automotive energy in quantity but also reasonably match automotive energy types. This is an attempt to avoid a surplus or deficiency of automotive energy types and ensure scientific use of automotive energy infrastructure in the processes of energy production, transport, and filling.
6. Competitiveness of automotive industry: although China's automotive industry began late and lacked competitiveness in traditional automotive and energy technologies, it has a greater possibility to excel in the area of new energy

vehicles. The country's new energy automotive market will provide the impetus for upgrading and introducing competition into the automotive industry once it reaches a certain scale.

The sustainable automotive energy system is an ideal policy target scenario. We believe there are six basic standards for estimating the sustainability of China's automotive energy system: the transportation economy, energy efficiency, greenhouse gas emissions, security of energy supply, supply and demand matching of fuel types, and automotive industry leadership.

1.3 Contents of This Book

China's automotive energy situations are very complex. They are related to energy, economy, environment, technology, society, industry leadership, and public policy. Systematic, in-depth multidisciplinary and comprehensive studies will provide a solution for the country's automotive energy challenges. Commissioned by the National Energy Administration, Ministry of Industry and Information Technology, Ministry of Science and Technology, and other government departments, Tsinghua University's China Automotive Energy Research Center (CAERC) has studied the sustainable development of the country's automotive energy industry. After studying influences such as energy, economy, environment, technology, society, industry leadership, and policy, plus the life cycle of automotive energy in several technology roadmaps, CAERC put forward five scenarios of automotive energy development in China. These are the Reference Scenario (RS), Electric Vehicles Development Scenario (EVDS), Fuel Cell Vehicles Development Scenario (FCVDS), Biofuels Development Scenario (BDS), and Integrated Policy Scenario (IPS). In scenario analyses, CAERC assessed automotive energy demand management, improvement of vehicle fuel economy, and technical specifications of BEVs, FCVs, and second-generation biofuels. CAERC made policy recommendations and institutional arrangements for promoting sustainable transformation of the national automotive energy system. The aforementioned research works and CAERC achievements are summarized in this 13-chapter book. Chapters 2 through 13 are described as follows.

Chapter **2** Motor Vehicle Development and Air Pollution Control illustrates environmental problems caused by the development of motor vehicles in China, focuses on the relationship between urban motor vehicles and air pollution, analyzes pollution control measures of motor vehicles, and summarizes relevant challenges and future prospects.

Chapter **3** Scenario Analyses of China's Vehicle Ownership and Vehicle Traffic Services focuses on future projections of vehicle ownership and vehicle traffic services; forecast of ownership of private passenger vehicles, public transport vehicles, and other vehicles, through creating predictive models related to population density, household income, and other factors; and also forecast of passenger and freight vehicle traffic services, combined with future projections of annual vehicle mileage and passenger/cargo rates.

Chapter **4** Vehicle Powertrain Technology: systematic assessment of the development status of China's motor vehicle power systems, which includes technologies of high-efficiency and clean vehicle power, hybrid and pure electric drive. Also, there is an introduction to RD&D of these technologies in the country.

Chapter **5** Petroleum-Derived Liquid Fuels: analysis of the country's oil supply chain around petroleum-based vehicle issues, as well as examination of oil production, imports, refining, and consumption. Strategic recommendations are provided for future development.

Chapter **6** Natural Gas: discussion of the future of China's natural gas supply and demand and technology development of natural gas vehicles. Also, there is the discussion of the potential for using natural gas in automotive energy.

Chapter **7** Coal-Derived Liquid Fuels: analysis of the supply chain of coal-based liquid fuel technology; comprehensive and objective assessment of technological developments of coal-based liquid fuel. Basic developmental recommendations are provided.

Chapter **8** Liquid Biofuels: assessment and commentary on various resources and technologies that would affect China's future biofuels. There is multi-scenario analysis of future automotive biofuels (fuel ethanol and biodiesel), combined with future key technological breakthrough points and policies.

Chapter **9** Energy for Electric Vehicles: analysis of electrical system development status and future development scenarios, with respect to development issues of automotive electrical energy. There is the discussion of key effects on electrical system development and assessment of the possible use of electricity as an automotive fossil-fuel alternative. An outlook for feasible directions is also provided.

Chapter **10** Hydrogen and Fuel Cell Vehicle Technology: analysis of China's hydrogen production and utilization, with respect to development issues of hydrogen and fuel cell vehicles. Exploration of sources of future automotive hydrogen, together with technological development of fuel cell vehicles, is also presented.

Chapter **11** Life-Cycle Energy Consumption and Greenhouse Gas Emissions of Automotive Energy Pathways: after introducing models and related parameters of different fuel and power technologies, the life cycle of the multi-fuel/vehicle line is comprehensively compared, based on the same computing platform as China's automotive energy pathways. There is also the forecast of energy consumption and carbon emissions under the life cycle of the key technical path.

Chapter **12** Scenario Analyses of Automotive Energy: five future development scenarios of China's automotive energy are provided, with respect to its sustainable transformation—the Reference Scenario (RS), Electric Vehicles Development Scenario (EVDS), Fuel Cell Vehicles Development Scenario (FCVDS), Biofuels Development Scenario (BDS), and Integrated Policy Scenario (IPS). Analysis and assessment of these five scenarios within the scope of energy, economy, environment, and technology, and industry leadership are also presented.

Chapter **13** Policy Recommendations Regarding Sustainable Development of China's Automotive Energy: quantitative analysis of sustainable transformation of the country's automotive energy system with policy recommendations.

Chapter 2
Motor Vehicle Development and Air Pollution Control

Huo Hong, Yao Zhiliang, and He Kebin

Abstract This chapter first introduces the general ambient environmental issues caused by vehicles in China and then simulates CO, HC, NO_x, and particulate matter (PM) emissions from vehicles in 12 selected typical Chinese cities during 1990–2009. The results show a decreasing trend in CO and HC emissions but an increasing trend in NO_x and PM emissions in the examined cities. Megacities (e.g., Beijing and Shanghai) have stricter emission standards than the national level, so their vehicle emissions decrease faster than those of other cities. Also, the ambient SO_2, NO_2, and PM_{10} concentrations in Beijing, Shanghai, and Guangzhou show a decreasing trend during the past decade. However, in cities where the emission measures are relatively lenient (e.g., Jinan, Ningbo, and Chongqing), the NO_x and PM emissions increased significantly. Therefore, vehicle pollution is no longer a problem that exists only in large cities. Local governments need to pay great attention to the fact that vehicle pollution is rapidly rising in provincial capitals and prefecture-level cities. This chapter finally discusses the measures implemented during recent 10 years to control vehicle emissions in China.

The rapid vehicle growth in China has caused various environmental issues, especially urban air pollution. Fortunately, the national government and local governments have implemented many measures to control vehicle emissions. It is important to emphasize that vehicle emission-control measures must be in accord with vehicle development in order to protect the urban ambient environment.

Keywords Vehicle emissions • Air pollution • Emission standards

---

Huo Hong
Institute of Energy, Environment and Economy, Tsinghua University, Beijing 100084, P.R. China

Yao Zhiliang • He Kebin (✉)
School of Environment, Tsinghua University, Beijing 100084, P.R. China
e-mail: hekb@tsinghua.edu.cn

CAERC, Tsinghua University, *Sustainable Automotive Energy System in China*,
DOI 10.1007/978-3-642-36847-9_2, © Springer-Verlag Berlin Heidelberg 2013

2.1 Environmental Problems Caused by Motor Vehicles

With the growth of China's economy and urbanization in addition to the continuous improvement in living standards in the country, the number of motor vehicles has rapidly risen over the past several decades. In 32 years from 1978 to 2010, China's vehicle population increased from 1.36 to 80 million (excluding motorcycles, low-speed trucks, and low-speed electric vehicles). The annual average growth rate was 14 %, though since 2000 this rate has reached 17 %. The rapid increase in the number of vehicles has brought greater convenience and a higher quality of life, but it has also resulted in serious environmental problems.

Owing to the relatively low levels of pollution control for motor vehicles as well as the slow development of infrastructure construction and transportation management in China, individual vehicle emission factors are generally higher than in developed countries. At the same time, because of the high density of urban traffic and population, vehicle emission density and pollutant concentrations are generally high, which are greatly injurious to health.

Toward controlling vehicle pollution, China is rapidly undertaking the production of unleaded gasoline, and it has implemented new vehicle emission standards. However, because China was late in starting vehicle pollution control, automobiles with relatively backward pollution-control mechanisms still account for a certain proportion of the vehicle population, and the average emission level is much higher than in the United States and European countries. Further, construction of the supporting infrastructure and transportation management has failed to keep pace with the rapid growth in the number of vehicles. Insufficient transport structure and composition have led to chronic traffic saturation on the major roads of many big cities. Low average running speeds and the high incidence of idling have aggravated the air pollution in China's cities (Sidebar 2.1).

Sidebar 2.1: Impact of Vehicle Pollutants on Health and the Environment

The main vehicle pollutants are carbon monoxide (CO), hydrocarbons (HCs), nitrogen oxides (NO_x), and fine particulate matter ($PM_{2.5}$). These pollutants can cause serious harm to human health. CO and HCs result from incomplete combustion. CO can lead to a decrease in oxygen transmission function in the blood. Low concentrations of CO cause headaches, dizziness, and intoxication; high concentrations are lethal. Among HCs, benzene and polycyclic aromatic substances are proven carcinogens. Among NO_x, NO_2, in particular, is highly toxic. It is a red-brown gas with a pungent smell, and it can greatly damage the human respiratory and immune systems at a concentration of 5 ppm. Following inhalation, $PM_{2.5}$ can become deposited in lungs and lead to diseases in the respiratory system. In addition, it can be very harmful through its surface adsorption of many toxic substances. Furthermore, photochemical

(continued)

(continued)
reactions occur between NO_x and HCs, generating low-altitude ozone and photochemical smog. NO_x are a major source of acid rain, which is also very harmful to human health. Black carbon (BC) in vehicle emissions not only causes decreased visibility but also impacts on the climate.

Table 2.1 Annual change of NO_2 concentration in Beijing (μg/m^3)

Year	2006	2007	2009
Central city	69.5	67.5	59.3
Outer city (central city excluded)	63.3	61.8	57
Suburbs	50.4	47	45.3

Note: The central city comprises Dongcheng, Xicheng, Xuanwu, and Chongwen; the outer city comprises Haidian, Chaoyang, Fengtai, and Shijingshan; and the suburbs comprise Fangshan, Daxing, Tongzhou, Mentougou, Changping, Shunyi, Pinggu, Yanqing, Huairou, and Miyun

Table 2.2 Contributions of pollution sources to total emissions in Beijing in 2008 (excluding fugitive emissions, %) (Wang et al. 2010)

Pollution source	NO_x	PM_{10}	HC
Power plants	8	2	5
Building materials	10	45	2
Chemical industry	1	4	1
Smelting	1	14	1
Industrial boilers	14	15	11
Mobile sources	66	20	79

Table 2.1 shows the annual change of NO_2 concentration in Beijing (Beijing Municipality's Bureau of Environmental Protection 2007, 2008, 2010). It is observed that the NO_2 concentrations in the central city were 31–44 % higher than in suburban areas. In the traffic-dense central city, the characteristics of vehicle pollution are very clear.

With increased auto emissions, the contribution of vehicle pollutants to air pollution is currently showing an upward trend in Chinese cities. The contributions of vehicles to total CO and HC emissions are more than 50 % in most cities, and it is even over 90 % in megacities. In population-dense city centers, the proportion of vehicle pollutants among both total emissions and emission concentrations is over 80 %, and this figure is continuously rising. Vehicle emissions have thus become a major source of air pollution in cities (Huo 2005; Guan et al. 2006; Guan and Yu 2007; Jin 2009; Li et al. 2010; Yang et al. 2009). In cities that have experienced a rapid growth in vehicle population, such as Beijing, Shanghai, and Guangzhou, automobiles have become the primary source of air pollution. Vehicle pollution has become the top priority in city air pollution control. Table 2.2 shows the contributions of major pollution sources to the total emissions in Beijing in 2008

(Wang et al. 2010). As shown, vehicles were the dominant source of NO_x and HCs, both of which are ozone precursors. Therefore, vehicles are the major cause of air quality deterioration in the city.

In 2006, the CO, NO_x, and HCs from vehicles accounted for over 20 % of the total emissions nationwide. Currently, black carbon (BC) is exerting a important impact both on climate and human health, and it has increasingly drawn attention in global academic community and among policy makers. The BC emissions from vehicles account for over 10 % of total BC emissions in China (Zhang et al. 2009).

2.2 Calculation of Vehicle Pollutant Emissions in Cities

Generally speaking, vehicles produce more pollution in China than in the United States and European countries. Figure 2.1 presents a comparison of per capita vehicle emissions between Beijing and some developed countries (Huo et al. 2011). At present, Beijing produces the same level of per capita vehicle emissions as the United States and European countries. However, the vehicle ownership in Beijing is much lower than in the developed countries (220 vehicles/1,000 people in Beijing versus 600–800 vehicles/1,000 people in the USA and European countries), which indicates that, individually, Chinese vehicles produce far more emissions. In addition, it is notable that the per capita vehicle pollution is decreasing in those

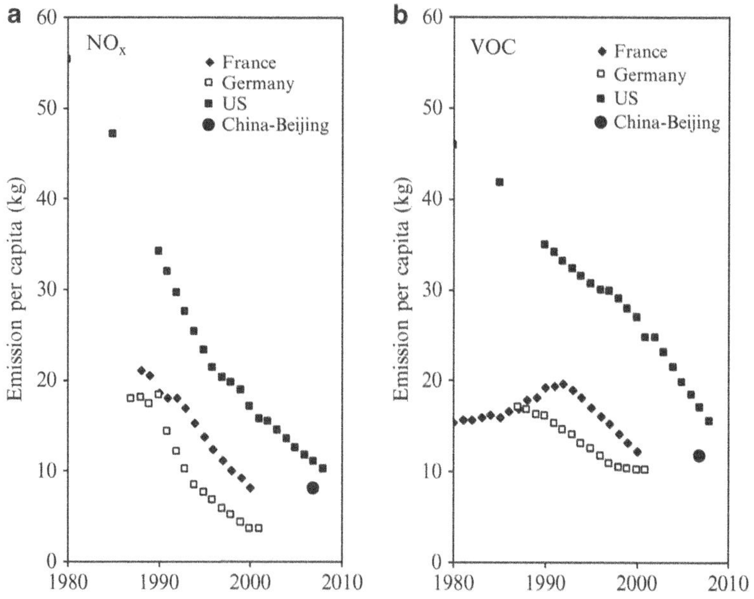

Fig. 2.1 Comparison of per capita vehicle emission in Beijing, the USA, Germany, and France (Revised from Huo et al. (2011), Copyright 2011, with permission from Elsevier)

countries, because the vehicle ownership has become saturated and vehicle emission controls are becoming more stringent. By contrast, China's vehicle ownership per thousand people is increasing. Therefore, it is necessary to impose stricter measures to control vehicle emissions. Because owners may drive their vehicles for about 15 years before buying a new one, the sooner the measures are implemented, the better benefit will be got from the measures. It is also important to put new-energy vehicles on the market as soon as possible to improve the air quality in cities.

To obtain a better understanding of the variations in vehicle emissions in different Chinese cities, this study selected 12 typical cities of various sizes. Vehicles were classified according to type, and the factors that affect vehicles' operation levels and emission factors were analyzed. We developed a vehicle emission inventory method on the basis of available data and examined the vehicle emission characteristics and variations.

The total emission can be calculated using Eq. (2.1):

$$Q_i = \sum_j \left(P_j \times \mathrm{VKT}_j \times \mathrm{EF}_{i,j}\right) \tag{2.1}$$

where i represents the pollutants, including CO, HCs, NO_x, and PM, and j represents the vehicle type, including light-duty gasoline buses, light-duty diesel trucks, heavy-duty diesel buses, heavy-duty diesel trucks, and motorcycles. Q is the total annual emission (tons). P is the vehicle population (million). The registered population of different types of vehicles can be calculated using the method proposed by Huo et al. (2011). In this method, the dynamic vehicle turnover was simulated using vehicle survival function rates, newly registered vehicles, and scrapped vehicles. Vehicle survival functions are variation curves that show, with the aging of vehicles, the changing proportion in the target type's current population with respect to the originally registered population. In this chapter, the motor vehicle survival rates are constant. The target type's population in the target year constitutes the current population that survived to the target year based on the variation in survival rate and the newly registered population. Since sales of China's motor vehicles are currently enjoying robust growth, there is a rapid renewal of automobiles, but there are major regional differences.

In Eq. (2.1), VKT_j is the average annual vehicle kilometers traveled of vehicle type j (km). As an important factor for vehicle operation level and energy consumption, VKT reflects the development of vehicles and transportation. Because there are no official statistics relating to VKT, previous researchers usually obtained VKT data from surveys. The chapter uses the following two methods to determine the traveled distance: (1) questionnaires to obtain data relating to vehicle driving characteristics (including VKT) in different regions (using this relatively easy and feasible method, we could obtain more detailed information, but we would also need a large dataset to ensure reliability) and (2) directly obtain VKT data from transportation management departments (in this way, the data's veracity can be guaranteed, though the information is hard to obtain).

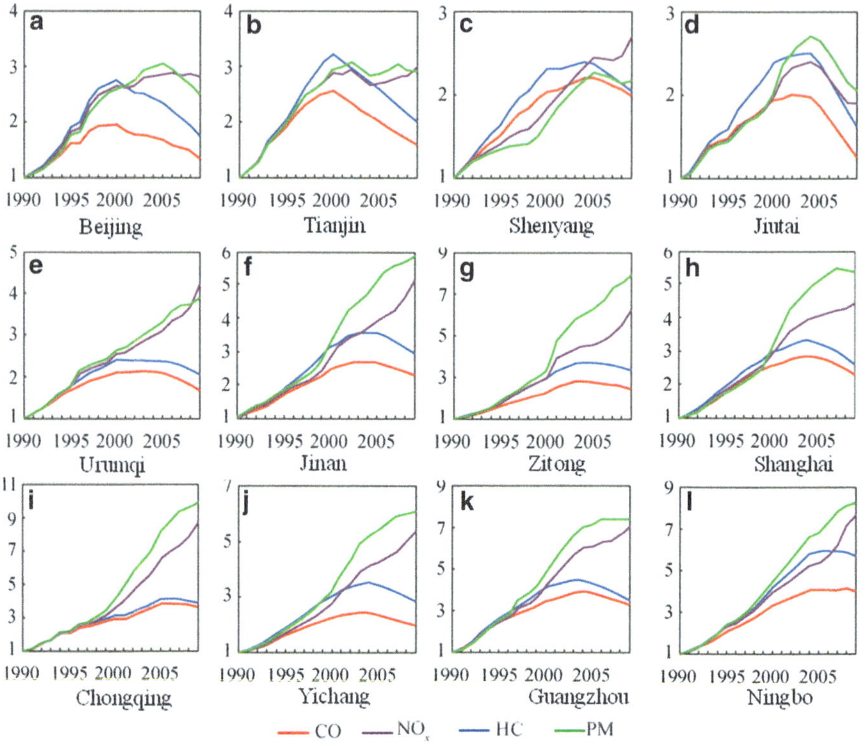

Fig. 2.2 Variation trends of common vehicle pollutants in 12 typical cities from 1990 to 2009. *Note*: The vertical axis represents the growth in emissions as a multiple of the emission level in 1990; the horizontal axis represents years

In Eq. (2.1), EF_i is the emission factor of vehicle pollutant i (g/km). Different emission standards require the different emission limits of vehicle pollutants. Therefore, the EF of the same type of vehicles differs greatly under different emission standards. In order to obtain an accurate vehicle inventory, it is important to simulate the number of vehicles under each emission standard. EF calculation is performed using the International Vehicle Emission (IVE) model.

Figure 2.2 shows the variation trends in vehicle emissions in 12 typical cities from 1990 to 2009. It is observed that there is a decreasing trend in CO and HC emissions but an increasing trend in NO_x and PM emissions. In megacities such as Beijing and Shanghai, owing to the stricter emission standards implemented, vehicle emissions decrease faster than those in other cities; however, in cities such as Jinan, Ningbo, and Chongqing, where the vehicle population is rapidly proliferating and which have been relatively slow to introduce emission measures, the emission of NO_x and PM has increased dramatically. This demonstrates that vehicle pollution is no longer a problem that exists only in large cities. Local governments need to pay great attention to the fact that vehicle pollution is rapidly rising in provincial capitals and prefecture-level cities. Since there are many of these small and medium-sized cities, they will have a greater impact on the overall variation trends in the nationwide emission of vehicle pollutants than the major cities will.

2.3 Air Pollution Status of Three Key Cities

2.3.1 *Beijing*

In 2008, Beijing had 274 days when the air quality reached the national Class II air quality standard or better, amounting to 74.5 % of the whole year. The average annual concentrations of SO_2, CO, NO_2, and PM_{10} in the atmosphere in Beijing were 0.036, 1.4, 0.049, and 0.122 mg/m^3, respectively, which were 23.4, 30.0, 25.8, and 17.6 % lower than 2007 levels. The concentrations of SO_2, CO, and NO_2 all met the national standards, and that of PM_{10} exceeded national standards by 22 %. In particular, during the 17 days of the 2008 Olympic Games (August 8–24), a series of temporary protection measures were implemented in Beijing and surrounding areas. As a result, there were 10 days with Class I and 7 days with Class II air quality. During the period of the Olympics, the concentrations of atmospheric SO_2, CO, NO_2, and PM_{10} were 0.008, 0.8, 0.023, and 0.057 mg/m^3, respectively, 46.7, 42.9, 57.4, and 53.7 % lower than the corresponding period of the previous year. The commitment toward holding a "Green Olympics" was achieved.

Changes in the concentrations of major pollutants in Beijing from 2000 to 2008 (Beijing Municipality's Bureau of Environmental Protection 2001, 2002, 2003, 2004, 2005, 2006, 2007, 2008, 2009) are presented in Fig. 2.3. As can be seen, the average annual concentrations of SO_2, CO, NO_2, and PM_{10} significantly decreased

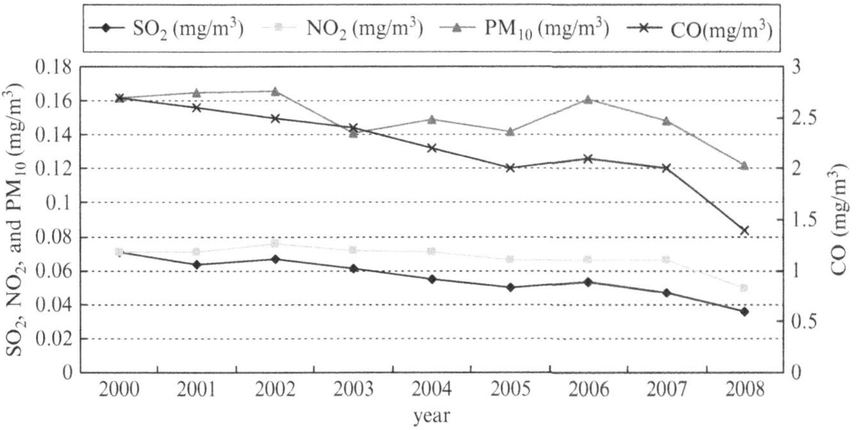

Fig. 2.3 Variation trends in SO_2, CO, NO_2, and PM_{10} concentrations in Beijing from 2000 to 2008

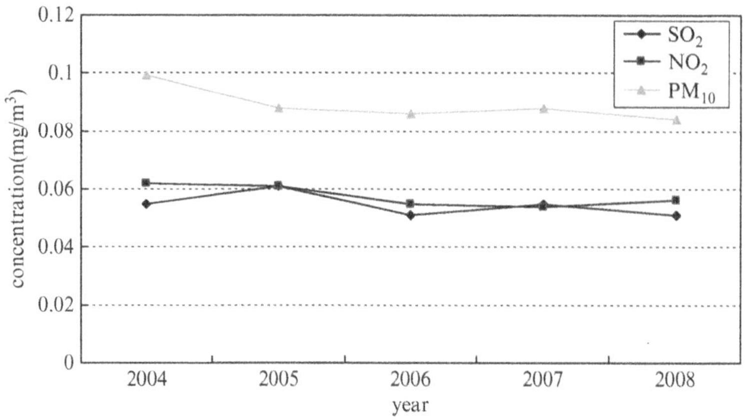

Fig. 2.4 Variation trends of SO_2, NO_2, and PM_{10} concentration in Shanghai from 2004 to 2008

by 49.3, 48.1, 30.9, and 24.7 %, respectively, in 2008, compared with 2000 levels. It can be concluded that air pollution control in Beijing in recent years has been fairly effective, especially regarding SO_2 and CO concentrations. However, it should be noted that the effect of controls on NO_2 and PM_{10} levels is still relatively weak, which indicates that motor vehicle pollution in Beijing remains a rather serious problem, and it has made an ever greater contribution to air pollution.

2.3.2 Shanghai

In 2008, there were 328 days in Shanghai when the air quality was excellent (meeting Class I or Class II), which was about the same number as the previous year. There were 313 days when PM_{10} was the primary pollutant, which amounted to 85.5 % of the entire year; the number of such days for SO_2 was 37, which accounted for 10.1 %; the number of such days for NO_2 was 9, accounting for 2.5 %. There were 5 days when both PM_{10} and SO_2 were the primary pollutants, which accounted for 1.4 %; there was 1 day when PM_{10} and NO_2 were the primary pollutants, accounting for 0.3 %; and there was 1 day when SO_2 and NO_2 were the primary pollutants, which accounted for 0.3 %.

The average annual concentrations of PM_{10}, NO_2, and SO_2 in Shanghai were lower than the Class II of the national air quality standard in 2008. The annual daily average PM_{10} concentration in Shanghai in 2008 was 0.084 mg/m^3, which was 0.004 mg/m^3 lower than in 2007; that of SO_2 was 0.051 mg/m^3, which was 0.004 mg/m^3 lower than 2007 levels; and that of NO_2 was 0.056 mg/m^3, which was 0.002 mg/m^3 higher than in 2007.

The variation trends of the concentrations of major pollutants in Shanghai from 2004 to 2008 are shown in Fig. 2.4 (Shanghai Municipality's Bureau of

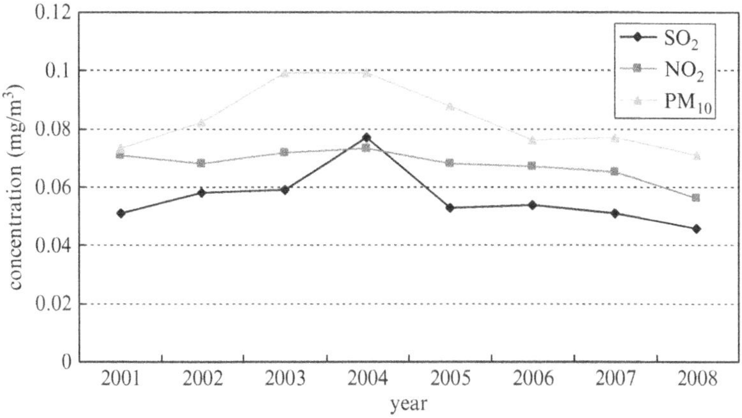

Fig. 2.5 Variation trends of SO_2, NO_2, and PM_{10} concentrations in Guangzhou from 2001 to 2008

Environmental Protection 2006, 2007, 2008, 2009, 2010). As is evident in the figure, the concentrations of SO_2, NO_2, and PM_{10} were reduced by 7.3, 9.7, and 15.2 %, respectively, in Shanghai in 2008 compared with 2004 levels. Unlike in Beijing, the concentration of SO_2 in Shanghai was reduced by a relatively small margin, which could be closely related to the strict control of Shanghai's vehicle population. As a result, air pollution in Shanghai is combined by soot and motor vehicle emissions.

2.3.3 Guangzhou

In 2008, air quality in Guangzhou showed an improvement over the previous year, which was the result of continuous improvement efforts. The annual average concentrations of SO_2, NO_2, and PM_{10} were 0.046, 0.056, and 0.071 mg/m^3, respectively, which met Class II of the National Ambient Air Quality Standard. In 2008, dustfall in Guangzhou reached the recommended standard for Guangdong Province. The annual average concentrations of SO_2, NO_2, and PM_{10} in 2008 showed a decrease of 9.8, 13.8, and 7.8 % compared to the 2007 levels, respectively; the monthly average concentrations of dustfall declined by 6.5 %, and the number of good air quality days increased by 3.03 %.

Variation trends in the concentrations of major pollutants in Guangzhou from 2001 to 2008 (Guangzhou Municipality's Bureau of Environmental Protection 2002, 2003, 2004, 2005, 2006, 2007, 2008, 2009) are shown in Fig. 2.5. As can be seen in the figure, the concentrations of SO_2, NO_2, and PM_{10} initially showed an increase before gradually declining; the highest values occurred in 2004. The annual average concentrations of SO_2, NO_2, and PM_{10} were 0.077, 0.073, and 0.099 mg/m^3, respectively, in 2004. The concentrations of the three pollutants in 2008 were, respectively, 40.3, 23.3, and 28.3 % lower than in 2004. It is evident that a series of pollution-control measures adopted in Guangzhou has achieved remarkable effects since 2004, particularly in terms of coal-burning pollution, though problems through motor vehicle pollution have become increasingly prominent. The situation is thus similar to that in Beijing. Motor vehicle pollution has become the focus of urban air pollution control.

2.4 Motor Vehicle Pollution-Control and Energy-Conservation Measures

2.4.1 Vehicle Emission Standards for New Vehicles

To deal with pollution problems caused by the rapid development of motor vehicles, developed countries have adopted more stringent new vehicle emission standards. Currently, the vehicle emission regulations of the United States, Japan, and Europe are the three main systems around the world. Many other countries have, to different extent, adopted these regulations and standards. The emission standards and regulations of China are based on the European system.

China's vehicle emission standard system can be roughly divided into four periods. In the first period, which was before 1989, China controlled the CO and HC emissions from gasoline vehicles based on the idling-based measurement and smoke emissions from diesel vehicles by means of free acceleration smoke. The second period started in 1989, when vehicle emission control was extended from vehicle exhaust emissions to crankcase emissions and the measurement was shifted from simple idling method to a preliminary driving cycle-based method. The third period began in 1993 with the establishment and implementation of the GB14761 series standards, and by then, China has established a comprehensive vehicle emission-control system including exhaust emissions, crankcase emissions, and evaporative emissions.

The fourth period began in 2001. This was when the State Environmental Protection Administration (the former Ministry of Environmental Protection) successively implemented national standards, such as the National Emission Standards and Emission Measurement Methods for Light-Duty Vehicles (I, II) (GB18352 series); this was also when the state's control policy of gradually tightening vehicle emission standards began to come into effect. The State Environmental Protection Administration and the State Administration of Quality Supervision, Inspection and Quarantine jointly established three national emissions standards for vehicle pollutant emission on April 16, 2001, and another one was issued on November 27, 2002. These four standards in conjunction with reserved standards together constitute the current vehicle emission standard system in China; they provide unified coverage of motor vehicle emission control in all major aspects of China's vehicle emission standard system.

On July 1, 2004, the phase II of the national light-duty vehicle emission standards (National II), which is equivalent to the Euro II Standards, was implemented nationwide. National II can reduce individual vehicle emissions by 40 % compared to National I vehicles. The implementation of the National II standards shows that the vehicle pollution control in China has entered a new stage. The State Environmental Protection Administration required that all new light-duty vehicles had to meet the emission limit. In July 1, 2008, National III Standard was officially implemented throughout the country, which further reduced the EFs of vehicles.

Because facing more serious vehicle pollution, large cities like Beijing, Shanghai, and Guangzhou have adopted stricter control measures and introduced the new standards ahead of the national level. Taking Beijing, for example, National I, which is equivalent to Euro I, was adopted in 1999; National IV was introduced on March 1, 2008, and National V in 2013. These measures have dramatically reduced the EFs of individual vehicle pollutants.

Current emission standards have produced a tremendous reduction in pollutant emissions. According to estimates by the International Council on Clean Transportation (ICCT), the results of measures adopted in China over the past decade have been very positive. If those control measures had not been introduced, there would have been 80 % more PM, 77 % more HC, 70 % more CO, and 55 % more NO_2 emissions (ICCT 2010).

2.4.2 *Fuel Quality*

The composition of fuel has a direct impact on vehicle emissions, and hence changing this composition will reduce emissions. The benefit in vehicle emission reduction from improving fuel quality is clear and instant. For example, vehicles using leaded gasoline were responsible for 90 % of lead concentration in many large cities. There is an obvious relationship between the lead content in gasoline and lead concentration in the air, so reducing the amount of lead in gasoline could significantly reduce ambient lead concentrations. As vehicle emission standards become stricter, fuel quality has become a key factor in deciding the success of vehicle emission-control actions.

2.4.2.1 Unleaded Gasoline

Lead is added to gasoline during production to increase the antiknock performance of gasoline. However, lead is an atmospheric pollutant and causes damage to the catalytic converter and oxygen sensor. To remove the hazard of lead in the environment and to ensure long-term normal operation of the catalytic converters and oxygen sensors, it is important to use unleaded gasoline.

Since 1996, the proportion of unleaded gasoline has increased annually in China. According to the annual vehicle pollution report prepared by Chinese Environmental

Science Academy and Tsinghua University in 2001, the proportion of unleaded gasoline among total gasoline was 51.82 % in 1990, 56.30 % in 1994, 65 % in 1997, and 80 % in 1998. Following the notice by the General Office of the State Council about banning production, sales, and use of leaded vehicle gasoline (State Council Document No. [1998] 129), the production of leaded gasoline was banned nationwide on January 1, 2000; from July 1, 2000, it was prohibited to sell or use leaded gasoline. In this way, the fuel quality in China has dramatically improved, allowing in-depth cleaning of vehicle exhausts.

2.4.2.2 Low-Sulfur Fuel

The sulfur content of fuel has a great impact on vehicle emissions, especially for the vehicles that meet the National III standards or higher. Sulfur can damage the catalytic converter, particulate trap, and other post-processing parts, which significantly reduces the effects of emission purification and means that vehicles fail to meet the emission standard.

Constrained by fuel-refining technology, China implemented the National II fuel standard (equivalent to Euro II) nationwide before 2009, in which the sulfur content of gasoline was 350 mg/kg and that of diesel 2,000 mg/kg. China planned to adopt fuel standards that were equivalent to Euro III in 2010 throughout the country. In 2010, the sulfur content of gasoline was limited to 150 mg/kg and that of diesel 350 mg/kg. In order to meet the new vehicle emission standard, some large cities such as Beijing have already adopted National IV, which calls for a further reduction in the sulfur content of gasoline to 50 mg/kg and that of diesel to 50 mg/kg. National IV was implemented on October 1, 2009 in Shanghai. In Beijing, the National V fuel standard was implemented in 2012, which requires the sulfur content to be lower than 10 mg/kg (Sidebar 2.2).

Sidebar 2.2: Effect of Improving Fuel Quality on Energy Saving

At present, fuel quality is a major issue for applying high-efficient diesel engine technologies. In China, fuel quality cannot meet the demands of the high-efficiency vehicle technologies. Therefore, improving fuel quality is beneficial not only to the successful implementation of emission standards but also to the development of high-efficient technologies.

2.4.3 Controlling Emissions from In-Use Vehicles

Emissions from in-use vehicles deteriorate as their accumulative mileages increase. In addition, vehicles with low accumulative mileages may have a high-emission level if they are improperly maintained. Therefore, timely measures need to be taken to detect improper maintenance to avoid a rise in vehicle emissions. Through routine inspections, the inspection/maintenance (I/M) programs are one means of identifying vehicles that are poor with respect to emissions; in this way, relevant parts can be cleaned, replaced, or properly adjusted so that they can return to a normal working condition in a timely fashion.

With routine inspection and maintenance, the automobile emission-control system is able to function properly. This plays a very important role in controlling in-use vehicle pollution for the following reasons. First, I/M programs can identify those vehicles that are producing high emissions through malfunction or other mechanical problems. Pollutants emitted from this small number of high-emission vehicles often account for a large proportion of total emissions. Related foreign studies indicate that high-emission vehicles, which account for 20 % of the entire vehicle population, contribute over 60 % of total vehicle emissions. Second, I/M programs can also be used to identify emission-control failure and prevent the removal of emission-control equipment on vehicles. If there is failure in the functions of catalytic converters or oxygen sensors, CO and HC emissions can be increased by over 20-fold, and NO_x emission can be increased by three- to fivefold. However, because malfunctions of emission-control equipment do not affect driving performance, drivers are usually not aware of this problem. The advantage of the I/M programs lies in their ability to identify vehicles with emission-control problems and require them to be repaired and undergo maintenance, which ensures that vehicle emissions remain at an appropriate level, thereby effectively preventing abnormal vehicle emissions and helping to control vehicle pollution.

Some large cities in China, such as Beijing and Shanghai, have implemented the I/M programs. In Beijing, for example, before 2003, the main method for gasoline vehicle inspection was the dual-idling method. Since March 2003, the method for inspecting vehicles has been under simple driving conditions. Three sets of emission limits were adopted according to the differences in vehicle type, technology, and vehicle weight. For diesel vehicles, the free acceleration test mode was replaced by the lug-down smoke test mode. The I/M programs currently being conducted in Beijing are different from the general type of programs that exist elsewhere in the country. Outside Beijing, there are centralized programs that are responsible only for testing and also decentralized programs that combine inspection and maintenance. The size of the inspection facilities is relatively small. Annual inspections of vehicles in Beijing are mainly carried out by collective and state-owned enterprises in addition to the police, the military, and other such institutions. At all testing stations, there is a combination of traffic-management and safety tests, and repairs can also be carried out at these testing stations. Vehicles that pass the emission test can proceed to security checks and registration. Vehicles that fail the emission test must be kept at an affiliated service station of the inspection facilities or at an independent service station and then returned for rechecking. I/M programs play an active role in controlling vehicle emissions in Beijing.

2.4.4 Eliminating Vehicles with Dated Technology

According to the new standards for vehicle scrappage issued in 1997, automobiles that exceed the national emission standards even after maintenance should be scrapped. The drivers of in-use vehicles that meet the national standards are encouraged to update their emission-control systems, though it is not mandatory requirement. Taxing old vehicles is also an effective and appropriate method of keeping these automobiles out of key cities. Based on these principles, local environmental government and vehicle management should introduce economic policies toward effectively eliminating old vehicles. It is therefore necessary to establish vehicle supervision and management sectors to eliminate old vehicles that do not meet national standards.

From the perspective of emission performance, renewing a car's engine can help it meet the emission standards, and this achieves the same effect as taking the vehicle off the road. Renewing car engines is a method that has been used in some developing countries to reduce urban vehicle emissions. In such places, efforts are made to remove old vehicles from key urban areas and to replace them with low-emission automobiles that use closed-loop three-way catalytic purification. Depending on the local situation, incentives such as partial vehicle tax exemption and offering preferential loans (such as for renewing taxis) may be provided.

In China, vehicles that fail to meet the National I standards are referred to as "yellow-label vehicles." The emission levels of these vehicles are quite high, and they can be several times higher than vehicles that conform to National II. The elimination of yellow-label vehicles will significantly reduce emissions. During the period of the Beijing Olympic Games, yellow-label vehicles were forbidden to travel in urban areas, and this produced positive results. Additional measures have since been taken to accelerate the elimination of yellow-label vehicles in Beijing. Since October 1, 2009, yellow-label vehicles have not been allowed within and on the Sixth Ring Road. This measure is playing an important role in reducing vehicle pollution in Beijing (Sidebar 2.3).

Sidebar 2.3: Effect of Eliminating Old Vehicles on Energy Saving

Eliminating old vehicles can also benefit energy saving. China introduced its fuel-economy standards in 2005, but older vehicles are free from such standards. If older vehicles retired quickly from the fleet, the average fuel-economy level of the entire fleet would be elevated. To speed up the fleet's technological renovation and stimulate the automobile market, United States and European countries have offered incentives to encourage drivers to dump their old vehicles. For example, in 2009 when the global economy was in recession, the United States Congress devised a scrappage program, referred

(continued)

(continued)
to as "cash for clunkers," under which consumers could trade in their old, "gas-guzzling" vehicles and receive vouchers worth up to $4,500 to help pay for new, more fuel-efficient cars and light-duty trucks. This measure improved the overall fuel economy of American vehicles. It also helped stimulate the American automobile market and helped it somehow recover in 2010.

2.4.5 *Promoting Alternative Fuels*

China launched the Clean Vehicle Action Program in 1999. Headed by the Ministry of Science and Technology, together with 13 ministries and commissions, including the former State Development Planning Commission, the former State Environmental Protection Administration, and the former Ministry of Construction, the National Clean Vehicle Action Coordination Group was established. This program was originally executed in 12 cities and regions, including Beijing, Shanghai, Chongqing, and Sichuan; it was later expanded to 19 cities and regions. In 2006, China initiated the energy-efficient and new-energy vehicle high-tech plan, which promoted the rapid development of natural gas vehicles. By the end of 2009, more than 1,000 natural gas stations had been constructed, most of which were concentrated in the vicinity of natural gas sources, such as in Sichuan, Chongqing, Urumqi, Xi'an, and Lanzhou. A convenient natural gas supply and low prices have been the main forces behind the development of vehicles using compressed natural gas. Currently, about one-fourth of taxis and one-sixth of buses in China are equipped with natural gas engines.

In addition, China began to promote ethanol gasoline in 2001. Bioethanol gasoline (E10, 10 % ethanol blended in gasoline by volume) is widely used in 6 provinces (Heilongjiang, Jilin, Liaoning, Henan, Anhui) and 27 cities of 4 provinces (Hebei, Shandong, Jiangsu, and Hubei) (Sidebar 2.4).

Sidebar 2.4: Effect of Alternative Fuels on Energy Saving

Whether alternative fuels are able to reduce pollutant emissions is still a matter of debate. Researchers in China and overseas are performing tests and simulations to study the differences in emissions between vehicles combusting alternative fuels and conventional gasoline. Nevertheless, because alternative fuels are mainly extracted from coal, natural gas, and biomass, so then can save significant amount of oil.

2.4.6 *Developing Electric Vehicles*

Electric vehicles (EVs) are powered by electricity, which is mainly generated from coal and hydraulic power in China. Compared with conventional vehicles, EVs can reduce oil use by over 90 %. Emissions from EVs are mainly determined by energy consumption and emission level of the power generation processes (Huo et al. 2010) (Sidebar 2.5).

Sidebar 2.5: Developing EVs and the Environment

Evaluating the total emission of an EV throughout its life cycle first demands an investigation of the pollutant emission level of China's thermal power plants. The EF of NO_x emitted from thermal power plants differs according to the sizes of boilers, control techniques for NO_x emission, and coal quality. In recent years, with an increasing proportion of large-scale boilers and widespread use of low nitrogen burning (LNB) technology, the average EF of NO_x emitted from thermal power plants has gradually reduced—from 89 g/kg of coal in 1995 to 74 g/kg in 2004. The usage of LNB amounted to 62 % in 2005, and it reached 77 % in 2010. Since newly built power plants are obliged to install LNB facilities, the LNB rate is expected to reach 90 % or higher. When it does so, the average EF will be reduced to 5 g/kg of coal. Selective catalytic reduction (SCR) technology can substantially reduce NO_x emissions. At present, the penetration rate of SCR technology in China is only about 10 %.

Regarding NO_x emissions from vehicles, China is currently implementing National III emission control standards (except for Beijing and other large cities where National IV/V has been introduced). National IV and National V standards are expected to be adopted within a decade. By 2030, Chinese vehicles will meet European IV standards or even stricter.

To make EVs more attractive in China, the government should introduce certain policies. The government can make low-carbon districts the main markets for EVs. Although some advanced power-generation techniques and pollution-control techniques (SCR) have high costs, there are no technical barriers. Therefore, the government should promote these techniques as quickly as possible through a series of policies, especially in view of the benefits to the regional environment and health. It is also notable that since power plants usually have a longer life cycle than vehicles, it takes longer to make a technical update of such plants; thus, the government needs to coordinate with the policies of power and transportation departments to promote the development of clean EVs.

2.4.7 Temporary Traffic-Control Measures

In addition to common control measures, temporary traffic-control measures can be adopted to ensure good air quality and traffic conditions during special periods. Beijing implemented the following temporary transportation-control measures during the period of the 2008 Olympic Games:

1. Private vehicles had to abide by the odd-and-even license plate rule, which requires that vehicles with certain last digit on plates (odd or even) not to run on roads on certain weekdays.
2. During the event, 70 % of governmental vehicles were banned from the roads.
3. Trucks were allowed to go within the Sixth Ring Road only between midnight and 6 a.m.
4. Yellow-label vehicles were banned from the roads of Beijing.

As a result, the total vehicle travelled distance was reduced by 32.0 %. The average driving speed was increased from 25 to 37 km/h. During the event, the HC, CO, NO_x, and PM_{10} emissions in Beijing were reduced by 55.5, 56.8, 45.7, and 51.6 %, respectively (Wu et al. 2011; Zhou et al. 2010).

2.5 Challenges and Prospects

As has been noted, although the air quality of China has been improved in recent years, the absolute concentrations of pollutants, especially NO_2 emissions, are still high. In most Chinese cities, vehicle exhausts are the major source of air pollution. Vehicle pollution control needs to be more rigorous. A series of regulations and measures has been taken to control vehicle pollution since the period of the Ninth Five-Year Plan (1995–2000), and some progress has been made. However, these gains have been mostly offset by the explosive growth of the vehicle population and other traffic problems. In addition, with the development of Chinese society, the public has greater expectations regarding environmental issues.

In the Twelfth Five-Year Plan (2010–2015), NO_x was for the first time listed as an object of total emission control. Motor vehicles are the second-largest source of NO_x emissions, and thus they are a key to achieving NO_x control objectives. Since the autumn of 2011, $PM_{2.5}$ pollution over a period of consecutive hazy days attracted the public's concern regarding air quality conditions. In the new Ambient Air Quality Standard (Second Exposure Draft) developed by the Ministry of Environmental Protection in 2011, $PM_{2.5}$ and ozone 8-h period concentrations were listed in the Routine Air Quality Evaluation System, and the standard limits for PM_{10} and NO_2 became stricter. Studies of many Asian cities show that motor vehicles contribute 20–35 % of the total $PM_{2.5}$ emission. Therefore, implementing new standards will offer a full range of strategic choices regarding gasoline and diesel vehicles, alternative fuels, new-energy vehicles, and exhaust gas purification for the future China.

In progressing toward vehicle pollution control, there are two key factors—activity level and EFs. To this end, the following strategies are proposed.

2.5.1 Developing Public Transportation

Since it is impossible to control the vehicle population at the current stage, the main measure for controlling the activity level is to guide the public's choice of transport. Therefore, clean, convenient, inexpensive public transportation should be developed, and the public transport network needs to be perfected. This would encourage more people to choose public transportation, which would reduce the use of private cars and ultimately control vehicle emissions.

2.5.2 Improving Traffic Management

Improving traffic management will have a significant effect on vehicle emissions. At present, traffic management in Chinese cities is still at a relatively low level. Supervision at traffic intersections should be enhanced to alleviate traffic pressure, improve traffic flow, and thereby reduce emissions. Economic measures, such as using electronic toll systems on congested roads and increasing parking fees in central city areas, could be adopted to control travel in urban areas and thus reduce the harm to the environment caused by vehicle pollution in densely populated city centers.

2.5.3 Implementing Stricter Emission and Fuel Standards

As noted above, measures have been undertaken in China to control vehicle pollution, and some progress has been made. Stricter emission and fuel standards, such as the Euro V and Euro VI, should be implemented as soon as possible to reduce the EFs of individual gasoline and diesel vehicles.

2.5.4 Faster Development of Alternative Fuels

In addition to reducing emissions by gasoline and diesel vehicles, new ways should be explored to control emissions. For example, alternative fuels could be used as substitutes for gasoline and diesel. Though some progress has been achieved in China, investment in scientific and technological research into alternative fuels should be increased to deal with problems in using such alternative fuels and reduce costs as much as possible. With stricter vehicle pollution control, improvements in new fuels technology, and the popularization and use of alternative fuels, the pressure that the vehicle growth places on oil resources and the environment can be significantly relieved.

References

Beijing Municipality's Bureau of Environmental Protection (2001) 2000 report on the state of environment in Beijing
Beijing Municipality's Bureau of Environmental Protection (2002) 2001 report on the state of environment in Beijing
Beijing Municipality's Bureau of Environmental Protection (2003) 2002 report on the state of environment in Beijing
Beijing Municipality's Bureau of Environmental Protection (2004) 2003 report on the state of environment in Beijing
Beijing Municipality's Bureau of Environmental Protection (2005) 2004 report on the state of environment in Beijing
Beijing Municipality's Bureau of Environmental Protection (2006) 2005 report on the state of environment in Beijing
Beijing Municipality's Bureau of Environmental Protection (2007) 2006 report on the state of environment in Beijing
Beijing Municipality's Bureau of Environmental Protection (2008) 2007 report on the state of environment in Beijing
Beijing Municipality's Bureau of Environmental Protection (2009) 2008 report on the state of environment in Beijing
Beijing Municipality's Bureau of Environmental Protection (2010) 2009 report on the state of environment in Beijing
ICCT (International Council on Clean Transportation) (2010) 2010 overview of China's vehicle emission control program: past successes and future prospect
Guan G, Yu Q (2007) Characteristics and evaluation of motor vehicle exhaust pollution in the city. Guangzhou Chem Ind 3:15–17
Guan GC et al (2006) A study on the share rate of motor vehicle exhaust pollution in Foshan. J Foshan Univ (Nat Sci Ed) 24(1):59–62
Guangzhou Municipality's Bureau of Environmental Protection (2002) 2001 report on the state of environment in Guangzhou
Guangzhou Municipality's Bureau of Environmental Protection (2003) 2002 report on the state of environment in Guangzhou
Guangzhou Municipality's Bureau of Environmental Protection (2004) 2003 report on the state of environment in Guangzhou
Guangzhou Municipality's Bureau of Environmental Protection (2005) 2004 report on the state of environment in Guangzhou
Guangzhou Municipality's Bureau of Environmental Protection (2006) 2005 report on the state of environment in Guangzhou
Guangzhou Municipality's Bureau of Environmental Protection (2007) 2006 report on the state of environment in Guangzhou
Guangzhou Municipality's Bureau of Environmental Protection (2008) 2007 report on the state of environment in Guangzhou
Guangzhou Municipality's Bureau of Environmental Protection (2009) 2008 report on the state of environment in Guangzhou
Huo H (2005) Study on link-based light-duty vehicle emissions based on traffic characteristic. Doctoral thesis, Tsinghua University, Beijing

Huo H et al (2010) Environmental implication of electric vehicles in China. Environ Sci Technol 44:4856–4861
Huo H et al (2011) Modeling vehicle emissions in different types of Chinese cities: importance of vehicle fleet and local features. Environ Pollut 159:2954–2960
Jin XS (2009) Motor vehicle exhaust pollution situation and control measures in Guangdong. Automob Energy-Sav 3:20–23
Li CQ, Bai K, Zhang P, Song DL, Zhou ZH (2010) Study on the share rate of motor vehicle exhaust pollution in Chendu. Proceeding of Chengdu technology annual meeting, pp 1–6
Shanghai Municipality's Bureau of Environmental Protection (2006) 2005 report on the state of environment in Shanghai
Shanghai Municipality's Bureau of Environmental Protection (2007) 2006 report on the state of environment in Shanghai
Shanghai Municipality's Bureau of Environmental Protection (2008) 2007 report on the state of environment in Shanghai
Shanghai Municipality's Bureau of Environmental Protection (2009) 2008 report on the state of environment in Shanghai
Shanghai Municipality's Bureau of Environmental Protection (2010) 2009 report on the state of environment in Shanghai
Wang S et al (2010) Quantifying the air pollutant emission reduction during the 2008 olympic games in Beijing. Environ Sci Technol 2010(44):2490–2496
Wu Y et al (2011) On road vehicle emission control in Beijing: past, present, and future. Environ Sci Technol 45(1):147–153
Yang Q et al (2009) Study on vehicle's rate of pollution sharing in Chongqing's central city. J Southwest China Norm Univ 34(4):173–177
Zhang Q et al (2009) Asian emissions in 2006 for the NASAINTEXB mission. Atmos Chem Phys 9:5131–5153
Zhou Y, Wu Y, Yang L, Fu LX, He KB, Wang SX, Hao JM, Chen JC, Li CY (2010) The impact of transportation control measures on emission reductions during the 2008 Olympic Games in Bejing, China. Atmos Environ 44:285–293

Chapter 3
Scenario Analyses of China's Vehicle Ownership and Vehicle Traffic Services

Wang Hewu, Hao Han, and Ouyang Minggao

Abstract A hybrid vehicle ownership model, which comprises three sub-models (private passenger vehicle population model, public traffic vehicle population model, and other vehicle population model), was established to simulate the growth of China's vehicle population.

The passenger vehicle population model links the vehicle population and residents' income distribution to forecast the growth of the private passenger vehicle population. The public traffic vehicle population model links the vehicle population, human population, and urbanization rate to forecast the growth of the vehicle population of urban public buses and taxis. The other vehicle population model links vehicle population growth and GDP growth to forecast the population of all other vehicles, including passenger vehicles for public affairs, other buses, and trucks. Passenger and freight traffic volume were projected based on the forecast results for the vehicle population.

The vehicle population and traffic volume were projected under two scenarios of reference scenario and comprehensive policy scenario. The reference scenarios and comprehensive policy scenarios employed in this section are the same as those in Chap. 12 for the vehicle energy development scenario. Chapter 12 presents the scenario definitions in detail, and the present section describes only the content related to vehicle population and vehicle traffic volume.

Keywords Vehicle ownership • Traffic volume • China

Wang Hewu (✉) • Hao Han • Ouyang Minggao
China Automotive Energy Research Center, State Key Laboratory of Automotive Safety and Energy, Tsinghua University, Beijing 100084, P.R. China
e-mail: wanghw@tsinghua.edu.cn

CAERC, Tsinghua University, *Sustainable Automotive Energy System in China*,
DOI 10.1007/978-3-642-36847-9_3, © Springer-Verlag Berlin Heidelberg 2013

3.1 History and Current Status

Vehicle population and traffic volume are affected by many different factors, such as economic development, population, geography, technological progress, and policies. China's economy has achieved rapid growth, with its gross domestic product (GDP) having increased from RMB 454.6 billion in 1980 to RMB 39,798.3 billion in 2010—an average annual growth rate of 19.1 %. By 2010, GDP per capita had reached RMB 29,900, with the annual average disposable income of urban residents and net income of rural residents having reached RMB 19,100 and RMB 5,900, respectively. The growth in GDP and personal income has been the economic basis for the rise in China's vehicle population and traffic volume. To promote the development of China's automotive industry, the government established its Automotive Industry Policies in 1994, which defined the automotive industry as one of the key industries in China's national economy for stimulating the development of other industries. In 2004, to address the current situation in the domestic and overseas automotive industry after China entered the World Trade Organization, the State Council implemented its Policies on Automotive Industry Development. These policies set numerous objectives relating to upgrading the industrial structure, increasing international competitiveness, meeting growing consumer demand, and promoting positive development of the industry. In 2009, in response to the international financial crisis, the State Council promulgated its Automotive Industry Revitalization Planning, which played an important role in stabilizing automobile consumption, accelerating structural adjustments, improving independent innovative capability, and promoting upgrading of the industry.

Propelled by these factors, China's vehicle population rose rapidly. As shown in Fig. 3.1, China's vehicle production rose from 0.51 million in 1990 to 18.26 million in 2010, with an average annual growth rate of 15.9 %. With the spread of household vehicles, both the production and sales of passenger vehicles achieved

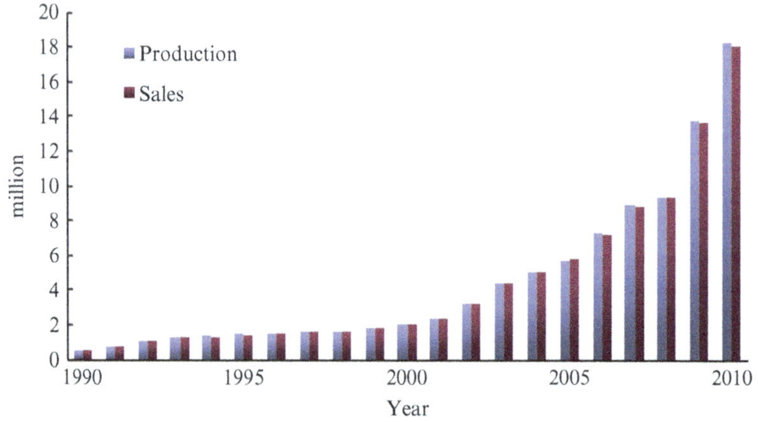

Fig. 3.1 Historical vehicle production and sales in China

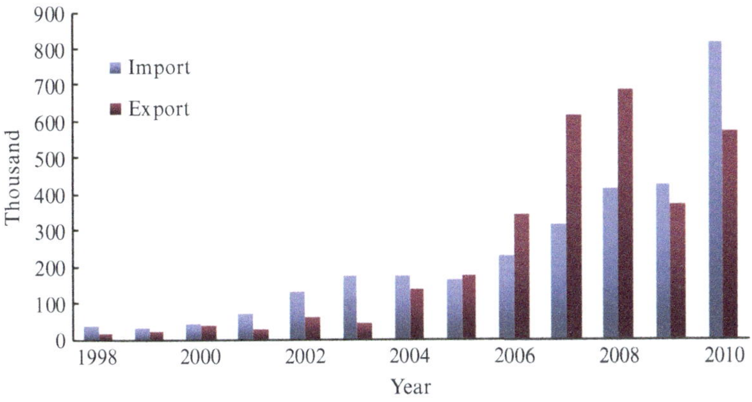

Fig. 3.2 Historical vehicle imports and exports for China

the fastest growth, amounting to 13.9 and 13.76 million, respectively, in 2010, and passenger vehicles that year accounted for 76 % of the total production and sales. With increasing demand for mass passenger and cargo transportation, the production and sales of buses and trucks also achieved rapid growth. By 2010, the production and sales of buses amounted to 0.447 and 0.443 million, respectively, while those of trucks were 3.92 and 3.86 million. China Association of Automotive Manufactures (2011).

Vehicle imports and exports in China have also increased constantly. As shown in Fig. 3.2, vehicle imports rose from 0.4 million in 1998 to 0.81 million in 2010; vehicle exports rose from 14,000 to 567,000 over the same period. China's vehicle exports exceeded imports in 2005. However, owing to the impact of the global financial crisis, vehicle exports experienced a fall in 2009 and 2010.

In the present study, the number of newly registered vehicles was estimated based on historical production, import, and export figures. As seen in Fig. 3.3, the number of newly registered vehicles rose from 2.16 million in 2000 to 18.57 million in 2010. Among those, the number of newly registered passenger vehicles increased from 1.23 million in 2000 to 14.44 million in 2010; the number of newly registered buses and trucks grew from 0.17 to 0.41 million and from 0.76 to 3.71 million, respectively, over the same period.

The vehicle population has increased with the growth in the number of newly registered vehicles. In our research, the total vehicle population was estimated based on the number of newly registered vehicles, and the results (from 2000 to 2010) are presented in Fig. 3.4. The total vehicle population at the end of 2010 amounted to 79.97 million (excluding low-speed vehicles; the same category of vehicles is excluded in the figures below)—an increase of 27 % from 2009. China's vehicle ownership in 2010 was 59.6 vehicles per 1,000 persons, which was approximately equivalent to the United States level in 1918 and the Japanese level in 1965. In 2010, the number of passenger vehicles was 59.57 million (accounting for 74.5 % of the total population), buses 2.78 million (3.5 %), and trucks 17.62 million (22 %). Compared with developed countries, the number of passenger vehicles in China

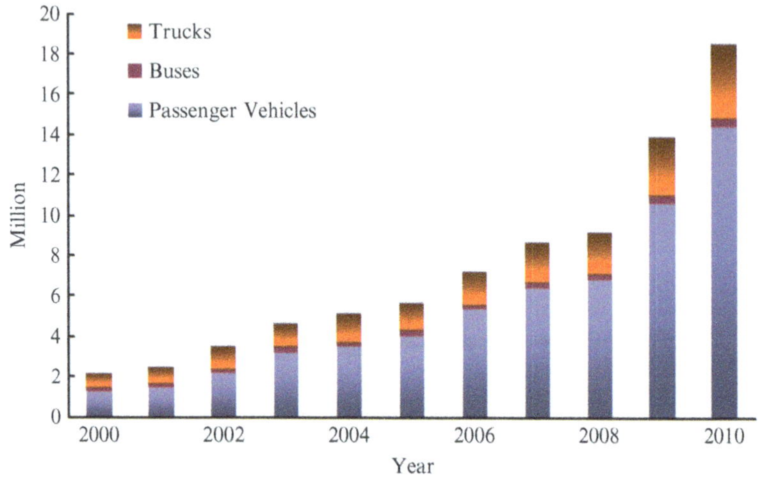

Fig. 3.3 New vehicle registrations in China between 2000 and 2010

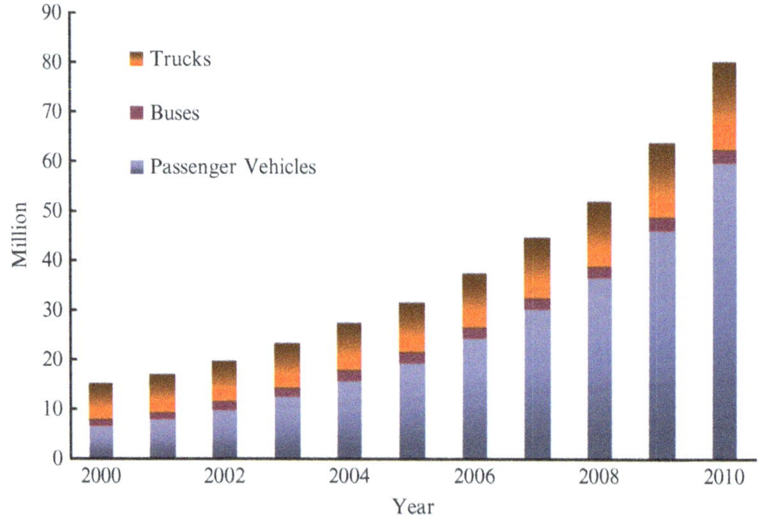

Fig. 3.4 Total vehicle registrations in China between 2000 and 2010

accounts for a lower percentage of the total vehicle population since the penetration rate of sedans is not yet that high in Chinese families.

China's vehicle population is expanding and it has great growth potential. Many institutions and individual scholars have conducted projections of China's vehicle population. In terms of domestic research, according to the Research on China's Energy Development Strategy and Policy led by the National Development and Reform Commission in 2004, China's vehicle population is expected to reach 110 million by 2020 (National Development and Reform Commission 2004). According to another estimate using income distribution figures for the year 2006, China's vehicle population will reach 228 million by 2030 (Shen Zhongyuan 2006). The 2050 China Energy and CO_2 Emissions Report published in 2009 projected that China's vehicle population will reach 363–397 million by 2030 and 558–605 million by 2050 (China Energy and Carbon Emission Research Group 2009). In terms of global research, Wang et al. of the Argonne National Laboratory using the Gompertz model in 2006 forecast that China's vehicle population would reach 247–287 million by 2030 and 486–662 million by 2050. In 2007, the International Energy Agency (IEA) in *World Energy Outlook 2007* forecast that China's vehicle population would reach 270–410 million by 2030. In 2007, Dargay et al. (2007) using the improved Gompertz model forecast that China's vehicle population would reach 390 million by 2030. Wang et al. (2011) of the University of California (UC), Davis, summarized the above results and concluded that all these studies generally underestimated the growth rate of China's vehicle population. Based on a comparison with the increasing vehicle population of the developed countries at a similar stage, estimated that China's vehicle population would reach 332–419 million by 2022 (Sidebar 3.1).

Sidebar 3.1: China's Vehicle Population Research Approaches

Wang et al. of UC Davis selected seven countries that at one time were at a similar economic level and increasing stage of vehicle population as China; the countries were Japan, the United States, Germany, Italy, Korea, Spain, and Brazil. The study defined the year when the vehicle population of those countries reached China's current level in 2008 (37–38 vehicles per 1,000 persons). Wang et al. then investigated the growth rate of those countries' vehicle population over 15 years from the base year of 2008, and they took the average growth rate as the growth rate of China's vehicle population rate from year 2008 to 2024. This method is simple and practical and made no assumptions regarding GDP or the growth in per capita income. The projected result was much higher than that of other studies and is closer to the current condition regarding current vehicle production and sales.

Wang et al. of Argonne National Laboratory adopted the Gompertz model (a kind of growth model) as the vehicle population growth model, and the input of the model was per capita GDP with three model parameters. Three scenarios were designed using a parameter representing the saturation level of the vehicle population: high, middle, and low saturation, corresponding to 600, 500, and 400 vehicles per 1,000 persons, respectively. The other two parameters were regressed based on per capita GDP and vehicle ownership from 1978 to 2004. Dargay et al. made a projection using a similar method. This approach is sensitive to the assumption of vehicle ownership saturation level, selection of historical data, and assumption of future GDP growth rate.

(continued)

(continued)

The higher the vehicle ownership saturation level is set, the more recent are the historical data employed, the faster is the assumed future GDP growth, and the higher the result. Thus, studies using the same approach may obtain quite different projection results.

Shen Zhongyuan of the Institute of Energy Economics of Japan built a vehicle population forecast model for China based on a distribution curve for residents' income. The vehicle population was estimated according to the integral of two results derived from a logarithmic normal distribution curve as the income distribution curve and from a logistic curve as the vehicle diffusion curve. With this approach, a parametric regression can be made with the income distribution and vehicle diffusion curves in accordance with the historical data. This study employed residents' income distribution as the forecast basis. The advantage with this approach is that the influence of income level and income difference on vehicle diffusion rate can be modeled. However, this also requires the assumption of other parameters, such as future average income of urban and rural residents, and Gini coefficient.

Source	Forecast method	Key assumption	Results (10^6)
Wang et al. (2011)	Used growth rate of vehicle ownership of developed countries at a similar stage to estimate vehicle ownership	Selection of similar countries	2022: 332–419
Huo et al. (2007)	Took growth curve as the vehicle population growth model, estimated the model parameters based on historical data	GDP growth rate, saturation of vehicle ownership	2020: 125–134 2050: 486–662
Shen Zhongyuan (2006)	Estimated vehicle ownership according to income distribution and vehicle diffusion curves	Growth of residents' income, change in Gini coefficient	2030: 228

3.2 Vehicle Ownership and Traffic-Volume Projection Model

3.2.1 Model Structure

3.2.1.1 Vehicle Classification and Model Function

Different types of vehicles vary remarkably in vehicle population growth pattern, vehicle-use characteristics, fuel economy, and other aspects. To build a corresponding analytical model, it is necessary to classify the vehicles into different categories. Currently, China Association of Automobile Manufacturers is responsible for issuing statistics concerning China's vehicle production and sales; China Customs issues data relating to vehicle imports and exports; the Ministry of Public Security Traffic Management issues figures relating to vehicle registration; statistics relating to commercial vehicles are surveyed and issued by the Ministry of Transport. The manner of vehicle classification adopted in those statistics is quite different. A vehicle-type-based classification can provide basic details regarding the fleet characteristics of vehicles. However, for the same vehicle type, different uses would affect the fleet characteristics. With passenger vehicles, for example, both the vehicle use and ownership growth patterns are quite different for private passenger vehicles, government and business passenger vehicles, and taxis. Therefore, it is necessary to classify vehicles in two ways—vehicle type and utility. In this study, we classified vehicles into 11 categories based on vehicle type and utility. Table 3.1 presents the vehicle classification and description employed in the present study. Passenger vehicles are classified into private passenger vehicles, government and business passenger vehicles, and taxis; buses are classified into urban transit buses, non-transit operating buses, nonoperating buses based on their utility, and large buses, medium-sized buses, and light buses based on their length; trucks are classified into semitrailers, heavy-duty trucks, medium-duty trucks, light-duty trucks, and mini-trucks.

The vehicle traffic-volume projection model comprises three sub-models: vehicle population projection model, passenger vehicle traffic-volume projection model, and commercial vehicle traffic-volume projection model. The model's input parameters include:

1. GDP growth, residents' income distribution, and other economic quantities
2. Population growth, increase in urbanization rate, and other population and geographic quantities
3. Vehicle population control, traffic-demand management, and other public policy quantities

The model has the following major functions:

1. Simulating the relationship between vehicle population and macroeconomic factors, population, geographic factors, and public policy factors
2. Projecting China's future vehicle population and traffic volume

Table 3.1 Vehicle classification

Major type	Description	Segmentation	Segmentation description
Passenger vehicles	Vehicles used for passenger transport with nine seats or fewer	Private passenger vehicles	Mainly used for private purposes
		Government and business passenger vehicles	Used by governments or enterprises
		Taxis	Used for urban taxis
Buses	Vehicles used for passenger transport, with over nine seats	Urban transit buses	Used for urban public bus systems
		Non-transit operating buses	Used for long-distance passenger traffic
		Nonoperating buses	Used as shuttle buses
Trucks	Vehicles used for freight transport	Semitrailer towing trucks	Towing vehicles used to tow semitrailers
		Heavy-duty trucks	Max. total weight over 14 tons
		Medium-duty trucks	Max. total weight over 6 tons and less than 14 tons
		Light-duty trucks	Max. total weight over 1.8 tons and less than 6 tons
		Mini-trucks	Max. total weight less than or equal to 1.8 tons

3. Making a sensitivity analysis of the factors that may affect vehicle population and traffic volume

3.2.1.2 Passenger Vehicle Traffic-Volume Model

Figure 3.5 shows the structure of the passenger vehicle traffic-volume forecast model. China's historical passenger vehicle registration figures are obtained from statistical yearbooks; the vehicle population is classified according to private passenger vehicles, government and business passenger vehicles, and urban taxis. Then, the total registration number and vehicle-age distribution of China's passenger vehicles are obtained based on the passenger vehicle survival pattern model. From the vehicle population forecast model, a forecast is made of the future number of total passenger vehicle registrations based on the historical number of total passenger vehicle registrations; a calculation is then made of future new vehicle registrations based on the vehicle survival function. The traffic volume of passenger vehicles is obtained by considering the annual mileage of passenger vehicles. Finally, the fuel consumption and greenhouse gas (GHG) emissions of passenger vehicle are obtained following assumptions regarding fuel economy and GHG emission factors of new vehicles, as described in Chap. 11 of this book. The structure and content of part of the integration and optimization models appear in Figs. 3.5 and 3.6.

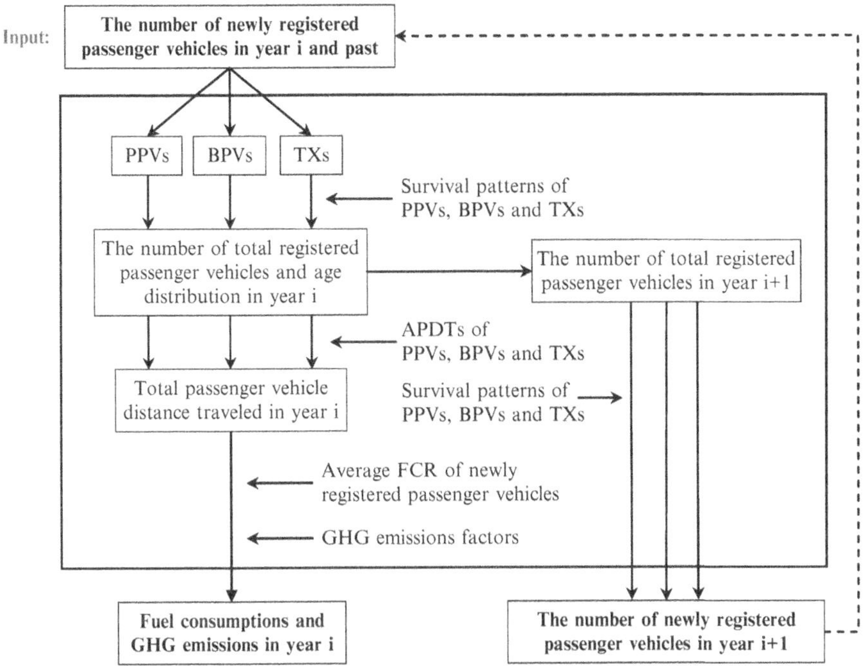

Fig. 3.5 Structure of passenger vehicle traffic-volume forecast model

3.2.1.3 Commercial Vehicle Traffic-Volume Model

The commercial vehicle traffic-volume model is shown in Fig. 3.6. The number of China's historical commercial vehicle new registrations is obtained from statistical yearbooks; the number of China's current total commercial vehicle registrations and vehicle-age distribution are calculated according to the survival function of commercial vehicles. Unlike with the passenger vehicle model, the commercial vehicle model first estimates the historical commercial vehicle traffic volume according to a single vehicle's annual mileage and passenger (cargo) capacity before forecasting the future growth of traffic volume. Then, the future number of total commercial vehicle registrations is calculated based on the estimated traffic-volume demand. Finally, the fuel consumption and GHG emissions of commercial vehicles are calculated in accordance with assumptions regarding the fuel economy and GHG emission factors of new vehicles.

3.2.1.4 Basic Assumptions of the Model

Table 3.2 presents the basic macro assumptions adopted in this study, including total GDP, population, and urbanization rate.

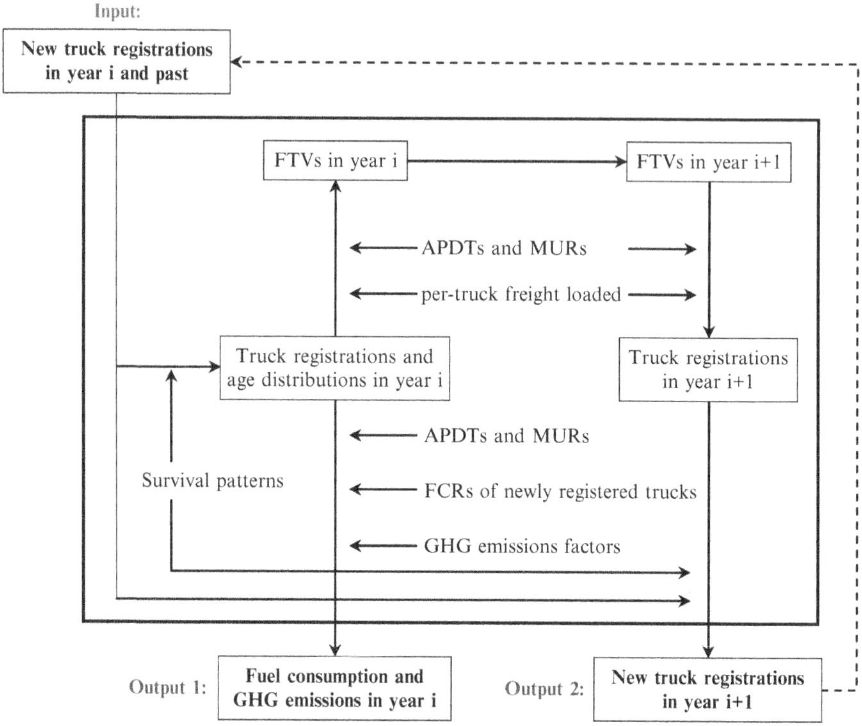

Fig. 3.6 Forecast model for commercial vehicle traffic volume

Table 3.2 Model parameter assumptions

Year	2010	2020	2030	2040	2050
Total population (0.1 billion)	13.6	14.4	14.7	14.7	14.6
Urbanization rate (%)	49	63	70	74	79
Total urban population (0.1 billion)	6.7	9.1	10.3	10.9	11.5
Average persons per urban household	2.88	2.8	2.75	2.7	2.65
Total number of urban households (0.1 billion)	2.3	3.2	3.7	4.0	4.4
Total rural population (0.1 billion)	6.9	5.3	4.4	3.8	3.1
Average persons per rural household	3.8	3.5	3.4	3.2	3
Total number of rural households (0.1 billion)	1.8	1.5	1.3	1.2	1.0
Total GDP (trillion RMB at 2010 level)	40.1	79.0	128.6	199.6	284.3

3.2.2 Vehicle Population Forecast Model

The vehicle population forecast model comprises three sub-models: private passenger vehicle population model, public traffic vehicle population model, and other vehicle population model. Among these, the passenger vehicle population model links the vehicle population and residents' income distribution to forecast the growth of the private passenger vehicle population. The public traffic vehicle population model links the vehicle population, human population, and urbanization rate to forecast the growth of the vehicle population of urban public buses and taxis. The other vehicle population model links vehicle population growth and GDP growth to forecast the population of all other vehicles, including passenger vehicles for public affairs, other buses, and trucks.

3.2.2.1 Private Passenger Vehicle Population Model

Model

For private passenger vehicles, the main driving factor of population growth is increasing residents' income level. This study investigates the relationship between China's household income and vehicle ownership according to the available historical data; it then analyzes the inherent laws and assumes vehicle ownership under the condition of increasing household income; and it forecasts the private passenger vehicle population through Eq. 3.1 according to residents' income distribution:

$$\mathrm{UPV}_i = \int \mathrm{UP}_i \cdot \mathrm{ULND}_i(x) \cdot \mathrm{UPVO}_i(x)\mathrm{d}x \qquad (3.1)$$

where UPV_i signifies the urban private passenger vehicle population in year i, UP_i the urban population in year i, $\mathrm{ULND}_i(x)$ the value of the urban household income distribution density function in year i corresponding to x, and $\mathrm{UPVO}_i(x)$ the average household's vehicle population rate when the household income in year i equals x. It should be noted that Eq. 3.1 is for the urban private passenger vehicle population; the rural private passenger vehicle population is estimated by a similar method.

To estimate the private passenger vehicle population using Eq. 3.1, it is necessary to obtain the household income distribution function. Related domestic and foreign studies have adopted many functions for the household income distribution function; of these, the logarithmic normal distribution has been the most widely used, and it is expressed in Eq. 3.2. In this study, we employ the logarithmic normal distribution as the urban household disposable income distribution function.

$$f(x;\mu,\sigma) = \frac{1}{x\sigma\sqrt{2\pi}} \exp\left(-\frac{[\ln(x)-\mu]^2}{2\sigma^2}\right) \qquad (3.2)$$

The relationship between the two parameters μ and σ in Eq. 3.2 and average household income α and Gini coefficient G calculated using the household income distribution can be expressed by Eqs. 3.3 and 3.4 (Aitchison and Brown 1963).

$$\alpha = \exp\left(\mu + \frac{1}{2}\sigma^2\right) \qquad (3.3)$$

$$G = \mathrm{erf}\left(\frac{\sigma}{2}\right) \qquad (3.4)$$

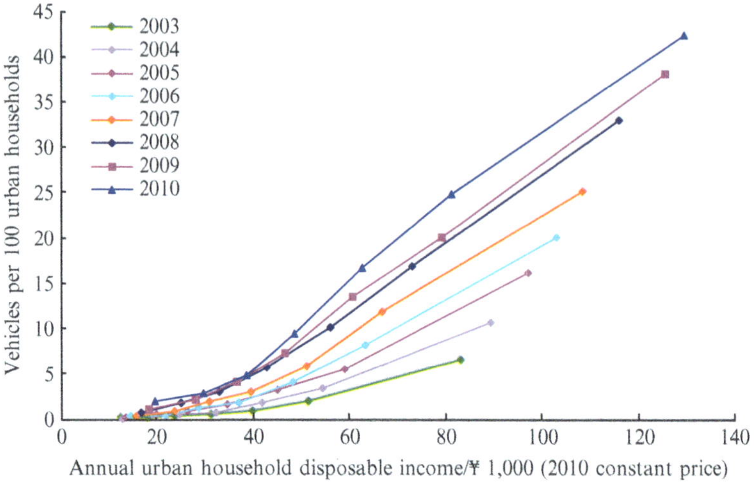

Fig. 3.7 Historical household vehicle population with rising income levels

Data and Scenario Assumption

The China Statistical Yearbook contains statistics for China's urban and rural resident household income and vehicle population level. Figure 3.7 presents passenger vehicle ownership under different household income levels obtained from the yearbook. It is evident that vehicle ownership rises with increasing income level, and there is a significantly linear relationship between vehicle ownership and household income when annual household income exceeds RMB 50,000. Additionally, with the same household income level, private vehicle ownership in more recent years is higher than in earlier years. There may be three reasons for this. First, the price of vehicles has decreased annually (Huo and Wang 2011). Since 2003, the price of vehicles in China, especially of household passenger vehicles, has been falling, which means that even with the same income level, households have more recently had greater ability to purchase vehicles. The second reason is the demonstration effect of vehicle purchase. Since China's vehicle ownership level is at present still very low, there is a strong demonstration effect for families with cars compared with families without cars. The final reason is higher income expectations. Since income levels in China have quickly increased over recent years, people have relatively high expectations regarding their future incomes, and this also stimulates vehicle purchases. It is also possible to see that the difference between the various years has decreased; for example, the curve for 2009 is not so different from that for 2008. This indicates that the results of falling vehicle prices, the demonstration effect, and residents' income expectations are constantly diminishing. Thus, in our opinion, before vehicle ownership reaches the level of one vehicle per household, there will be a linear increase from the historical curve as household income changes. As the level of one vehicle per household approaches, vehicle ownership will decrease with

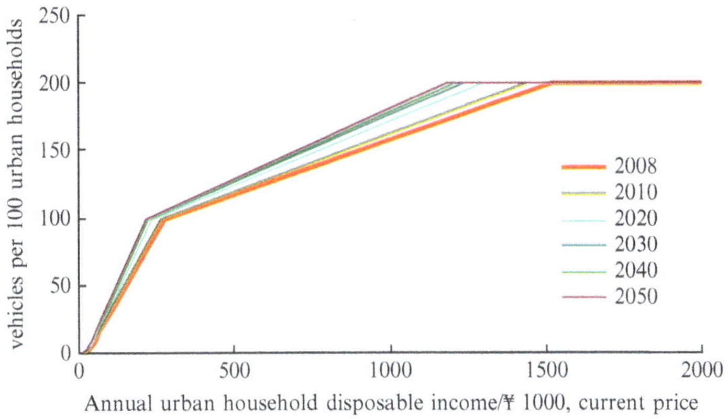

Fig. 3.8 Assumption of vehicle population with different household income

increasing household income, and it will reach a state of saturation after the level of two vehicles per household is attained.

Following the above assumptions, Fig. 3.8 presents the vehicle population rate with higher average household disposable income. From this, it is evident that in 2010, when the average household disposable income reached RMB 276,000, the ownership level would have reached one vehicle per household; when the average household disposable income reaches RMB 1,520,000, the population level would rise to two vehicles per household, and thereafter it would not rise with increasing average household income level. In view of the fact that vehicle ownership under the same household income rises annually owing to the fall in vehicle prices, we assume that vehicle ownership after 2010 will be higher than in 2010 with same household income level.

The average urban household disposable income and Gini coefficient were calculated based on the statistics of urban residents' income given in the China Statistical Yearbook; it was assumed that the annual growth rate in China's average household disposable income every 10 years from the year 2010 to 2050 would be 8, 6.8, 4.7, and 3.4 % (in 2010 prices) and that the Gini coefficient would be reduced by 5 % every 10 years. Calculating the characteristic parameters of the logarithmic normal distribution in Eqs. 3.3 and 3.4, we obtain the future distribution of urban household disposable income in China as shown in Fig. 3.9.

3.2.2.2 Public Traffic Vehicle Population Model

Model

The public bus and taxi population is affected by such factors as human population, area, and a city's economic development level; it is also closely related to a city's transport planning. This study investigated the current public bus population of cities

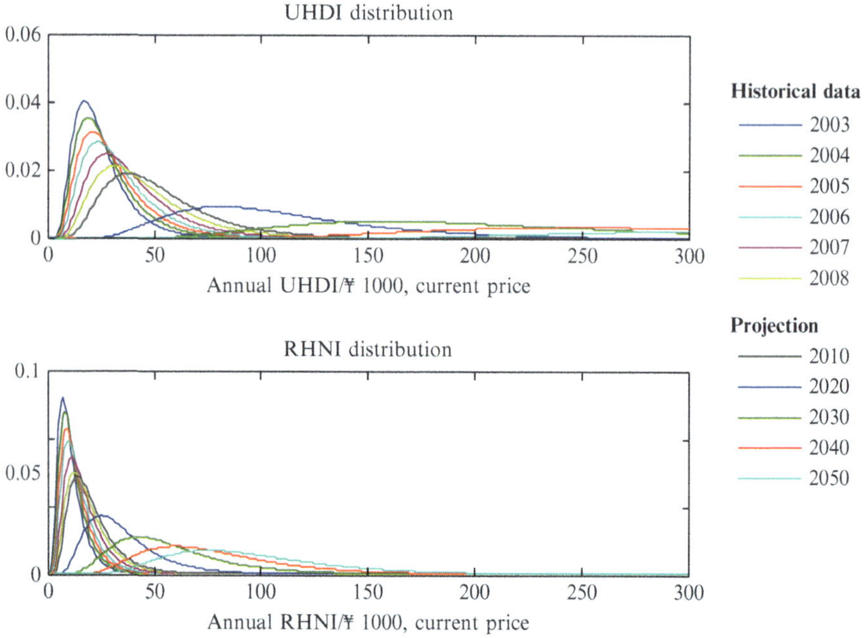

Fig. 3.9 Forecast of distribution of China's future urban household income. *Note*: data for 2005 and 2010 are historical data; the others are forecasts

at prefecture level or above in China (National Bureau of Statistics of the People's Republic of China 2010); we analyzed the inherent relationship between the public bus population and the city's human population; we then made a forecast of China's future public bus population based on the conclusions concerning China's future population and urbanization rate. The urban taxi and bus population is forecast using Eq. 3.5:

$$\mathrm{UPTV}_i = \sum_n \mathrm{UP}_{i,n} \cdot \mathrm{VD}_{i,n} \tag{3.5}$$

where UPTV_i signifies the urban public bus or taxi population in year i, $\mathrm{UP}_{i,n}$ the population of the city of category n in year i (1,000 persons), and $\mathrm{VD}_{i,n}$ the public bus or taxi density of the city of category n in year i (vehicles/1,000 persons).

Data and Scenario Assumption

Figure 3.10 shows the relationship between the population of selected cities at prefecture level and the number of taxis. From this, it is evident that the taxi population generally rises with increasing population; however, in cities with a lower human population, the taxi population is somewhat low. There may be two reasons for this.

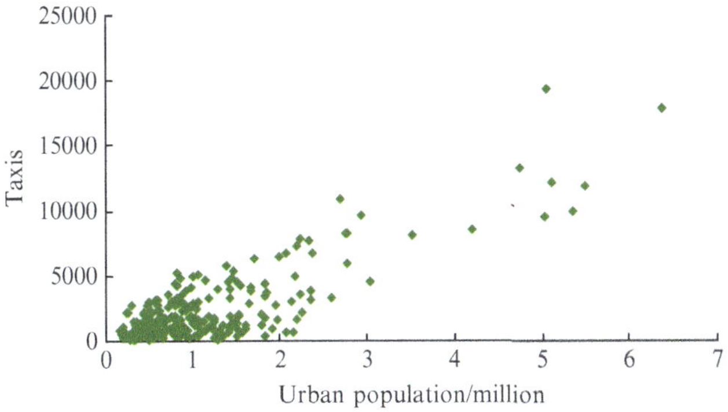

Fig. 3.10 Relationship between taxis and residential population in cities

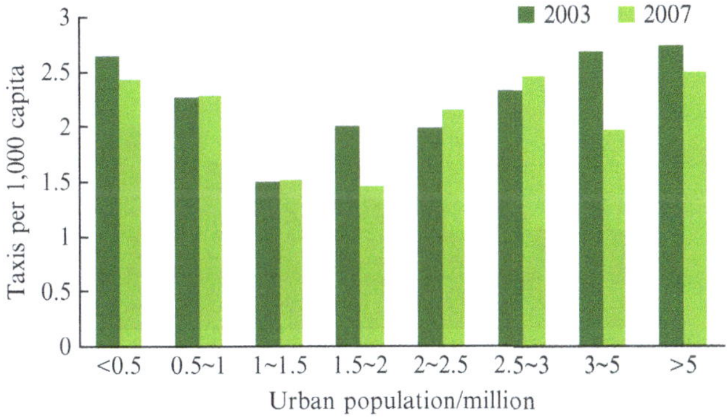

Fig. 3.11 Taxi density in Chinese cities of different scales

First, the urban taxi population is highly subject to related transport planning by local governments, and different local policies may result in large differences in the taxi population; thus, the urban taxi population is to a large extent affected by local factors. Second, motorcycle taxis are widely used in many cities as an alternative to regular taxis; thus, the taxi population does not necessarily reflect the size of a city.

From Fig. 3.10, it is evident that China's cities can be divided into eight levels according to the residential population; a calculation of each level of taxi density is presented in Fig. 3.11. There, it can be seen that the taxi density is relatively high in a small city with a population of under one million residents; with increasing population, the taxi density first falls before rising. The main reason for a small city with a population of under one million people having a relatively high taxi population density is that these cities lack the large-scale public transport systems, such as subway and public buses, and the private passenger vehicle population is relatively low; thus, there is mostly a dependency on taxis for middle- and long-distance travel. Moreover, from a comparison of the data between 2003 and 2007, it can be seen that the taxi density in the same city does not change greatly

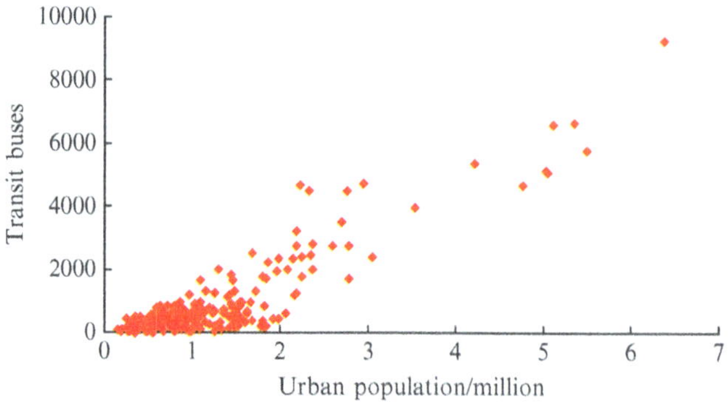

Fig. 3.12 Relationship between transit buses and population in cities

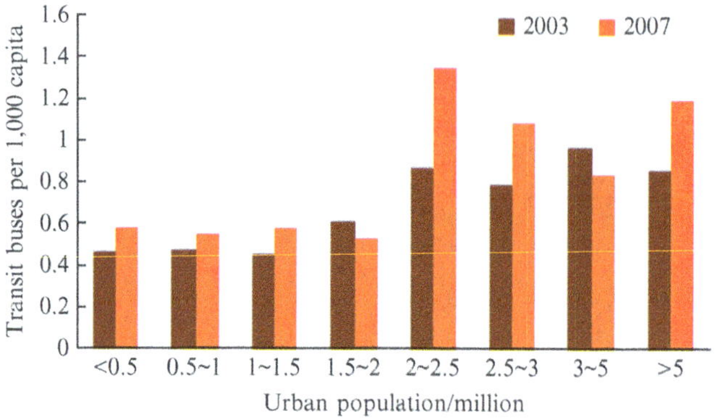

Fig. 3.13 Public bus density of cities with different populations

with time under the same population scale. There may be two reasons for this. First, taxis are a relatively mature type of public transport. Second, since the energy consumption per traffic unit with urban taxis is higher than with buses, local governments implement policies to control the total taxi population.

Next, we analyzed urban public buses using the same method. Figure 3.12 shows the relationship between the human population and the number of public buses in China's cities at prefecture level in 2007. From Fig. 3.12, it can be seen that the bus population generally rises with an increase in the number of city residents; however, there is divergence in the data distribution in the case of cities with smaller populations. The main reason for this is that the number of public buses is closely dependent on a city's public bus system planning and thus is very much subject to local variation.

The results of our analysis are shown in Fig. 3.13 based on the data presented in Fig. 3.12 using the same method as that used for taxis. From this, it is evident that, unlike taxis, the number of transit buses in cities with over two million residents shows a dramatic leap; this indicates that the demand for public transport systems

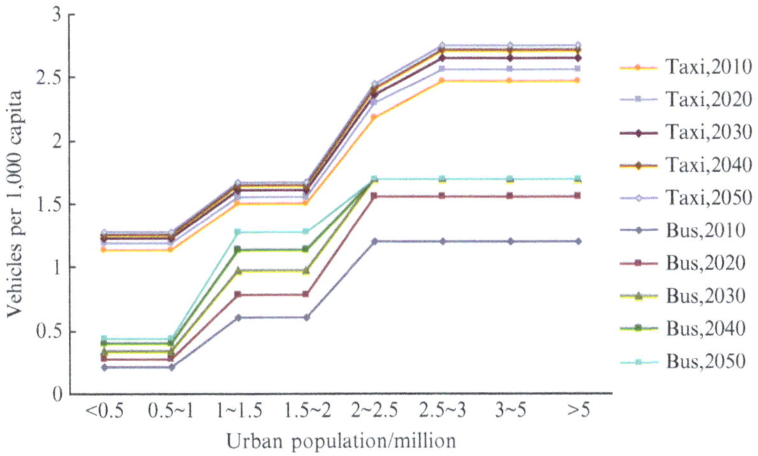

Fig. 3.14 Taxi and public bus density of cities with different populations

rises quickly with increasing city size. At the same time, when comparing data for 2003 and 2007, it can be seen that the public bus density for 2007 is significantly higher than for 2003 with the same urban population size. This indicates that China's rapid development greatly supports the development of the public bus system and that the density of China's public buses will increase further.

Our analysis indicates the relationship between the density of urban taxis and public buses and urban population size. Based on this, our research concludes that China's urban taxi and public bus densities will continue to rise in the future.

As evident in Fig. 3.14, according to our analysis, there is no large potential for growth in the urban taxi density with a constant urban population size, so we conclude that taxis have a low growth rate. However, with public buses, we take the highest public bus density among the population-based city groups to be the average public bus density for 2050 since this city group reflects the largest development potential for public buses. Therefore, we can conclude that all cities with different population sizes have large growth potential for public buses.

Figure 3.15 shows China's future population distribution in cities with different sizes; it assumes the rates for China's future population growth and urbanization presented in Sect. 3.2. Based on the above assumptions, we can forecast China's urban taxi and public bus population using Eq. 3.5.

3.2.2.3 Other Vehicle Population Model

Model

For categories of vehicles other than private passenger vehicles, public buses, and taxis, we investigated the relationship between the population of passenger vehicles

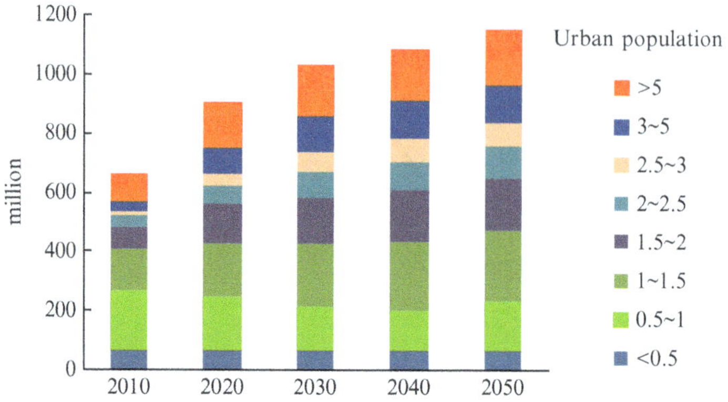

Fig. 3.15 Forecast for China's future urban population distribution

for public affairs, coaches (excluding public buses), and trucks and GDP growth. We propose an index of vehicle population growth elasticity; we then forecast the population of other categories of vehicles based on judging the vehicle population growth elasticity and future macroeconomic trends. In this sub-model, we forecast the population of passenger vehicles for public affairs, coaches, and trucks using Eq. 3.6:

$$\mathrm{EUV}_{i,k} = \mathrm{EUV}_{i-1,k,} \cdot \frac{\mathrm{GDP}_i}{\mathrm{GDP}_{i-1}} \cdot E_{i,k} \tag{3.6}$$

where $\mathrm{EUV}_{i,k}$ signifies the population of the vehicles of category k in year i, GDP_i the GDP in year i, and $E_{i,k}$ the elasticity coefficient of growth of vehicles of category k to GDP growth in year i.

Data and Scenario Setting

Figure 3.16 shows the relationship between the population of categories of vehicles and China's total GDP for 2003–2008. From this, we can see that the population of all categories of vehicles increases as total GDP rises with different growth rates. Among these vehicles, passenger vehicles for public affairs present the fastest growth rate, followed by light and mini-passenger vehicles; other categories of vehicles have a relatively slow growth rate.

To forecast China's future vehicle population using Eq. 3.6, we shall assume vehicle population growth elasticity.

In this study, we assume growth elasticity by considering the history of vehicle population growth in developed countries. Figure 3.17 shows the relationship between coach population growth and GDP growth in developed countries. It is evident in this figure that the coach population growth rate falls with a decreasing

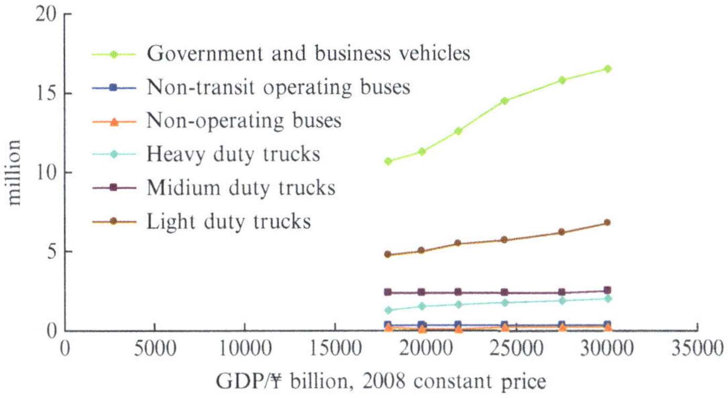

Fig. 3.16 Population growth curves of China's passenger vehicles for public affairs, coaches, and trucks and GDP

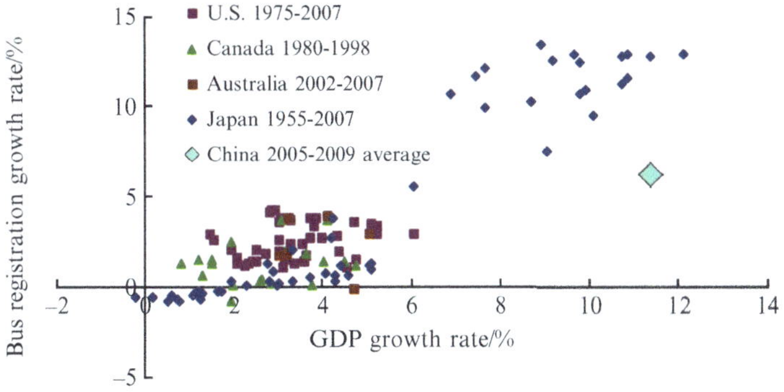

Fig. 3.17 Coach population growth and GDP growth in developed countries

GDP growth: it is less than 5 % when the GDP growth rate is reduced to less than 6 %, and it may fluctuate within a certain range. The average growth rate of the coach population from 2005 to 2009 has been 7 %, and from the experience of developed countries, it may show a further increase with decreasing GDP growth rate.

Figure 3.18 shows the relationship between truck population growth and GDP growth in developed countries. It is evident that the truck population growth rate follows the same pattern as that for coaches. The average growth rate of the truck population from 2005 to 2009 has been 12 %; from the experience of developed countries, China's future truck population will show a further increase with slower GDP growth rate. Generally speaking, China is currently in a period with the biggest population growth elasticity for both coaches and trucks. In the future, with a changing economic structure and saturated demand for passenger and cargo transportation, the population growth elasticity of both coaches and trucks will gradually reduce.

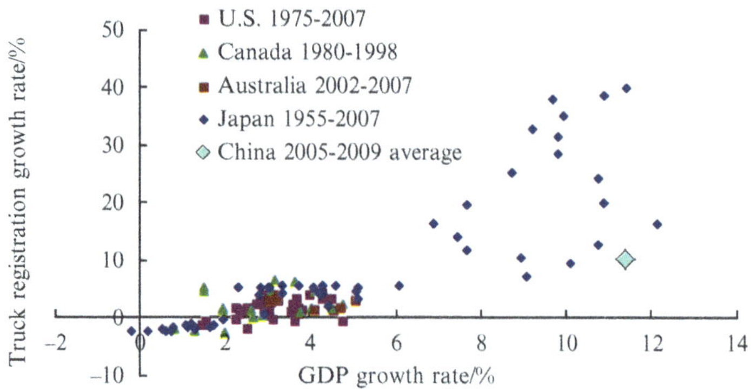

Fig. 3.18 Truck population growth and GDP growth in developed countries

3.2.3 *Vehicle Traffic-Volume Forecast Model*

Vehicle traffic volume is mainly propelled by growing social demand for passenger and cargo transportation. This factor is reflected in the forecast for the vehicle population; accordingly, this study will make a forecast of the vehicle traffic volume using the forecast results for the vehicle population. Equation 3.7 is used to calculate China's vehicle traffic volume:

$$\mathrm{RTS}_i = \sum_k \mathrm{VP}_{i,k} \cdot \mathrm{KPV}_{i,k} \cdot \mathrm{LC}_{i,k} \tag{3.7}$$

where RTS_i signifies the vehicle traffic volume (passenger or cargo transportation) in year i (the passenger transportation's traffic volume unit is in person kilometers, while the cargo transportation's is in ton kilometers), $\mathrm{VP}_{i,k}$ the population of vehicles of category k in year i, $\mathrm{KPV}_{i,k}$ the mileage of each vehicle of category k in year i, and $\mathrm{LC}_{i,k}$ the average passenger capacity of each vehicle of category k in year i.

3.2.3.1 Annual Vehicle Mileage

Surveys relating to the mileage achieved by various categories of vehicles are mainly conducted by the related national departments and research institutions; the survey results may vary depending on the survey region and target groups. Table 3.3 summarizes the estimations of China's annual vehicle mileage made by the main domestic research institutions. The Vehicle Emission Control Center of the State Environmental Protection Administration conducted a vehicle mileage survey on 20 Chinese cities in 2007. In that survey, the number of light buses exceeded 650,000 vehicles, and the survey showed that the annual average mileage for light buses in 2007 in China was 26,900 km. According to related studies, the annual mileage of

Table 3.3 Estimation of annual vehicle mileage in China by various studies (unit: 1,000 km)

Vehicle type	2000	2002	2007	2008
Sedan	24	26	26.9	
Large bus	40	48.6		
Medium bus	35	47.3		
Small bus		33.6		
Minibus		34		
Heavy-duty truck	40	50		65
Medium-duty truck	25	24		40
Light-duty truck	21	20		25
Mini-truck	20	38.4		

Data source: Lin Xiuli et al. (2009), China Academy of Transportation Sciences, MOT, 2009

urban sedans in China before 2005 was 24,000–27,000 km; among those vehicles, the annual mileage of taxis was 90,000–115,000 km and that of other categories of vehicles was 16,000–18,000 km. According to the national road transportation census in 2008 carried out by the Ministry of Transport, the annual mileage of heavy trucks was 65,000 km and that of light trucks was 25,000 km. These results may be higher than the average annual mileage of all trucks since the survey targets were all commuter vehicles. Compared with developed countries, China's annual mileage for private passenger vehicles is relatively high, and this may gradually reduce with the country's rapidly developing public traffic systems. The annual mileage for urban taxis mainly depends on daily working hours and average speeds; the annual mileage for urban buses and long-distance coaches largely depends on driving routes; the annual mileage of passenger vehicles for public affairs and trucks mainly depends on the purposes for which they are used, but the annual mileage for these vehicles does not in general change significantly. The present study intends to forecast the change in the future mileage of private passenger vehicles.

3.2.3.2 Forecast of Private Passenger Vehicle Mileage

Since there is a high population of private passenger vehicles and they have larger growth potential than other vehicles, the mileage of this category of vehicle is an important factor in terms of vehicle energy demand. The annual mileage of private passenger vehicles can be divided into short-distance and long-distance mileage according to purpose. The main purpose of urban short-distance travel is daily travel (such as commuting to work) and weekend suburban travel; this mileage is mainly affected by the development of public transport. Mileage in long-distance travel relates to intercity travel, and it is mainly affected by the development of long-distance coaches and high-speed railways. Driven by such economic factors as increasing oil prices and such policy factors as fuel taxes, the cost of using private passenger vehicles has increased with decreasing competitive power. In addition, China's urban public transport and intercity high-speed railways will undergo rapid development and achieve speed and comfort levels comparable with those of private passenger vehicles; thus, public transport should take over some of the mileage from the latter category. It can be anticipated that the annual mileage of China's private passenger vehicles will gradually undergo a reduction. The present study adopts a model based on residents' travel to forecast the annual mileage for private passenger vehicles. The annual mileage for urban or rural private passenger vehicles can be expressed by Eq. 3.8:

$$\mathrm{KPV} = \mathrm{UTF} \cdot \mathrm{UTD} \cdot \mathrm{US} + \mathrm{ITF} \cdot \mathrm{ITD} \cdot \mathrm{IS} \tag{3.8}$$

where KPV signifies the annual mileage of each private passenger vehicle, UTF the urban travel frequency of car owners, UTD the average distance traveled by car owners, US the percentage of traveling by car owners in private passenger vehicles, ITF the intercity travel frequency of car owners, ITD the average intercity travel distance of car owners, and IS the percentage of intercity traveling by car owners in private passenger vehicles.

According to the survey results on travel by the residents of some cities, the average number of trips made by urban residents on a travel day was two to three, while the equivalent figure for rural residents was 0.5. Considering that the average travel time for car owners is usually higher than that for non-car owners, the present study assumes the daily number of trips made by car-owning urban residents on a travel day to be four, while for rural car owners it is two. The average urban travel distance depends on several factors, such as city structure and size. We have made assumptions on the average travel distances in China's large, medium-sized, and small cities and in rural areas for the year 2010 based on survey results on travel by the residents of some cities. We also assume that the average urban travel distance will increase with greater city size. Since the current public transport systems in China are unable to compete with private passenger vehicles in terms of speed and comfort, most car owners travel in their vehicles. However, with the development of high-quality public systems, such as subways, these can be expected to provide alternatives for the urban use of private passenger vehicles.

For intercity travel, we can make approximate estimates based on the use of commuter coaches. According to the China Statistical Yearbook, the country's road passenger capacity by commuter coaches in 2009 was 27.8 billion person trips—about 20 trips per person—and the average distance traveled was 49 km. This figure mainly reflects the passenger transport conditions of intercity commuter coaches, and most of these passengers are people without cars. The present study assumes that residents with and without cars have the same intercity travel characteristics; the intercity travel frequency of Chinese people with cars in 2010 was 20 trips per year with an average distance traveled of 50 km. Table 3.4 shows the assumptions of the present study in detail.

The change in the percentage of people traveling by private passenger vehicles is the main reason for the change in the mileage of such vehicles. In this study, we provide reference scenarios and comprehensive substitution policy scenarios to reflect the different percentages of residents' traveling by private passenger vehicles. Under the reference scenario, 98 % of travel by urban residents with cars in 2010

Table 3.4 Travel frequency and average distance traveled by Chinese people with cars in 2010

Item	Large city	Medium-sized city	Small city	Rural area
Average urban trips per day	4	4	4	2
Average urban travel distance per day (km)	12	6	2	1
Average intercity trips per day	10	10	10	10
Average intercity travel distance per day (km)	50	50	50	50

was by private passenger vehicles, and the equivalent figure for intercity travel was 90 %. By 2050, the proportion of residents with cars in large, medium-sized, and small cities traveling by private passenger vehicles will be reduced to 60, 60, and 80 %, respectively; the percentage of their intercity travel in such vehicles will be reduced to 50 %. Under the comprehensive substitution policy scenario, the percentage of intercity travel in private passenger vehicles by residents with cars is reduced at a rate 1.5-fold faster than under the reference scenario. The annual mileage of private passenger vehicles in large, medium-sized, and small cities and rural areas can be calculated using Eq. 3.8, and then the weighted average value of the annual mileage of private passenger vehicles can be determined based on the vehicle population.

3.2.3.3 Average Passenger (Cargo) Capacity of Vehicles

According to Eq. 3.7, the vehicle traffic volume is the product of the vehicle population, the average mileage of each vehicle, and the average passenger (cargo) capacity of each vehicle. Based on China's historical vehicle traffic volume concluded from the China Statistical Yearbook, the average passenger (cargo) capacity of each vehicle can be deduced using Eq. 3.7. Since the passenger (cargo) capacity of each vehicle changes slowly, and the change in the average passenger (cargo) capacity of each vehicle and that of the vehicle population are linked to each other in terms of the use of commuter coaches, we assume there is no change.

3.3 Scenario Analysis

3.3.1 Scenario Definition

The reference scenarios and comprehensive policy scenarios employed in this section are the same as those in Chap. 12 for the vehicle energy development scenario. Chapter 12 presents the scenario definitions in detail, and the present section describes only the content related to vehicle population and vehicle traffic volume.

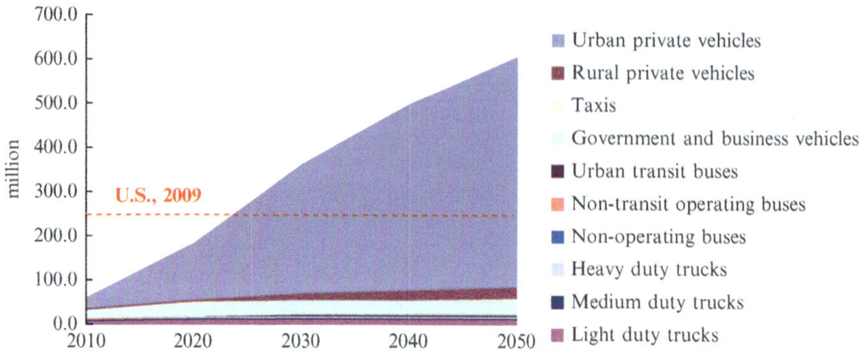

Fig. 3.19 Forecast results for China's vehicle population

Reference scenarios form the basis for the comprehensive policy scenarios. Under the reference scenario, the vehicle population, increase, and mileage depend on the macroeconomic conditions and population and geographic factors; they include GDP growth rate, growth and structural change in the population, and urbanization rate. These factors vary with changes in macroeconomic conditions, population, and geographic factors without regard to restrictions of resources, the environment, security, fairness, and vehicle-use management policy.

Comprehensive policy scenarios are the target of reference scenarios. Under comprehensive policy scenarios, private passenger vehicles are subject to development of public transport and intercity transport systems; thus, the traffic volume of such vehicles becomes significantly reduced. With the rapid development of China's high-speed railway network, the original low-speed railways will be used mainly for cargo transport. Railway cargo transport will compete with long-distance truck transport by virtue of the low transport costs and high reliability. Thus, the development of the railway cargo transport system will lead to a reduction in the cargo service volume of trucks.

3.3.2 Vehicle Population

Figure 3.19 shows the projected changes in China's vehicle population.

The period from 2010 to 2020 is the one that displays the fastest growth in China's vehicle population; China's vehicle population may reach 270 million vehicles by 2020 at an average annual growth rate of 13 %. Among these, passenger vehicle will show the fastest growth rate: the figure will reach 240 million vehicles by 2020, accounting for 89.1 % of the total vehicle population. The population of coaches and trucks will amount to 3.28 and 26.27 million vehicles, respectively, by 2020, accounting for 1.2 and 9.7 % of the total population. The vehicle population

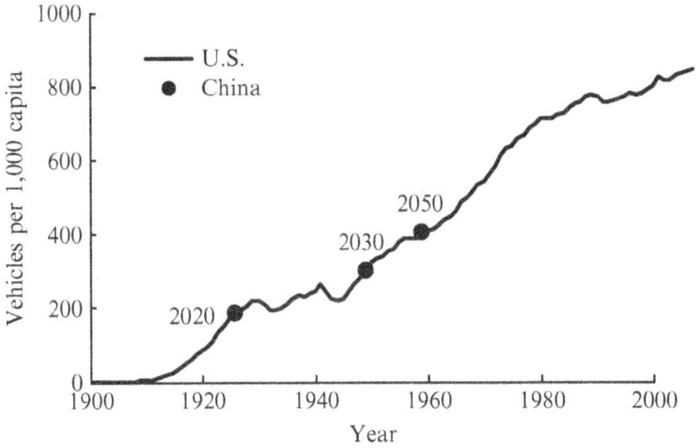

Fig. 3.20 Comparison of vehicle ownership between United States and China

growth rate will slow down over the period 2020–2030, and the total vehicle population will reach 440 million vehicles by 2030 at an average annual growth rate of 5 %. The passenger vehicle population will amount to 400 million vehicles by 2030, representing an increase to 91 % of the total vehicle population. The population of coaches and trucks will become 4.03 and 36.74 million vehicles, respectively, accounting for 0.9 and 8.3 % of the total population. The vehicle population growth rate will further diminish over the period 2030–2050. By 2050, the total vehicle population may attain 588 million vehicles—an approximately 7.4-fold increase over the figure for 2010; this would comprise 550 million passenger vehicles, 4.1 million coaches, and 36.8 million trucks.

According to the assumptions of this study relating to vehicle population forecasts and population growth, as shown in Fig. 3.20 (Davis et al. 2010), the vehicle population per 1,000 persons in China by the years 2020, 2030, and 2050 will amount to 189, 300, and 403 vehicles, respectively—equivalent to the levels for the United States in 1926, 1949, and 1959. Therefore, we can conclude that China's vehicle population rate will contain to maintain a large gap relative to that of developed countries owing to limitations of resources, the environment, and other factors in addition to the large population base.

Figure 3.21 shows a comparison of forecast results for China's vehicle population as determined in domestic and foreign studies. From that figure, it is evident that the forecast value predicted by the present study for 2020 is lower than that of UC Davis, though it is higher than the value found by other studies. The various studies are significantly different in their forecast values for 2030; among them, the present study reported the highest value. This study's forecast for 2050 is equivalent to that of the US Argonne National Laboratory and the 2050 China Energy and Carbon Emission Research Group. Since each investigation has its own particular forecast method and basic data, we need to compare the forecast results according to

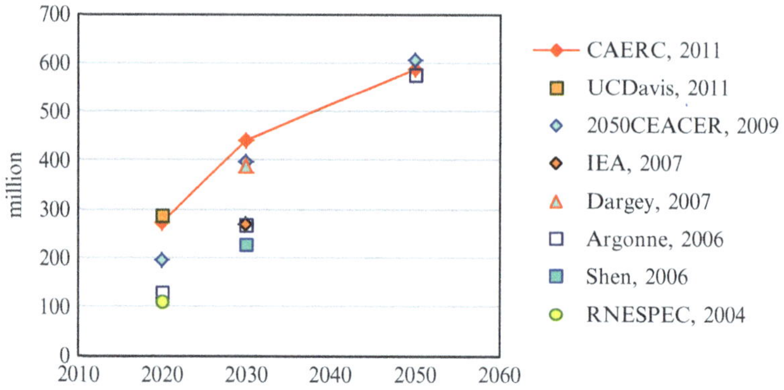

Fig. 3.21 Comparison of forecast results among domestic and foreign studies on China's vehicle population

the methods employed and the basic assumptions. Generally speaking, the forecast results of earlier studies are lower because of their relatively conservative estimates with regard to China's vehicle output and sales volume and population growth rate. Using more up-to-date data, later studies are better able to reflect the high-speed growth of China's vehicle population in recent years, and they thus derive higher forecast values.

It should be noted that the UC Davis study adopted forecast results based on the average growth rate of seven countries; the study by the 2050 CEACER (China Energy and CO_2 Emissions Report) group adopted forecast results using vehicle population figures from a basic scenario; the IEA study adopted forecast results using vehicle population figures from a reference scenario; the Argonne National Laboratory study adopted forecast results using a scenario with a saturated vehicle population. It should also be stated that the forecast results regarding vehicle population provided by the present study are closely related to a specific scenario assumption; thus, the forecast values will change with changes in the scenario assumption.

Figure 3.22 indicates the influence of changes in household income on the private vehicle population under the scenario assumption. From this figure, it is evident that the forecast results on the private passenger vehicle population up to the year 2020 change by ±15 % when China's household income changes by ±20 %. With time, the change in household income will have an increasingly weaker influence on the forecast results for the private passenger vehicle population: by 2050, with a household income change within the range of ±15 %, the forecast results on private passenger vehicle population will change by only ±5 %. Since there is uncertainty with respect to the growth of China's vehicle population, the forecast results of the present study reflect the influence of various factors on vehicle population.

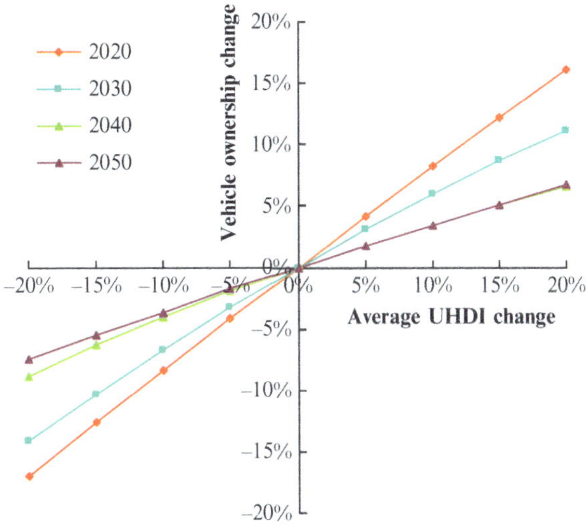

Fig. 3.22 Influence of changes in household income on the private vehicle population in the scenario assumption

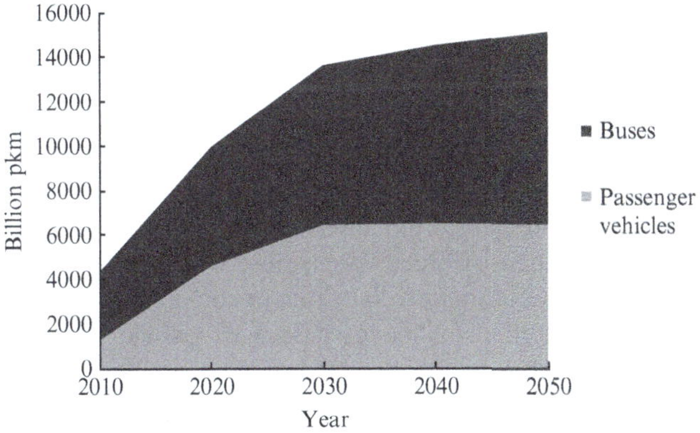

Fig. 3.23 Vehicle passenger transport under reference scenarios

3.3.3 *Traffic Volume of Vehicle Passenger Transport*

Figure 3.23 presents the change in China's traffic volume of vehicle passenger transport under reference scenarios. The period from 2010 to 2020 is that with the fastest growth in the traffic volume of passenger transport. The traffic volume of vehicle passenger transport will rise to 9,914 billion person km by 2020—a rise from 4,310 billion person km in 2010; among that volume, passenger vehicles will account for 5,322 billion person km by 2020, accounting for 53.7 % of the total traffic volume. From 2020 to 2030, the growth rate in the traffic volume of passenger

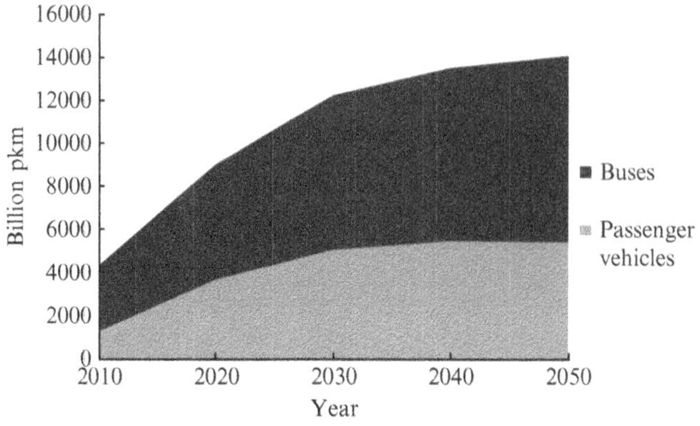

Fig. 3.24 Passenger vehicle transport under comprehensive policy scenarios

vehicle transport will slow down: it will reach 13,605 billion person km by 2030, of which 47.4 % of the traffic volume will be accounted for by passenger vehicles and 52.6 % by buses. From 2030 to 2050, the growth rate in the traffic volume of passenger vehicle transport will slow down further: it will reach 15,096 billion person km by 2050, of which 42.8 % of the traffic volume will be accounted for by passenger vehicles and 57.2 % by buses. The percentage of passenger vehicles in the traffic volume will show an increase from 2010 to 2050, reflecting the scenario assumption of a rapidly increasing passenger vehicle population.

Figure 3.24 shows the change in China's traffic volume of vehicle passenger transport under comprehensive policy scenarios. Compared with the reference scenarios, the growth rate in traffic volume under comprehensive policy scenarios is relatively low. The traffic volume of passenger vehicle transport will reach 8,995; 12,224; and 14,053 billion person km by year 2020, 2030, and 2050, respectively; this is equivalent to 90.7, 89.9, and 93.1 % of the volume under the reference scenarios during the same periods. The decrease results from the traffic volume of passenger vehicles, which will reach 3,673; 5,067; and 5,425 billion person km by 2020, 2030, and 2050, respectively; this is equivalent to 80.0, 78.6, and 83.9 % of the volume under the reference scenarios during the same periods.

From a comparison of the traffic volume of passenger vehicle transport under the reference and comprehensive policy scenarios, it is evident that if vehicle population growth is to be maintained, the traffic volume—especially the traffic volume of private passenger vehicles—will have to undergo a significant reduction. The decrease in the traffic volume for passenger vehicle transport will be mainly compensated by urban mass transit, based on the subway; this will result in reduced average travel times and distances with private passenger vehicles.

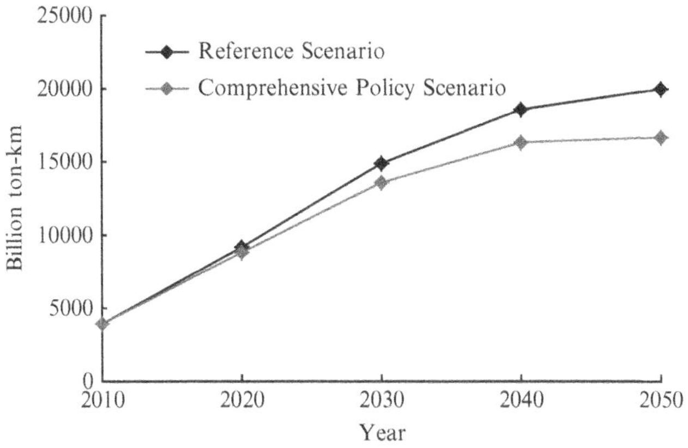

Fig. 3.25 China's traffic volume for cargo vehicle transport under reference and comprehensive policy scenarios

3.3.4 Traffic Volume of Vehicle Cargo Transport

Figure 3.25 shows China's traffic volume for cargo vehicle transport under reference and comprehensive policy scenarios. Under reference scenarios, the traffic volume of cargo vehicle transport shows a rapidly rising trend: it will reach 9,077 billion ton km, 14,765 billion ton km, and 19,924 billion ton km by 2020, 2030, and 2050, respectively.

Under comprehensive policy scenarios, the traffic volume of cargo vehicle transport will basically become saturated after 2040. It will amount to 87,420; 13,526; and 16,571 billion ton km by 2020, 2030, and 2050, respectively; this is equivalent to 96.3, 91.6, and 83.2 % of the same periods under reference scenarios.

From a comparison of the traffic volume for vehicle cargo transport under reference and comprehensive policy scenarios, the traffic volume of cargo vehicle transport will also undergo a large reduction. The decrease in the traffic volume of cargo vehicle transport will be mainly compensated by the traffic volume of railway cargo transport.

3.4 Conclusions

1. The growth in China's vehicle population and traffic volume is subject to various complex factors, including macroeconomic development, urbanization, population change, fluctuations in vehicle prices, and traffic policy. With a gradually rising income level of the population, increasing demand for passenger

and cargo transport, and a national policy to stimulate the development of the auto industry, China's vehicle population has undergone a rapid increase over the past 10 years. In coming decades, macroeconomic development will still be the internal motivating force for the growth of China's vehicle population and traffic volume. However, in the meantime, the rapidly growing vehicle population and traffic volume will also put great pressure on China's urban traffic and energy supply in addition to other aspects. Some cities, such as Beijing and Shanghai, have adopted administrative measures to control the vehicle population in an effort to relieve the pressure from urban traffic. Such measures will exert an obvious impact on the growth of China's vehicle population. On the whole, while maintaining current macroeconomic and national policy trends, China's vehicle population will continue its rapid increase for quite some time in the future.

2. According to the forecasts of the current study, China's vehicle population will amount to 270, 440, and 590 million vehicles by 2020, 2030, and 2050, respectively. Private passenger vehicles will be the main force for the growth of the total vehicle population: the private passenger vehicle population may reach 240, 400, and 550 million vehicles in 2020, 2030, and 2050, respectively; its proportion of the total population will increase in the future. In addition, China's bus and truck population will also undergo a major increase. The bus population will amount to 3.28, 4.03, and 4.1 million vehicles by 2020, 2030, and 2050; the truck population will reach 26.27, 36.74, and 36.8 million vehicles by 2020, 2030, and 2050. Though China's vehicle population will rise rapidly in the future—influenced by population, resource, environmental, and other factors—China's vehicle population level will remain far behind that of developed countries in Europe and the United States; it will therefore be difficult for China to attain the current vehicle population level in those countries.
3. In accordance with the rapidly growing vehicle population, China's vehicle traffic volume will also undergo rapid growth. Under reference scenarios, the traffic volume for passenger vehicle transport will reach 9,914, 13,605, and 15,096 billion person km by 2020, 2030, and 2050, respectively; under comprehensive policy scenarios, the values will amount to 8,995; 12,224; and 14,053 billion person km, respectively. Under reference scenarios, the traffic volume for cargo vehicle transport will reach 9,077; 14,765; and 19,924 billion ton km by 2020, 2030, and 2050, respectively; under comprehensive policy scenarios, the values will amount to 8,742; 13,526; and 16,571 billion ton km, respectively. Rapid growth in the vehicle traffic volume signifies rapid growth in vehicle energy demands and increasing emissions of GHGs and general contaminants. Therefore, it is necessary to reduce unit consumption and introduce low-carbon alternative fuels to cope with such challenges.
4. Compared with reference scenarios, China's vehicle population under comprehensive policy scenarios shows no change. The reason for this is that one of the basic principles of comprehensive policy scenarios is to protect rapid development of the domestic auto industry to maintain the stimulating effect of that industry on the national economy; however, under comprehensive policy scenarios, the growth of the vehicle population is supposed to be the same

as under reference scenarios. However, in terms of vehicle traffic volume, the population under comprehensive policy scenarios is lower than under reference scenarios. Under reference scenarios, the traffic volume of passenger vehicles and cargo transport by 2050 will be 3.5- and 5.2-fold higher, respectively, than in 2010, and there will be a 3.3- and 4.3-fold respective increase in the traffic volume of passenger vehicles and cargo transport. The reason for the vehicle traffic volume under comprehensive policy scenarios being lower than under reference scenarios is that the demand for vehicle traffic volume is supposed to be diminished by reduced vehicle use through traffic-management measures, such as policies, regulations, and alternative modes of transport. The decreased vehicle traffic volume is transferred to other transport modes through various methods, for example, urban traveling by bus instead of by private passenger vehicles, intercity travel by rail instead of by private passenger vehicle, and long-distance cargo transport by truck being transferred to rail. By transferring the traffic volume from private passenger vehicle traffic, with relatively high energy consumption, to public transport and rail traffic, with relatively low energy consumption, it should be possible to greatly conserve energy and reduce emissions.

5. From a comparison of reference and comprehensive policy scenarios, we can conclude that traffic-demand management is an important measure for reducing vehicle traffic volume, thereby reducing vehicle energy demand and carbon emissions. To ensure a normal increase in the vehicle population, the vehicle traffic volume can be radically decreased using various traffic-management measures. Such measures may include the following: limiting the use of private passenger vehicles, increasing the use of public transport, promoting the development of urban and intercity high-speed rail systems as alternatives for travel by passenger vehicles, and developing rail cargo transport, especially long-distance rail cargo transport, as an alternative to long-distance truck transport.

References

Aitchison J, Brown JAC (1963) The lognormal distribution. Cambridge University Press, Cambridge

Australian Bureau of Statistics (2009) Motor vehicle census 2009. Australian Bureau of Statistics, Canberra

China Academy of Transportation Sciences (2009) Investigation of vehicle travel of operating vehicles. Presentation on the Energy Foundation conference, Beijing

China Association of Automotive Manufactures (2011) China automotive industry yearbook 2011. China Automotive Industry Yearbook Press, Beijing

China Energy and Carbon Emission Research Group (2009) 2050 China energy and carbon emission report. Science Press, Beijing

Dargay J, Gately D, Sommer M (2007) Vehicle population and income growth, worldwide: 1960–2030. Energy J 28(4):143–170

Davis SC, Diegel SW, Boundy RG (2010) Transportation energy data book, 29th edn. Oak Ridge National Laboratory, Oak Ridge

Heston A, Summers R, Aten B (2009) Penn world table version 6.3. Center for International Comparisons of Production, Income and Prices at the University of Pennsylvania, Philadelphia

Huo H, Wang M (2011) Modeling future vehicle sales and stock in China. Energy Policy 43(4):17–29

Huo H et al (2007) Projection of China's motor vehicle growth, oil demand, and CO_2 emissions through 2050. Transportation Res Rec 2038:69–77

International Energy Agency (2007) World energy outlook 2007. IEA, Paris

Japan Automobile Manufacturers Association (2009) Motor vehicle statistics of Japan 2009. Japan Automobile Manufacturers Association, Inc., Tokyo

Lin Xiuli et al (2009) China's motor vehicle mileage distribution rule. Environ Sci Res 22(3): 377–380

Ministry of Transport of the People's Republic of China (2008) Compilation of national highway and waterway transportation volume special survey data. China Economic Publishing House, Beijing

National Bureau of Statistics of the People's Republic of China (2010) China city statistics yearbook 2010. China Statistics Press, Beijing

National Bureau of Statistics of the People's Republic of China (2011) China city statistics yearbook 2011. China Statistics Press, Beijing

National Development and Reform Commission (2004) China energy comprehensive development strategy and policy research. http://www.efchina.org/CSEPCN/FReports.do. Accessed 17 Jan 2013

North American Transportation Statistics (2010) Transportation vehicle. http://nats.sct.gob.mx. Accessed 17 Jan 2013

Shen Zhongyuan (2006) Forecasting China's vehicle population by income distribution curve. Environ Sci Res 28(8):11–15

Wang Y, Teter J, Sperling D (2011) China's soaring vehicle population: even greater than forecasted? Energy Policy 39(6):3296–3306

Chapter 4
Vehicle Powertrain Technology

Wang Hewu, Du Jiuyu, and Ouyang Minggao

Abstract In China, the internal-combustion engine accounts for over 99 % of vehicle power technology, and over 99 % passenger cars and 73 % commercial vehicles are powered by gasoline and diesel engines, respectively; the diesel-powered passenger cars are mainly SUVs and gasoline commercial vehicles mainly consist of mini-trucks, light trucks, and buses. The vehicle's power system is in the initial stage of electrification, but over 50 % of two-wheeled motor are electrified.

There is an upward tendency in gasoline and diesel consumption from Chinese vehicles, but the increase in gasoline consumption is lower than with diesel; vehicles consumed 85 % gasoline and 40 % diesel. There is a bottleneck in the development of diesel-powered passenger car, that is, the poor diesel quality.

Start-stop and ISG technologies can make 5–12 % and 20 % reductions in fuel consumption, respectively, and meet regulations of Chinese phase III fuel economy level. HEV technology is more widely used in commercial vehicles than in passenger cars in China. PHEV vehicle and extended-range electric vehicle technology have been applied only in a few vehicle models, still at the stage of technical verification. Hydrogen fuel-cell technology has been technically verified for urban buses and other passenger cars.

Wang Hewu (✉) • Du Jiuyu • Ouyang Minggao
China Automotive Energy Research Center, State Key Laboratory of Automotive Safety and Energy, Tsinghua University, Beijing 100084, P.R. China
e-mail: wanghw@tsinghua.edu.cn

CAERC, Tsinghua University, *Sustainable Automotive Energy System in China*,
DOI 10.1007/978-3-642-36847-9_4,

4.1 Current Status of Chinese Vehicle Powertrains

4.1.1 Development Status of Chinese Vehicles and Powertrain Constitution

The civilian vehicle population of China in a broad sense, including 12.84 million low-speed vehicles (three-wheeled vehicles and low-speed trucks, Sidebar 4.1; National Bureau of Statistics of China 2011a, c), was 90.86 million vehicles at the end of 2010. The human population was 1.37 billion (National Bureau of Statistics of China 2011d); the number of vehicles amounted to 67.3 per 1,000 people, which is half the global average. Figure 4.1 presents an overview of the development process. In the period of over 20 years from 1978, when China began to implement its reform and opening-up policy, to 2000, the growth rate in the number of vehicles showed a relatively stable trend, with an average annual growth rate of 11.6 %. However, during the first 10 years of the twenty-first century, the growth rate in the vehicle population increased substantially. From year 2000 to 2005, the average annual growth rate in the number of vehicles was 14.5 %; the vehicle population during those 5 years doubled. From 2005 to 2010, the average annual growth rate rose to 19.8 %; the vehicle population during those 5 years showed a 1.77-fold increase. Though the number of low-speed vehicles during that period decreased to a certain degree, the overall vehicle population increased from 51 to 90.46 million from 2005 to 2010—an increase of 77.4 %.

Sidebar 4.1: Low-Speed Vehicles

Low-speed vehicles are those whose maximum speed is below 45 km/h. Before 2004, they were referred to as agricultural vehicles, and they are widely used for transporting materials in rural areas. They include both three-wheeled and four-wheeled agricultural vehicles. In 2004, a national standard "Safety Specifications for Power-Driven Vehicles Operating on Roads" was implemented: three-wheeled agricultural vehicles were renamed three-wheeled vehicles and four-wheeled agricultural vehicles were renamed low-speed trucks—the two categories were together referred to as low-speed vehicles. Low-speed vehicles, passenger cars, and trucks were listed as the three major vehicle categories. The administrative authority of low-speed vehicles was changed from the Agricultural Machinery Department to the Public Security Department. Owing to the changes in statistical coverage, there are no official statistics for the number of low-speed vehicles before 2004.

It is evident in Fig. 4.1 that the composition of China's vehicle population has undergone significant change. The quantity and proportion of passenger cars has

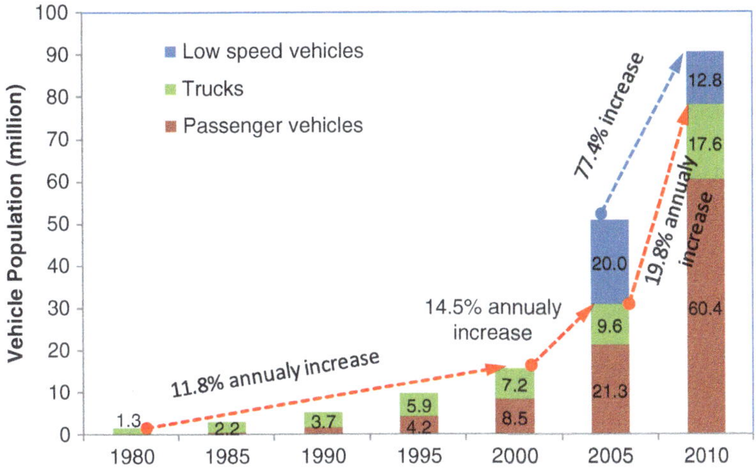

Fig. 4.1 Chinese vehicle population and growth over the past 30 years

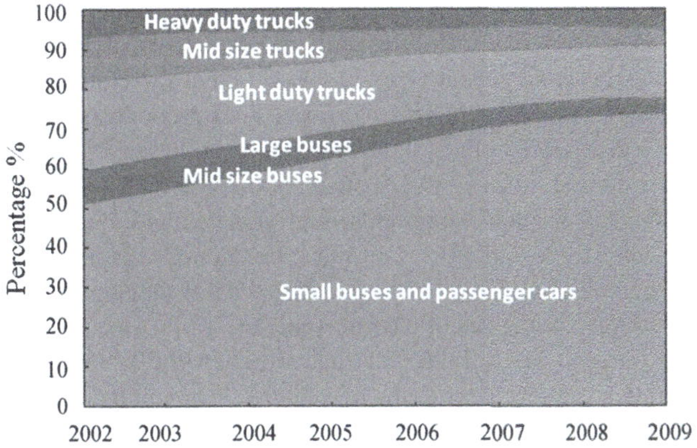

Fig. 4.2 Composition of Chinese vehicle population from 2002 to 2009

increased substantially. This has been mainly due to the large increase of passenger car sales through purchases by families. Figure 4.2 shows the changing composition of the vehicle population in China from 2002 to 2009 (National Bureau of Statistics of China 2011c). A significant characteristic is that the proportion of passenger cars (namely, passenger cars, including private passenger cars, commercial and service passenger cars, taxis) increased rapidly—from 50.7 % in 2002 to 73.9 % in 2009.

The situation in China with respect to vehicle powertrain is the same as that elsewhere in the world. Although various fuels are used, energy conversion is achieved almost entirely through the internal-combustion engine. Statistics from

Table 4.1 Vehicle powertrain technology in China

		2005	2006	2007	2008	2009	2010
Calculated in-vehicle sales (%)	Diesel vehicles	24.91	23.03	23.19	23.15	20.33	20.06
	Gasoline vehicles	75.06	76.91	76.78	76.83	79.73	79.83
	Other fuels vehicles	0.02	0.05	0.02	0.03	0.04	0.11
Calculated in power output of engine installed (%)	Power output of diesel engine	–	–	33.67	32.73	28.84	30.72
	Power output of gasoline vehicles	–	–	66.28	67.20	71.10	69.22
	Others engines	–	–	0.05	0.07	0.06	0.05

the China Association of Automobile Manufacturers (2011) show that the number of natural gas vehicles in China between 2005 and 2009 (including liquefied and compressed natural gas vehicles) increased from 127,000 to 400,000; the number of liquefied petroleum gas vehicles decreased from 110,000 to 70,000. The total number of gas-fueled vehicles increased from 237,000 to 470,000. The proportion of gas-fueled vehicles among the total vehicle population (31.6 million in 2005 and 62.81 million in 2009—both excluding low-speed vehicles) remained fairly constant at about 0.75 %; the remaining 99.25 % of vehicles used gasoline and diesel.

It can be seen from vehicle sales data (China Association of Automobile Manufacturers 2011) from recent years (Table 4.1) that there has been an increase in the proportion of gasoline vehicles in terms of sales proportion—from 75.06 % in 2005 to 79.83 % in 2010. By contrast, the proportion of diesel vehicles has shown an overall downward trend—from 24.91 % in 2005 to 20.06 % in 2010. The proportion of vehicles using other fuels (alternative-fuel vehicles and electric vehicles) has remained low—mostly at under 0.1 % of sales. It is notable that the sales proportion of vehicles using other fuels in 2010 exceeded 0.1 % for the first time amounting to 0.11 % owing to increased sales of electric vehicles.

It can also be seen from Table 4.1 that although the proportion of gasoline vehicles amounts to 75–80 % in terms of vehicle sales volume, the results appear very differently when analyzed with respect to the power of the vehicle's engine. From 2007 to 2010, the total power output of gasoline engines accounted for 66–71 % of the total vehicle engine power; the total power output of diesel engines accounted for 28.8–33.6 %; the installed power of engines using other fuels accounted for only 0.05–0.07 %.

4.1.2 Powertrain Technology of Chinese Passenger Cars

Passenger cars include basic passenger cars, multifunction vehicles, multipurpose passenger cars, and cross-type vehicles (Sidebar 4.2). Gasoline vehicles are overwhelmingly dominant among the passenger cars produced and sold in China. As evident in Table 4.2 (China Association of Automobile Manufacturers 2010), after 2006, the proportion of gasoline vehicles among all passenger cars exceeded 99 %;

Table 4.2 Type of powertrain used in China's passenger cars produced and sold

Type	Production									
	Quantity (thousand)					Proportion (%)				
Type	2006	2007	2008	2009	2010	2006	2007	2008	2009	2010
Gasoline	5,189	6,336	6,697	10,299	13,757	99.17	99.29	99.40	99.18	98.99
Diesel	41	44	39	81	126	0.79	0.70	0.58	0.78	0.91
Alternative	2	0.3	0.8	3	13	0.043	0.005	0.013	0.03	0.096
Total	5,233	6,381	6,737	10,383	13,897					
	Sales									
	Quantity (units)					Proportion (%)				
Type	2006	2007	2008	2009	2010	2006	2007	2008	2009	2010
Gasoline	5,134	6,253	6,712	10,247	13,619	99.20	99.30	99.36	99.19	98.99
Diesel	39	43	42	80	125	0.75	0.69	0.63	0.78	0.91
Alternative	2	0.4	0.9	3	12	0.042	0.007	0.013	0.03	0.093
Total	5,175	6,297	6,755	10,331	13,757					

the proportion of diesel vehicles was less than 1 %; the proportion of vehicles powered by other means—including natural gas vehicles, liquefied petroleum gas (LPG) vehicles, and battery electric vehicles—was even lower. It is notable that there was a rise in the number of battery electric vehicles in 2010; they pushed the number of passenger cars using other forms of power over 13,000—an increase of 10,000 compared with 2009—and the proportion rose to almost 0.1 %.

Sidebar 4.2: Cross-Type Passenger Cars

Cross-type passenger cars are passenger car types in addition to general vehicles, multipurpose vehicles (MPVs), and sports utility vehicles (SUVs). These vehicles were mainly referred to as mini passenger cars in the old classification.

Diesel passenger cars are mainly concentrated in SUV types. As shown in Fig. 4.3, the number of diesel passenger cars in 2005 was 41,000, and this figure fluctuated within a small range of 40,000–43,000 over the next 3 years. However, in 2009 and 2010, the sales volume showed a sudden increase to 79,000 and 113,000, respectively. Compared with the previous years, the growth rates were both above 60 %. This growth was due to the increase in sales of diesel SUVs: the sales volume of diesel SUVs in 2008 was 25,000, though the sales volumes in 2009 and 2010 increased, respectively, to 49,000 and 95,000. The increase in sales of diesel MPVs played a certain role in promoting the increase in the population of diesel passenger cars: the number of diesel MPVs was 2,400 in 2008, though the figures for 2009 and 2010 amounted to, respectively, 14,000 and 15,000 vehicles. It can also be seen in Fig. 4.3 that the number of diesel-powered basic passenger cars from 2005 to

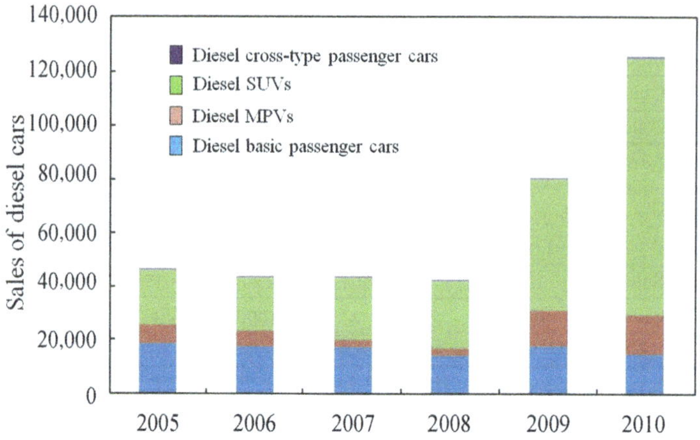

Fig. 4.3 Changes in sales of diesel passenger cars in China (2005–2010)

Table 4.3 Sales of diesel passenger cars in China by types

Year	2005	2006	2007	2008	2009	2010
Total sales (units)	41,380	40,198	42,965	42,089	68,880	112,655
Diesel basic passenger cars (%)	45.8	44.5	41.1	34.5	25.9	13.1
Diesel MPVs (%)	16.8	13.8	6.0	5.8	19.7	13.2
Diesel SUVs (%)	43.6	46.0	52.7	59.1	60.6	75.8
Diesel cross passenger cars (%)	0.6	0.5	0.8	1.1	0.4	0.7

2010 was between 19,000 and 15,000, and it showed a slight downward trend. The number of diesel cross-type passenger cars was very small; annual sales were under 1,000.

Table 4.3 shows the proportion of diesel passenger cars sold from 2005 to 2010 based on type. The data in the table indicate that the proportion of SUVs among all diesel passenger cars sold showed an annual increase from 43.6 % in 2006 to 75.8 % in 2010; in contrast, the proportion of diesel basic passenger cars decreased from 45.8 to 13.1 % over the same period. The proportion of diesel MPVs showed a large fluctuation, but it exceeded 13 % in 2009 and 2010. The proportion of diesel cross-type passenger cars has remained basically constant at around 1 % or under.

Over the past 5 years, most passenger cars sold in China have been basic passenger cars. Cross-type passenger cars have accounted for about 20 %; the proportion of SUVs has increased to some extent. Table 4.4 and Fig. 4.4 present the changes in the population of passenger cars sold in China from 2005 to 2010 in terms of quantity and type. That table and figure indicate that basic passenger cars have been dominant, having a maximum market share of 75 % in 2007 and minimum of 69 % in 2010. From 2007 to 2010, the proportion of basic passenger cars showed a slow downward trend. The proportion of cross-type passenger cars has increased: their share of the Chinese passenger car market rose from 15.7 % in 2007 to over 18 % in 2009 and 2010. SUVs have developed rapidly in recent

Table 4.4 Sales of Chinese passenger cars by types

Type	2005	2006	2007	2008	2009	2010
Total sales (million)	3.98	5.18	6.30	6.76	10.33	13.76
Basic passenger cars (%)	70.3	74.0	75.1	74.7	72.3	69.0
MPVs (%)	3.9	3.7	3.6	2.9	2.4	3.2
SUVs (%)	4.9	4.6	5.7	6.6	6.4	9.6
Cross passenger car (%)	20.9	17.7	15.7	15.7	18.9	18.1

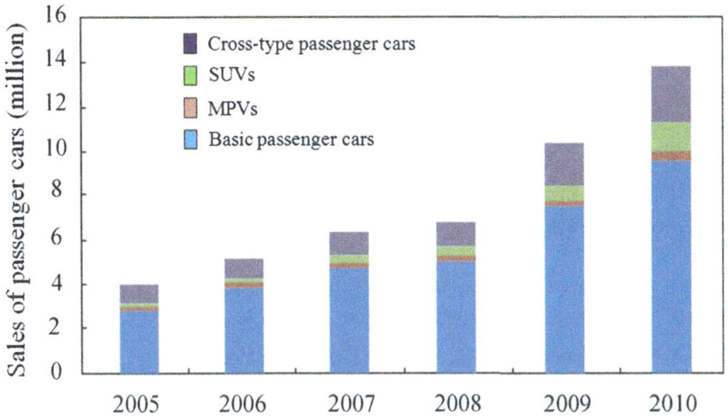

Fig. 4.4 Sales of Chinese passenger cars by vehicle type

years; their proportion rose continuously from 5.7 % in 2007 to 9.6 % in 2010, which amounts to an increase of 70 %. Regarding engine size, sales of Chinese passenger cars were mainly concentrated in vehicle types with a displacement of less 2 L; among these, compact cars with a displacement of 1–1.6 L accounted for the greatest quantity, and their proportion exceeded 50 %. However, since the sales of vehicles with an engine displacement of over 2 L have increased, the market share of compact cars has fallen. Changes in the population of passenger cars by vehicle type are presented in Fig. 4.5. The number of vehicles with an engine displacement of 1.0–1.6 L in 2005 was 1.68 million; that number rose to 3.49 million in 2008. Over that 3-year period, the sales volume doubled (106 % increase). Over the next 2 years, the sales volume again doubled (113 % increase) and amounted to 7.45 million. The main reason behind this rise was the recognition on the part of consumers of the functions of compact passenger cars and the preferential fiscal and taxation policies implemented by the government (Sidebar 4.3). At the same time, small passenger cars with a displacement of under 1 L showed a rapid increase from 1.27 million in 2009 to 2,010,000 in 2010, after having decreased from 0.97 million in 2005 to 0.71 million in 2008. It is also evident in Fig. 4.5 that the sales of midsize passenger cars with an engine displacement of 1.6–2.0 L also rapidly increased—from 0.97 million in 2005 to 1.65 million in 2008 and 2.85 million in 2010; their sales volume showed a 2.85-fold increase over the 5 years from 2005 to 2010. Above a displacement of 2.0 L, increasing displacement resulted in a lower proportion of sales.

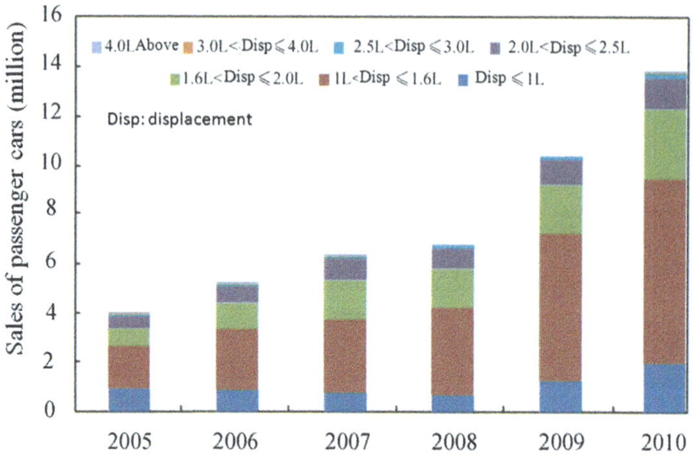

Fig. 4.5 Sales volume of Chinese passenger cars by displacement

Sidebar 4.3: Preferential Fiscal and Taxation Policies for Small-Displacement Vehicles

Vehicle purchase tax is a kind of tax levied on specified vehicles purchased by organizations and individuals in China. The tax rate is 10 % for vehicles with an engine displacement of 1.6 L. However, to expand domestic demand, the Chinese government in January 2009 began implementing preferential fiscal and taxation policies to promote the development of vehicles with a small displacement. Passenger cars with a displacement of 1.6 L or less were taxed at the rate of 5 % from January 1, 2009, to December 31, 2009, and at 7.5 % from January 1, 2010, to December 31, 2010. This preferential policy was discontinued in 2011.

Figure 4.6 shows an annual comparison of Chinese passenger cars by sales volume. It is evident in Fig. 4.6 that compact cars with a displacement of 1.0–1.6 L have dominated the passenger car market: their proportion from 2005 to 2010 was within the range of 42.5–57.3 %, and overall, they showed an increasing trend. The market share of compact cars was 42.5 % in 2005; this increased to 51.7 % in 2008 and further rose to 57.3 % in 2009 and 54.2 % in 2010. They were 20–30 percentage points higher than the vehicle type with the second-largest market share. The great market share of these vehicles shows that Chinese passenger cars are being developed with increased energy saving and emission reduction. It can also be seen in Fig. 4.6 that midsize vehicles with a displacement of 1.6–2.0 L exceeded the number of small vehicles in 2006 and became the main type of passenger car in China. Small vehicles with a displacement of under 1 L experienced a rapid

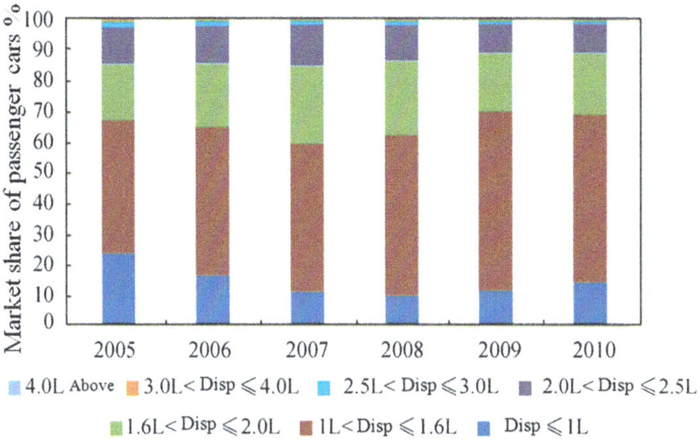

Fig. 4.6 Proportion of Chinese passenger cars sold by displacement

decline in market share. However, through the support of national preferential agricultural policies for rural vehicles (Sidebar 4.4), these mini-vehicles showed rapid development in the rural market and reversed the downward trend; the proportion of small-displacement vehicles rose in 2009 and 2010. The market share of mini-vehicle in 2005 was 24.3 %, but the market share then showed a reduction of 16.8, 11.9, and 10.5 % in the next 3 years and turned to 12.3 and 14.6 % in 2009 and 2010, respectively.

Sidebar 4.4: Policy for Rural Vehicles

Rural vehicles were subject to a preferential agricultural policy—the Vehicle Industry Restructuring and Rejuvenation Program—announced on January 14, 2009. Subsidies extended to some cross-type passenger cars (minibuses). The subsidies were implemented from March 1, 2009 to December 31, 2010. The subsidies amounted to 10 % of the vehicle price. Across the country, the amount of subsidies paid out amounted to RMB 5 billion.

4.1.3 Powertrain Technology of Chinese Commercial Vehicles

The annual sales of Chinese commercial vehicles (including buses and trucks, but excluding semitrailer tractors and vehicles without power or where the type of power could not be confirmed, such as partly built buses and trucks) are shown in Fig. 4.7 (China Association of Automobile Manufacturers 2010). The figure shows that the sales volume of diesel vehicles among Chinese commercial vehicles continuously

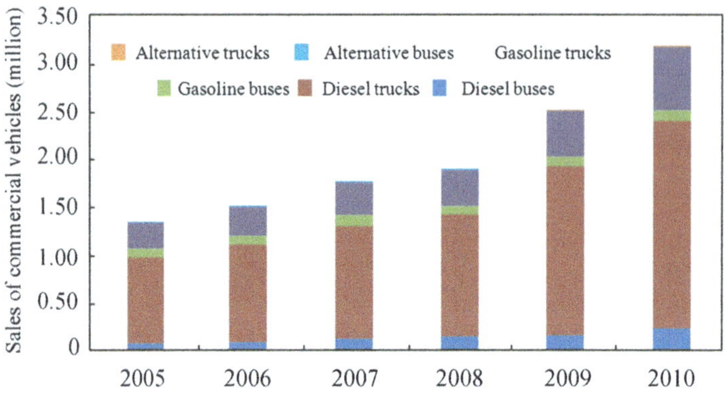

Fig. 4.7 Annual sales volume of Chinese commercial vehicles

Table 4.5 Sales volume of Chinese commercial vehicles

Year		2005	2006	2007	2008	2009	2010
Sales of commercial vehicles (million)		1.34	1.51	1.76	1.89	2.52	3.19
Diesel commercial vehicles (%)		73.1	74.3	74.5	75.3	76.7	75.8
In which	Buses (%)	6.6	6.9	7.6	8.3	6.8	7.5
	Trucks (%)	66.5	67.4	66.9	67.0	69.9	68.3
Gasoline commercial vehicles (%)		26.8	25.6	25.5	24.6	23.3	24.1
In which	Buses (%)	6.6	5.6	6.3	5.0	3.8	3.5
	Trucks (%)	20.1	20.0	19.1	19.6	19.3	20.5
Commercial vehicles using other fuels (%)		0.1	0.1	0.1	0.1	0.1	0.2
In which	Buses (%)	0.1	0.1	0.1	0.1	0.1	0.2
	Trucks (%)	0.0	0.0	0.0	0.0	0.0	0.0

rose from 0.89 million to 2.18 million between 2005 and 2010—a 1.44-fold increase over 5 years. The sales volume of diesel vehicles also rose from 88,000 to 238,000 over the same period—a 2.7-fold increase. Meanwhile, the sales volume of gasoline trucks and buses increased, respectively, from 270,000 to 640,000 and from 89,000 to 113,000 units—a 1.4- and 0.3-fold increase. The increase in sales of diesel and gasoline trucks was similarly consistent, but the increase in gasoline buses was far lower than that of diesel buses. Commercial vehicles using other fuels were mainly concentrated among buses: their volume showed a threefold increase between 2005 and 2010. Trucks using other fuels were sold only in 2009 and 2010, with their number being less than 100 vehicles.

Table 4.5 and Fig. 4.8 show the proportions of diesel trucks, diesel buses, gasoline trucks, gasoline buses and trucks, and buses using other fuels since 2005. A couple of points should be noted here. First, at present commercial vehicles mainly use diesel; the proportion of diesel commercial vehicles has exceeded 73 % since 2005 and maintained a slow growth trend. The dieselization of commercial vehicles has basically been achieved. Among commercial vehicles in 2005, the proportion of

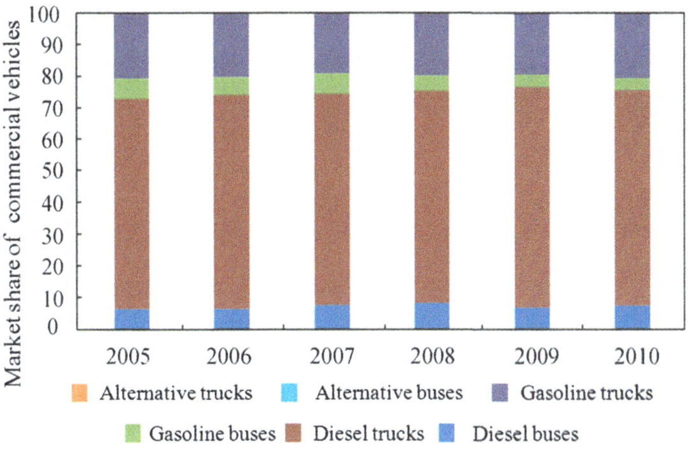

Fig. 4.8 Power used in Chinese commercial vehicles

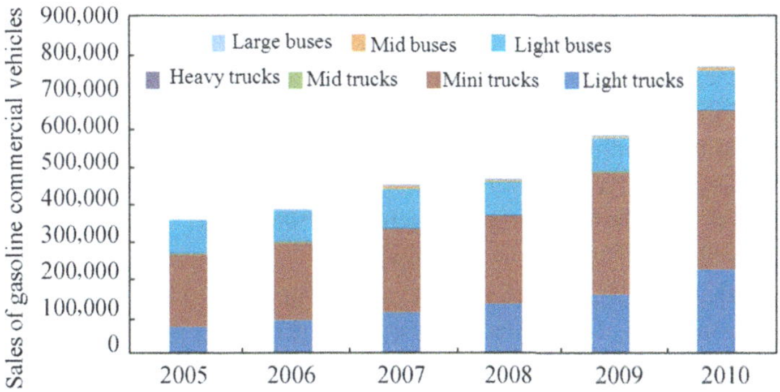

Fig. 4.9 Sales volume of gasoline commercial vehicles in China

diesel vehicles was 73.1 % and increased to 75.8 % in 2010; within that figure, the proportion of diesel trucks sold exceeded 66 % and accounted for the largest proportion in sales volume of Chinese commercial vehicles. Second, the proportion of gasoline vehicles among commercial vehicles slowly declined from 26.8 % in 2005 to 24.1 % in 2010. This was mainly due to the reduction in the proportion of gasoline buses (which dropped from 6.6 % in 2005 to 3.5 % in 2010); however, the proportion of gasoline trucks was largely constant at about 20 %. The proportion of vehicles using other fuels was still very low (0.2 %), and these were mainly natural gas city buses.

Among Chinese commercial vehicles, gasoline vehicles are mainly mini-trucks, light trucks, and light buses. The proportion of gasoline vehicles among midsize and heavy commercial vehicles is very low. As shown in Fig. 4.9, among 359,000 gasoline commercial vehicles sold in 2005, the numbers of mini-trucks, light trucks,

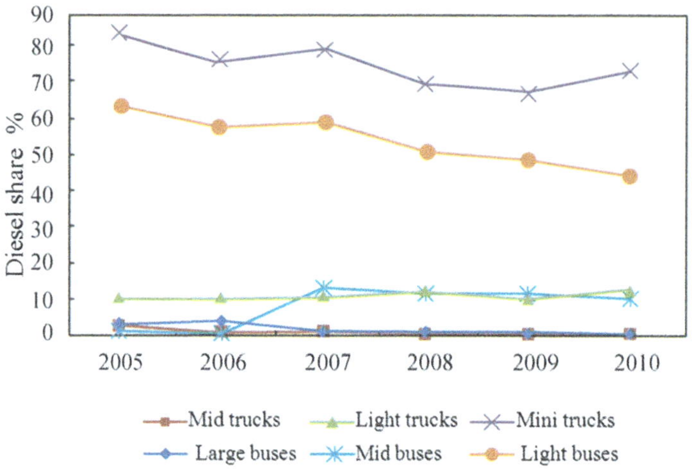

Fig. 4.10 Proportion of gasoline vehicles in commercial vehicles

and light buses were, respectively, 189,000, 79,000, and 88,000; their total was 356,000 and their proportion exceeded 99 %. In 2010, among the sales volume of 767,000 gasoline commercial vehicles, the numbers of mini-trucks, light trucks, and light buses were, respectively, 423,000, 229,000, and 106,000; their total was 768,000 and their proportion was still as high as 98.8 %.

As seen in Fig. 4.10, the current proportions of mini-trucks and light buses are still above 70 and 45 %, respectively, but their proportions have shown a downward trend. The proportion of mini-trucks with gasoline engines was 84.4 % in 2005, and thereafter it gradually declined; it further decreased to 67.7 % in 2009 and rose again to 73.9 % in 2010. The proportion of gasoline engines among light buses has shown a constant decrease from 63.9 % in 2005 to 49.8 % in 2009; it further declined to 44.8 % in 2010. The proportion of gasoline vehicles has been constantly lower than that of diesel vehicles. The proportion of gasoline vehicles over midsize buses and light trucks has remained fairly constant at 10 % since 2007, with dieselization being at 90 %. The proportion of large buses and midsize trucks powered by gasoline were both lower than 1 %. The market share of gasoline heavy duty trucks is almost zero.

4.1.4 Increased Proportion of Electrified Two-Wheeled Motor Vehicles

In China's two-wheeled motor vehicle industry, the proportion of electrically driven motor vehicles using traction battery and electric motor has rapidly increased. Figure 4.11 presents the annual output and power composition of two-wheeled motor vehicles since 2000 (Sun Li 2010). It can be seen in this figure that in 2000, the output of electric two-wheeled vehicles in China was only 0.15 million, whereas

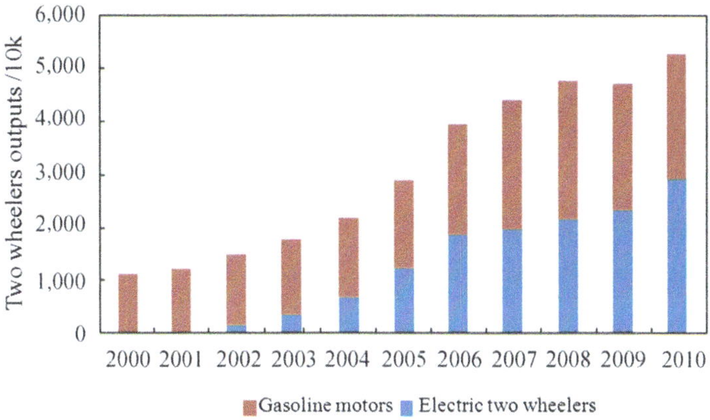

Fig. 4.11 Annual output and power of two-wheeled motor vehicles in China

Table 4.6 Output and power of two-wheeled vehicles in China

	Output (million)		
Year	Electric drive	Gasoline motor	Electrification rate (%)
2000	0.15	11.08	1.3
2001	0.5	11.8	4.1
2002	2	13.	13.3
2003	4	14.	22.2
2004	7	15.	31.8
2005	12.5	16.5	43.1
2006	19	20.4	48.2
2007	20	24.4	45.0
2008	21.89	25.94	45.8
2009	23.69	23.59	50.1
2010	29.54	23.43	55.8

the output of two-wheeled gasoline vehicles was 1.108 million; the proportion of electrification was just 1.3 %. In 2005, the output of two-wheeled electric vehicles amounted to 12.5 million, which was an 82-fold increase over the output in 2000; during the same period, the output of two-wheeled gasoline vehicles increased by only 48 % and amounted to 16.5 million. Over the next 5 years, the output of two-wheeled electric vehicles still underwent rapid development, but the growth rate of gasoline two-wheeled vehicles was slow. As seen in Table 4.6, the output of two-wheeled electric vehicles in 2009 exceeded that of two-wheeled gasoline vehicles for the first time; the electrification rate of two-wheeled vehicles amounted to 50.1 %. The output of two-wheeled electric vehicles in 2009 showed an increase of 5.85 million over the 2009 figure, with the total output amounting to 29.54 million; at the same time, the output of two-wheeled gasoline vehicles decreased to 23.43 million, which was 0.17 million fewer than the output in 2009. In 2010, the electrification rate of two-wheeled vehicles reached 55.8 %. Thus, two-wheeled motor vehicles have truly entered a period of electrification.

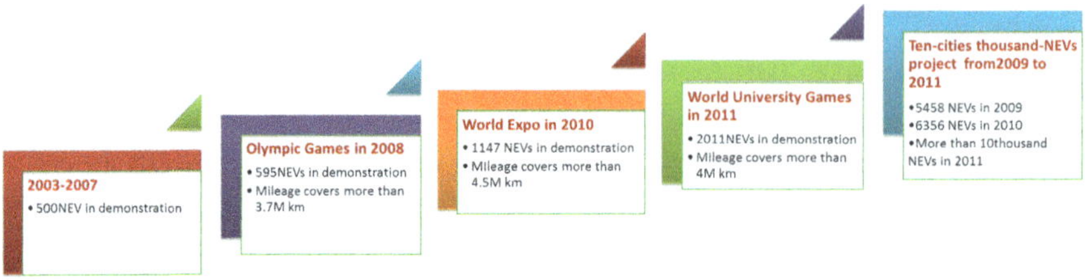

Fig. 4.12 Main demonstration activities for electric vehicles in China

4.1.5 Early Stage of Powertrain Electrification

Under the support of the 863 Program for electric vehicles sponsored by the Ministry of Science and Technology, demonstrations of electric vehicles have been conducted in cities. Hybrid electric vehicles and battery electric vehicles have been verified technically, and extensive tests relating to their operation have been carried out by various research organizations. Fuel-cell electric vehicles were extensively used for various functions. The main demonstration activities for Chinese electric vehicles are presented in Fig. 4.12. During the Beijing Olympic Games in 2008, almost 600 electric vehicles in service cover several million kilometers during that period. In 2009, the Ministry of Science and Technology, Ministry of Finance, Ministry of Industry and Information Technology, and the National Development and Reform Commission jointly launched the "Ten Cities, Thousand Vehicles" program, which promoted the large-scale commercialization of new-energy vehicles. During the 2010 Shanghai World Expo, 1,147 electric vehicles including the super capacity electric vehicles, fuel-cell electric vehicles, battery electric vehicles, and hybrid electric vehicles operated in demonstrate cities for over 4 million kilometers.

In 2010, the "Ten Cities, Thousand Vehicles" program was expanded to 25 cities, and the field of application was expanded to private cars in six cities. Through central financial subsidies, incentives were offered to promote the use of 10,000 electric vehicles for public traffic in 13 pilot cities, including Beijing, Shanghai, and Dalian (Fig. 4.13). As part of the program, about 5,000 vehicles were used for demonstration purposes in 2009. In 2010, 7,181 electric vehicles were put into operation. The number of electric vehicles used for demonstration purposes in 2011 amounted to 10,036. Among these, the number of battery electric buses exceeded 1,000, and there were 1,076 private electric vehicles.

In 2011, the public service sector accounted for 89 % of the electric vehicles in the Ten Cities, Thousand Vehicles program, whereas the private sector accounted for 11 %. In terms of vehicle type, 77 % were electric commercial vehicles, and electric passenger cars accounted for 23 %. As of April 2011, the program had employed 21 batches of battery electric vehicles, and 255 types of vehicles could be sold and operated in the selected cities; up to 47 production enterprises were involved in the program.

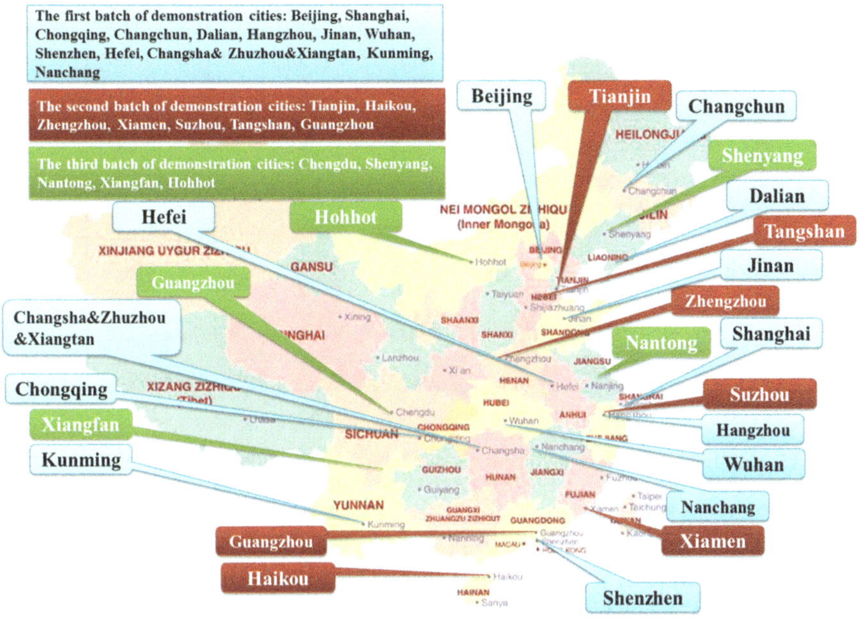

Fig. 4.13 Distribution of demo cities of Ten Cities Thousand Vehicles program

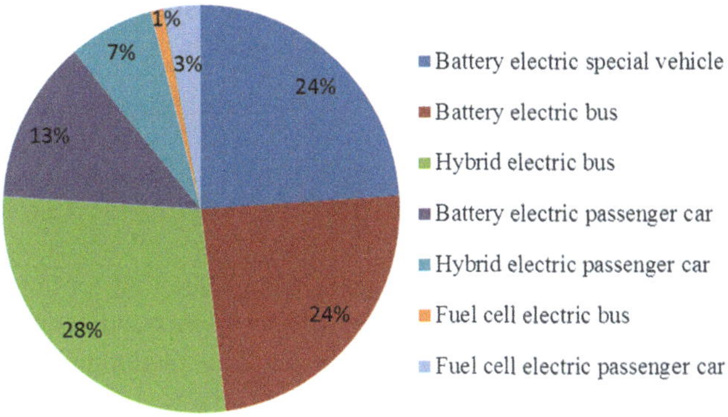

Fig. 4.14 Vehicles used in the Ten Cities, Thousand Vehicles program (April 2011)

Figure 4.14 shows the type of vehicles used in the first 21 batches recommended by the Ministry of Industry and Information Technology to participate in the program (Ministry of Industry and Information Technology). As evident in the figure, the proportion of battery electric vehicles was very high; among those, there were 55 types of battery electric city buses, which accounted for 24 % of all vehicles; there were 15 types of battery electric passenger cars, accounting for 13 % of all electric vehicles. There were up to 57 types of other special vehicles, such as ones for engineering and postal services, which accounted for 24 % of all vehicles. There

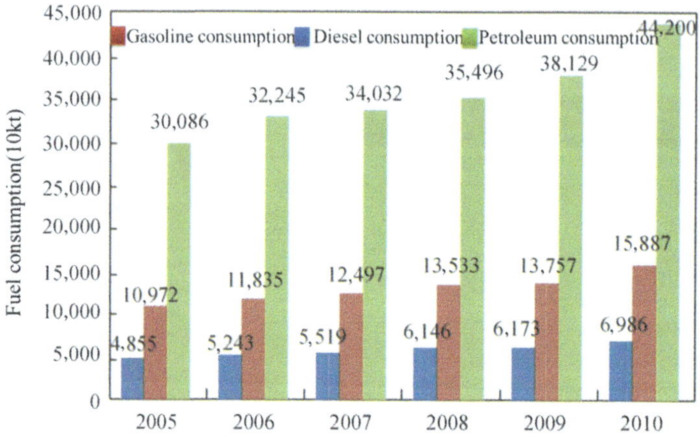

Fig. 4.15 Gasoline, diesel, and petroleum consumption in China (2005–2010)

were 134 types of battery electric vehicles, which accounted for up to 61 % of all vehicles. There were 61 types of hybrid electric buses that accounted for 20 % of all electric vehicles, which was the largest proportion of vehicles categorized according to type. However, the number of hybrid electric vehicle types was very low: there were only 14 types, and they accounted for 7 % of all the vehicles. In total, there were only nine models of fuel-cell electric vehicles, of which there were six fuel-cell passenger car models and three fuel-cell city bus models.

4.1.6 Current Status of Vehicle Fuel Consumption

Road transportation is gradually becoming the main field of oil consumption growth in China. With the increase in the number of family cars and economic stimulus measures by the government to encourage the purchase of automobiles, the vehicle population in China has risen rapidly in recent years. This has resulted in a continuous increase in fuel consumption. Figure 4.15 indicates the consumption of petroleum, diesel, and gasoline in China since 2005 based on data from the National Bureau of Statistics of China (2011d), the 2010 National Economic and Social Development Statistics Bulletin, and monthly data. The consumption of gasoline and diesel in 2005 was, respectively, 48.55 and 109.72 million tons—158 million tons in total. In 2010, these data increased, respectively, to 69.86, 158.87, and 228 million tons. The total consumption of gasoline and diesel showed an average annual growth rate of 7.5 % over this 5-year period. Correspondingly, the vehicle population increased from 31.59 million in 2005 (National Bureau of Statistics of China 2011b) to 78.02 million in 2010 (National Bureau of Statistics of China

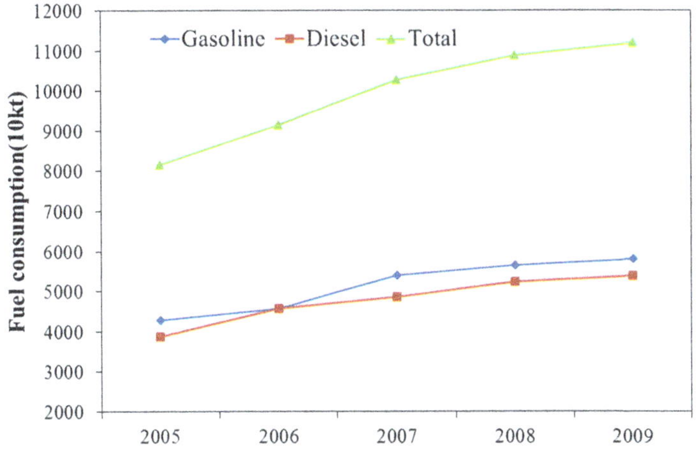

Fig. 4.16 Changes in gasoline and diesel consumption in China in recent years

2011b), which amounts to an average annual growth rate of 19.8 %. This indicates that the total consumption of gasoline and diesel rose as a result of the increase in the number of vehicles.

The China vehicle gasoline and diesel fuel consumptions increase in recent years, both the amount and growth rate of diesel consumption are higher than those of gasoline consumption. Figure 4.16 shows the gasoline and diesel consumed by vehicles from 2005 to 2009 (China Association of Automobile Manufacturers 2010). It can be seen in that figure that the total consumption of the two kinds of fuel increased from 81.97 million tons in 2005 to 112.22 million tons in 2009; the increase over that 4-year period was 36.9 %, with an average increase of 8.1 % annually. Gasoline consumption increased from 43.11 million tons in 2005 to 54 million tons in 2009—a 25.3 % increase, with an average increase of 5.8 % annually. Over the same period, diesel consumption by vehicles increased from 38.86 million tons in 2005 to 58.22 million metric tons in 2009—a 49.8 % increase, with an annual average increase of 10.6 %. Since 2006, the diesel consumption by vehicles has been higher than that of gasoline.

Gasoline consumption of Chinese vehicles accounted for over 85 % of total gasoline consumption. Table 4.7 (China Association of Automobile Manufacturers 2010) shows the total Chinese gasoline and diesel consumption and vehicle consumption since 2005. It can be seen in Table 4.7 that gasoline consumption by vehicles in 2005 was 43.11 million tons, which accounted for over 88.8 % of the total gasoline consumption in that year. Over the subsequent 4 years, that proportion remained at 85.5–88.5 %; thus, vehicles are the main gasoline consumers in China.

With the rising dieselization among commercial vehicles and their increased use in road transport, the proportion of diesel consumed by Chinese vehicles has risen to over 40 %. It can be seen from Table 4.7 that vehicles consumed 38,860,000 metric

Table 4.7 Gasoline and diesel consumption by vehicles in China (million tons)

	Gasoline fuel			Diesel fuel		
Year	Total consumption	Vehicle consumption	Proportion %	Total consumption	Vehicle consumption	Proportion %
2005	4.85	4.31	88.8	10.97	3.88	35.4
2006	5.24	4.59	87.6	11.83	4.59	38.8
2007	5.51	4.88	88.5	12.49	5.41	43.4
2008	6.14	5.25	85.5	13.53	5.67	41.9
2009	6.17	5.40	87.5	13.75	5.82	42.3

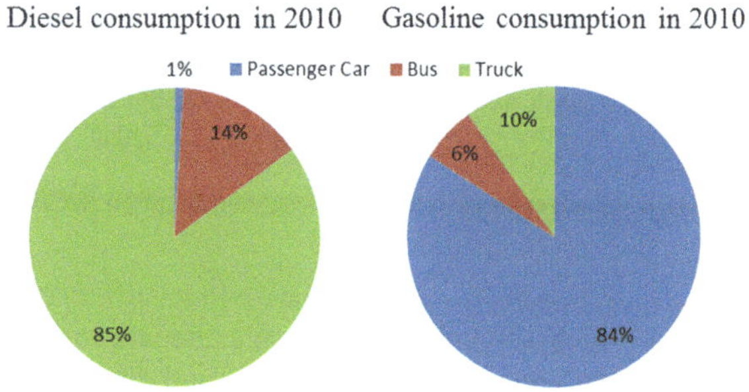

Fig. 4.17 Gasoline and diesel consumption according to vehicle type in China in 2010

tons of diesel in 2005, which accounted for 35.4 % of the diesel consumption in that year (109,720,000 metric tons); consumption of diesel by vehicles in 2007, 2008, and 2009 was, respectively, 43.3, 41.9, and 42.3 %.

The proportion of gasoline and diesel fuel consumed by different types of vehicles is shown in Fig. 4.17. In 2010, Chinese passenger cars consumed 60.29 million tons of gasoline, accounting for 84 % of total gasoline consumption; trucks consumed 7.06 million tons of gasoline, which accounted for 10 %. Passenger cars therefore account for the main consumption of gasoline. In the same year, passenger cars consumed 0.51 million tons of diesel, which accounted for 1 %. In 2010, buses consumed 10.54 million tons of diesel, which accounted for 14 % of total consumption; trucks consumed 66.62 million tons of diesel, which accounted for 85 % of total consumption; therefore, commercial vehicles, including buses and trucks, are the main field of diesel consumption.

However, owing to the development of alternative-energy and new-energy vehicles, there has also been annual consumption of natural gas and ethanol. Table 4.8 (Zhang Guobao 2010) shows that natural gas vehicles increased from 0.127 million to 0.45 million from 2005 to 2009; correspondingly, natural gas consumption by Chinese vehicles also increased from 1.6 billion m^3 to over 4 billion m^3 over the same period. In addition, because of policies encouraging the use of ethanol, consumption of ethanol by vehicles increased from 1.02 million tons in 2005 to 1.71 million tons in 2009. Chinese vehicles also consumed liquefied petroleum gas, methanol, electricity, and other fuels and energy, but there are no data for such quantities.

Table 4.8 Alternative vehicles and alternative-fuel consumption in China

Year	2005	2006	2007	2008	2009
Natural gas vehicle (thousand)	127	214	265	307	450
Natural gas consumption (billion m^3)	1.6	1.9	2.3	2.7	>4.0
Ethanol fuel (million tons)	1.02	1.32	1.45	–	1.71

4.2 Efficient and Clean Vehicle Powertrain Technology

In the face of challenges with respect to energy security and environmental protection, high-efficiency vehicle powertrain technology is playing an important role in the efficient use of energy and the reduction in greenhouse gas emissions. From the perspective of the full life cycle of fuels and taking conventional gasoline passenger cars as an example, 78 % of the fuel energy is consumed in the driving stage. From a further analysis of vehicle energy flow, it has been shown that only about 13 % of the fuel energy is actually used to drive the vehicle to overcome the air resistance and wheel friction resistance. Energy lost in fuel combustion process, idling, braking, and powering the vehicle accessories is 62, 17, 5.8, and 2.2 %, respectively. Vehicle energy-loss reduction technologies focus on researching and developing products that tackle these energy-loss processes. These technologies mainly include efficient engine technology, smart start-stop technology, recovering of braking energy, efficient transmission and air-resistance-reduction technology, and lightweight vehicle technology.

4.2.1 Efficient Engine Technology

4.2.1.1 Gasoline Direct Injection

The spark-ignition homogeneous charge gasoline engine has many merits, such as high specific power, lower vibration noise, and ease in controlling emission, and thus, it is widely used in passenger cars. However, the conventional thermal efficiency of an intake manifold injection gasoline engine is at least 10 % lower than that of a compression-ignition engine since it adopts a lower compression ratio and throttles to adjust the load, which result in more energy losses (Sidebar 4.5). As a result, gasoline direct injection (GDI), which adopts the higher compression ratio like diesel engine and reduces throttle loss by cancelling the intake throttle valve, was created to improve thermal efficiency. When combined with supercharging technology, high-compression-ratio, and multivalve technology, GDI technology can achieve lean gasoline combustion with high thermal efficiency, which can reduce fuel loss by 8–15 %. Compared with an intake manifold injection gasoline engine, the GDI gasoline engine has the following merits: high adiabatic index, low pumping loss and heat loss, good dynamic response, fast cold start, and strong deceleration fuel cutoff ability. Moreover, fuel evaporation in the cylinder can lower the temperature of the charge to enhance the intake efficiency and antiknock of the gasoline engine, thereby further increasing the compression ratio. Typical technologies in this area used around the world include Mitsubishi's GDI gasoline engine technology and Volkswagen's fuel stratified injection (FSI) gasoline engine technology. GDI technology has already been adopted by many carmakers as the main technology to achieve low-carbon emission target, and it will be one of the main technologies for gasoline engines in the future. Among the newly developed car gasoline engines in Europe in 2006–2007, GDI accounted for over 70 % (Wang Jianxin and Wang Zhi 2010). At present, GDI is in the technological development stage in China. The following engines have been successfully developed: the 1.6 TGDI of Cher y Company and the GW4G15-GDI supercharged direct-injection gasoline engine of Great Wall Motor. However, the promotion of GDI engines in China faces a number of problems, such as unsatisfactory fuel quality and the dependence on imports for key components.

Sidebar 4.5: Working Process of Gasoline and Diesel Engines

The mixed gases in a conventional internal-combustion engine cylinder include those that can be ignited by compression and those that can be ignited by a spark. Diesel engines adopt compression ignition with a compression ratio of about 12:1, whereas gasoline engines adopt spark ignition with a compression ratio of about 8:1. The load of the diesel engine can be adjusted by controlling the volume of fuel injected into the cylinder and at the same time keeping the volume of air entering the cylinder unchanged. This kind of adjustment is called constant flow control. The load of a gasoline engine can be adjusted by controlling the volume of gasoline going into the cylinder and also by controlling the corresponding volume of air to achieve effective combustion. This kind of adjustment is called variable flow control. Throttle valves have to be installed in the intake system to control the intake air volume, which results in intake loss.

4.2.1.2 Homogeneous Charge Compression Ignition

HCCI is a new concept of internal-combustion engine that differs from both diesel engines (nonhomogeneous charge compression ignition) and gasoline engines (homogeneous charge spark ignition). HCCI is a combination of traditional spark-ignition and compression-ignition engines. It mixes fuel and air into a lean mixture, which it charges into the engine cylinder; combustion is achieved under the higher pressure and temperature. By adopting higher compression ratio, exhaust gas recycling (EGR), intake heating, and supercharging, the homogeneous charge achieves the self-combustion by compression ignition. In this way, HCCI can ensure the stability of ignition and combustion; it can reduce the spread and duration of combustion.

From the perspective of thermal efficiency, HCCI combustion possesses the following merits. First, because it can combust at low temperature, the heat it transfers to the walls of the combustion chamber and cylinder is low, and the smokeless combustion reduces the loss of radiant heat transfer; it is characterized by low thermal loss. Second, it possesses the advantage of a short combustion duration because the combustion process is mainly controlled by the chemical reaction process instead of the mixing process. Therefore, its combustion duration is shorter than that of a conventional diesel engine. As a result, it is possible to make HCCI cycle very close to the Otto cycle. HCCI can achieve indicated thermal efficiency as high as 50 %. The particular merits of HCCI, such as homogeneous combustion at low temperature, can effectively reduce fuel consumption and solve emission problems, which are a concern with conventional combustion engines. HCCI can further improve thermal efficiency, and thus, it represents a major breakthrough in engine combustion technology.

HCCI technology uses aspects of the diesel-like compression-ignition engine in its reduced emission of nitrogen oxides and particulate matter. And HCCI technology utilizes aspects of spark-ignition engine technology in its reduced fuel consumption under partial load, which reduces pumping loss.

With the support of state science and technology development projects, such as the 973 (fundamental research) project, colleges and universities—for example, Tianjin University, Shanghai Jiao Tong University, and Tsinghua University—have already carried out preliminary research on gasoline and diesel HCCI, and they have made many patents in HCCI key technologies (such as HCCI ignition control, spark-ignition and compression-ignition switch, and HCCI combustion control strategy).

4.2.1.3 Engine Supercharging Technology

The purpose of a supercharger is to increase the air intake capacity of the engine and improve the engine characteristics, such as in power per liter, fuel economy, and emissions characteristics (Sidebar 4.6). Engine supercharging technology includes both exhaust gas turbo superchargers and mechanical superchargers. Exhaust gas turbo superchargers utilize the energy of the engine exhaust. They are economical in terms of fuel, but there is some delay in the operation. Mechanical superchargers are directly driven by the engine crankshaft and they possess the following merits: they can be easily matched to the engine; they have a compact structure and show a good response. Engine supercharging technology can effectively improve engine power at low speed, but energy is consumed in the process and there is limited chance to improve fuel economy. It is possible that the fuel consumption rate with mechanical superchargers will be greater than with non-supercharged engines.

Sidebar 4.6: Supercharging Technology

Supercharging technology adopts a specialized compressor to compress the air in the cylinder, thereby improving the density of air in the cylinder and increasing the volume of intake air so as to further improve engine power. It can greatly enhance the power and torque of the engine without increasing the engine displacement.

The adoption of supercharging technology to reduce the engine displacement can improve fuel economy with unchanged power. Fuel consumption with an intake manifold, injection turbo-supercharged engine is 13–18 % lower than that of a gasoline engine of the same power owing to the reduced size and movement of the corresponding load. If turbo-supercharging technology is combined with GDI technology and the characteristics of fuel injection in the cylinder are utilized, the compression ratio can be increased for advantageous control of the residual gas; in addition, the effect of turbo-charging delay can be reduced. By integrating these two technical measures, the fuel consumption of a supercharged engine can be further reduced by 6 %.

Engines that adopt supercharging technology in China include the following: the 1.0T and 1.5T turbo-supercharged engines and the 2.0 TGDI turbo-supercharged direct-injection engine developed by Brilliance; the GW4G15T turbo-supercharged engine and GW4G15-GDI-TC2 mechanical-supercharged direct-injection engine developed by Great Wall Motor; the 1.3S, 1.6S, and 1.8S mechanical supercharged engine and 1.6 TGDI turbo-supercharged direct-injection engine developed by Chery; and the 4G13T turbo-supercharged engines developed by Geely Automobile. However, in comparison with the mature industrialized engine supercharging technology abroad, China still has a long way to go in engine supercharging technology.

4.2.1.4 Variable Valve Technology

Variable valve technology includes variable valve timing (VVT; Sidebar 4.7) and variable valve lift technology. The adoption of VVT technology can enable the opening and closing time of the intake and outtake valves and the corresponding degree of valve lifting to satisfy the intake requirements of the engine under different operating conditions; for example, the torque and power of the engine can be enhanced by increasing the intake charge; the efficiency and emissions can be improved under the good combustion process through an appropriate intake flow (swirl ratio). An advanced variable valve and lift electronic control system can create an appropriate swirl ratio to achieve good performance when the engine is at low speed through a series of control systems and the lift difference between the main intake valve and secondary intake valve. When the engine is at high speed, VVT technology has the merit of a higher power output than a conventional four-valve engine. This technology can achieve both high power output and low fuel consumption; thus, it greatly improves fuel economy and increases engine power output.

Sidebar 4.7: Variable Valve Timing Technology

When combined with an advanced engine control strategy, variable valve timing technology can achieve a variable compression ratio. Under a large load, it can avoid knock by delaying the closing time of the intake valve to reduce the compression ratio; under moderate and small loads, it can improve their thermal efficiency by adjusting the closing time of the valves to increase the compression ratio.

Research indicates that intake VVT technology can reduce the fuel consumed by 6 % and the exhaust VVT technology can reduce the fuel consumed by 1–2 %. Auto enterprises outside China usually adopt two-way VVT, namely, intake VVT and exhaust VVT, whereas domestic auto companies often adopt just intake VVT; only 30 % of vehicles in China are manufactured using this technology. Therefore, there is great room for expansion of this technology in China. With the development and spread of domestic VVT technology, production costs will gradually decrease. VVT will be a basic configuration for engines in the near future.

4.2.1.5 Diesel Technology in Passenger Cars

In contrast to gasoline engines, diesel engines adopt compression ignition; they possess a higher compression ratio and are more thermal efficient than gasoline engines. The fuel saved is 30–40 % higher than with gasoline engines (Sidebar 4.8). Diesel engines are widely used in commercial vehicles (trucks and buses) in China, but their share among passenger cars is very low because of the high emission of particulates, reduced comfort (they are not as comfortable as gasoline engines mainly due to noise and vibration), high cost, and particular requirements of fuels. Among passenger cars in China, diesel engines are found only in SUVs and MPVs. With the application of advanced technologies—such as intercooling exhaust gas recirculation technology, electronically controlled common rail direct-injection (CRDI) technology, variable geometry turbo-charging technology, and exhaust gas catalytic converters and particulate filters—and the adoption of lightweight aluminum materials in the engine body and cylinder head, modern cars with diesel engines can completely meet the comfort and emission levels required of passenger cars. Diesel cars have been widely adopted in Europe and have gained increasing attention in the United States in recent years. For example, 77,900 diesel cars were sold in the United States in 2010, which was a 36.9 % increase over the number of diesel cars sold there in 2009.

Sidebar 4.8: Efficiency of Gasoline and Diesel Engines

The reason for diesel engines being more fuel economical than gasoline engine is as follows: first, they do not have throttle valves and thus can reduce intake loss; second, they adopt lean combustion and thus can reduce thermal loss during the spreading process in the cylinder; third, their compression ratio is as high as 16:1, and thus, their thermal power conversion efficiency is theoretically higher than that of gasoline engines, whose compression ratio is only 11:1.

Diesel cars have long been imported in China, for example, Volkswagen sells its Jetta diesel cars in China. Independently developed diesel engines for passenger cars include the Yuchai Group 4 W engine, the energy-saving and environmentally friendly D16TCI developed by the Yunnei Group, and the clean-car diesel engine D01 developed by Changfeng Motor. Among them, the displacement of 4 W engine, which adopts CRDI technology, variable turbo-charging technology, and an aluminum engine body and cylinder head, is 1.2 L and its power output is 55 kW.

The results of driving tests in actual road conditions indicate that in big cities, which are subject to frequent traffic jams, the fuel consumed by diesel cars is 30 % lower than with gasoline cars of the same type. Clearly, diesel cars save more energy (Sidebar 4.9).

Sidebar 4.9: Diesel Passenger Car Demonstration in Shanghai

After 1-year operation 50 Passat diesel taxis provided by Shanghai Volkswagen performed excellently in tests in terms of fuel economy and environment friendliness. Compared with gasoline cars of the same type, the fuel consumed by diesel cars over a distance of 100 miles was 35–42 % lower than gasoline cars.

4.2.2 Smart Start-Stop Technology

When a car encounters traffic lights or parks for a short period of time, the engine goes into idling mode and consumes energy unnecessarily. By mean of control technology, it is very easy to stop the engine instead of using the idling mode, thereby reducing fuel consumption and the emission of pollutants. At present, the main technologies that achieve idling stop include the following: belt-driven starter and generator (BSG) technology (also called slightly hybrid technology) and integrated starter and generator (ISG) technology (also called mild-hybrid technology).

4.2.2.1 BSG Technology

With BSG technology, the starter motor of the conventional internal-combustion engine becomes a motor that possesses two functions—electricity generation and power driving. BSG technology increases the power or drive torque to meet the required starting power when the engine runs from a stop to high speed. It therefore eliminates the conventional starter motor by starting the motor through engagement with the flywheel. To some extent, BSG technology can be considered a combination of a conventional starter motor and a generator, and it can achieve the functions of both in the original generator.

BSG technology is very simple and it results in fuel economy. Because of the low cost, this technology is now becoming the basic configuration for vehicles. The BSG consists of a motor, battery, and control unit. Typically, the power of the motor is 50–100 % higher than with conventional vehicles, the capacity of the power battery needs to be doubled, and the control unit can be independent or integrated into the engine control unit. Because of its low power, a BSG motor does not have the ability to drive or assist the drive of a vehicle, but its electricity-generation function can be used to partially recycle the braking energy.

The process by which BSG technology achieves the smart start-stop function is as follows. If the car is in empty load for a long period, for example, when encountering a traffic light, the internal-combustion engine enters idling mode, and the control unit disables the combustion engine. When the car is ready to start again, the control unit recognizes the intent of the driver (through the movement of the clutch to the driving position and movement of the accelerator), and the BSG quickly increases the rotating speed of the engine from zero to normal idle speed (about 800 rpm) to complete the starting of the engine and movement of the vehicle.

At present, the following BSG products are relatively mature and can be supplied independently: StARS (first generation, 2006) and i-StARS (second-generation, 2010) of the Valeo Group, STT of Robert Bosch GmbH, and Drop-in of Delphi Corp. There is no supply of mature, independent BSG products in China. Currently, BSG products are mainly developed by automobile manufacturers of a fixed direction. Companies that are equipped with BSG systems include the following: Chery Automobile Company (A5 sedan, self-developed), Changan Automobile (CX30 cars, Bosch STT Systems), SAIC-GM (LaCrosse, General Motors' BAS system), Dongfeng Motor (Fengshen S30), and Brilliance China Automotive (Junjie FSV, through independent research and development). Table 4.9 presents a comparison of the energy-saving rates of typical BSG vehicles in China.

The Changan Automobile CX30 low-carbon passenger car (model SC7166D) is equipped with a Bosch STT Systems BSG. On urban roads, the gasoline consumption per 100 km is 7.5 L, which is 22.7 % lower than that of comparison baseline model (SC7166C) which is 9.7 L/100 km. But it consumes 8.3 % more fuel under suburban cycle. As a result, the combined fuel consumption is only reduced by 5.5 %.

The Chery A5 BSC passenger car (model SQR7160A217/B) is equipped with a self-developed smart start-stop system. Its gasoline consumption under urban cycle

Table 4.9 Parameters of domestic vehicles using BSG

Vehicle maker and model	Fuel consumption of vehicles equipped with BSG (L/100 km)	Urban cycle	Suburban cycle	Combined cycle
Changan CX30	Fuel consumption	7.5	6.5	6.9
SC7166D vs	(L/100 km) and	9.7	6.0	7.3
SC7166C	change(%)	−22.7 %	8.3 %	−5.5 %
Chery A5		9.4	6.4	7.3
SQR7160A217/B vs		10.7	6.9	8.3
SQR7160A217		−12.1 %	−7.2 %	−12.0 %
Dongfeng S30		9.0	5.6	6.8
DFM 716 B1DHEV vs		10.5	5.7	7.5
DFM7160B1A		−14.3 %	−1.8 %	−9.3 %
Zhonghua		8.0	5.6	6.5
SY 7150X1SHEV vs		9.2	5.8	7.1
SY7150X1SBAB		−13.0 %	−3.4	−8.5
SAIC-GM LaCrosse		12.3	6.1	8.3
SGM7240HAT vs		13.0	6.8	9.1
SGM7240ATA		−5.4 %	−10.3 %	−8.8 %

is 12.2 % lower than baseline car (model SQR7160A217). Under the suburban cycle and combined cycle, it can save 7.2 and 12 % fuel, respectively.

The Dongfeng Fengshen S30 BSG car consumes 14.3 % less fuel on urban road cycle and 1.8 % less fuel on suburban road cycle. The reduction of gasoline consumption on combine road cycle is 9.3 %.

The SAIC-GM LaCrosse is equipped with the BAS start-stop device developed by General Motors Corporation. Compared with the same type of vehicles not equipped with the BAS start-stop device (model SGM7240ATA), it can reduce gasoline consumption by 5.4 % on urban roads cycle and 10.3 % on suburban roads cycle. The gasoline consumption on combined roads cycle is reduced by 8.8 %.

In addition, the Geely stop-go system, which is used in the Dihao EC7, can reduce gasoline consumption by 5–10 % on combined road cycle and by 5–15 % in heavy traffic condition.

The application of BSG can reduce the gasoline consumption of some vehicles to meet the level required by regulations at the second stage and also to meet the threshold level required by regulations at the third stage (Table 4.10). From the comparison in the table, it is evident that the Changan Automobile CX30 has already met the threshold level for passenger car gasoline consumption at the third stage. When equipped with BSG, its gasoline consumption is further reduced by 2.2 %. Although the Chery A5, Dongfeng S30, Junjie, and SAIC-GM LaCrosse meet the requirements for gasoline-consumption regulations at the second stage, none of them meet the requirements for gasoline-consumption regulations at the

Table 4.10 Fuel consumption with and w/o BSG and national stage limits

Vehicle type	Curb weight (kg)	Fuel consumption (L/100 km) Second stage limit	Third stage limit	w/o BSG	With BSG	Gap with second stage (%)	Gap with third stage (%)
Changan CX30	1,325	9.2	7.3	7.3	6.9	−5.5	0.0
Chery A5	1,350	9.2	7.3	8.3	7.3	0.0	13.7
Dongfeng S30	1,215	8.6	6.9	7.5	6.8	−1.4	8.7
Zhonghua Junjie	1,240	8.6	6.9	7.1	6.5	−5.8	2.9
SAIC-GM LaCrosse	1,630	10.7	8.5	9.1	8.3	−2.4	7.1

third stage: their gasoline consumption is, respectively, 13.7, 8.7, 2.9, and 7.1 % higher than the gasoline consumption required at the third stage. After adopting smart start-stop technology, their gasoline consumption is reduced and all of them are able to meet the requirements for gasoline-consumption regulations at the third stage. Moreover, the gasoline consumption of the Dongfeng S30, Junjie, and SAIC-GM LaCrosse is 1.7, 5.8, and 2.4 % lower than the threshold level. Therefore, the gasoline consumption of some vehicles can be reduced by adopting a smart start-stop system to meet the threshold level at the third stage.

4.2.2.2 ISG Technology

ISG is different from BSG in how it achieves its smart start-stop function. The difference lies in how ISG concentrates the function of the generator and motor into the starter of conventional vehicles instead of into the generator as in BSG technology. In addition to the smart start-stop function, ISG can also recycle braking energy and be used as auxiliary power. It can reduce gasoline consumption by 15–25 % (Sidebars 4.10 and 4.11).

Sidebar 4.10: Working Principles of ISG

ISG technology replaces the flywheel. Through the inertia of its own movement and its switch between the motor and generator, ISG can balance the fluctuation of the crankshaft of the internal-combustion engine and thus becomes an active flywheel and shock absorber. All the accessory movements of the internal-combustion engine are driven electrically. Toothed belts and gear sets can all be eliminated. At the same time, ISG technology can eliminate the need of traditional generators and motors. The internal-combustion engine attachment can thus be arranged more flexibly.

Sidebar 4.11: Comparison Between ISG and BSC Technologies

ISG technology is more energy saving than BSG technology for the following reasons:

1. It has an energy-saving smart start-stop function like BSG.
2. It has auxiliary power. ISG can switch to the electric motor mode when the internal-combustion engine is at low speed. In this way, it can increase the output of the power system and lower the displacement requirement required by engine.
3. It offers braking-energy regeneration. Under deceleration or braking conditions, the inertia of the vehicle enables the electric motor to operate in the generator mode and convert part of the braking energy of the vehicle into electric power, which is stored in the power battery, and at the same time produce braking torque.

The ISG system typically consists of a motor (with dual functions—electricity generation and driving), battery, and control system. With ISG, the motor is changed from the conventional cylindrical starter motor, which engages the flywheel of the engine through gears, into a disk-type motor. The disk-type motor, which often adopts a permanent-magnet synchronous motor, replaces the conventional flywheel of the engine and is directly integrated into the engine crankshaft. The battery capacity of the ISG system is about 2 kWh and is larger than with the BSG system. It is difficult for lead–acid batteries to meet the technical requirements for ISG, so high-performance nickel-metal hydride or lithium-ion batteries are employed. The control unit has to control the motor according to the driver's intention, the status of the vehicle, and the status of the battery capacity to achieve a fast start, power acceleration (motor drive), recycling of brake energy (by storing the braking energy in the power battery as electricity), and other different functions.

The difference of function between ISG and BSG is in that starter of engine acts as both electric generator and motor in ISG system, and in BSG system, that starter of engine only acts as electric generator. ISG has the functions of power assisting and brake energy recuperation to obtain 15–15 % fuel-saving ratio (shown as Sidebars 4.10 and 4.11).

The battery capacity of ISG is larger than that of BSG, being more than 2 kWh. Base on the vehicle operation status, the driver intention, and SOC of battery, the ECU sends control signals to determine the powertrain operating modes, including power assisting and braking-energy recuperation. The representative of internationally advanced technology of ISG is IMA of Honda Company, which has been successfully applied on Insight and Civic. In China, the key automotive companies all have their ISG HEV prototype. Chang-An and SAIC produce hybrid electric vehicle with ISG technology, for example, Jiexun, Zhixiang, and Roewe 750 hybrid electric vehicles, and the parameters of performance are shown in Table 4.11.

Table 4.11 Comparison of domestic vehicles equipped with ISG technology

Vehicle maker and Model		Urban cycle	Suburban cycle	Combined cycle
Changan Jiexun	Fuel consumption	8.8/12.1	5.5/7.5	6.8/8.8
SC6442H vs. SC6442B4	(L/100 km) and	−32.3 %	−26.7	−22.7
Changan Zhixiang	change (%)	8.4/10.8	5.4/6.6	6.4/8.1
SC7155HEV vs. SC7200D4		−22.2	−18.2	−21.0
Roewe 750		9.9/12.4	6.2/7.4	7.5/9.4
CSA7180ACHEV vs. CSA7180AC		−20.2	−16.2	−20.2

Table 4.12 Fuel consumption with and w/o ISG and national stage limits

Vehicle type	Curb weight (kg)	Fuel consumption (L/100 km) Second stage limit	Third stage limit	w/o BSG	With BSG	Gap with second stage (%)	Gap with third stage (%)
Changan Jiexun	1,501	9.7	7.7	8.9	6.8	15.6	−11.7
Changan Zhixiang	1,390	9.2	7.3	8.1	6.4	11.0	−12.3
Roewe 750	1,602	11.3	10.2	9.4	7.5	−7.8	−26.5

From Table 4.12, it can be seen that with ISG, the fuel economy can be improved greatly to meet the requirement of the third phase limit regulation of vehicle fuel consumption. For example, the fuel economy of equivalent ICE models is respectively 8 L/100 km and 9 L/100 km, which cannot meet the third phase limit regulation of vehicle fuel consumption of 7.78 L/100 km and 7.38 L/100 km. The fuel consumption of two kinds of vehicles is 15.6 and 11 % higher than the regulation value. However, if the ISG technology is applied, the fuel economy can been increased to 6.8 L/100 km and 6.4 L/100 km, 11.7 and 12.3 % lower than the regulation value. Targeting to the fourth phase regulation of fuel consumption limit, the ISG technology will play important role.

4.3 Hybrid Electric Vehicle Technologies

Hybrid electric vehicles (HEVs) are compound vehicles that are powered by a primary device—an internal-combustion engine and an auxiliary device—an electric drive system. Hybrid electric vehicle technology combines and controls a vehicle's internal-combustion engine, motor, and energy-storage units (for example, the traction battery and super capacitor) in an optimal fashion. In addition, automatic start-stop functions, auxiliary drive, regenerative braking energy, optimization of the engine's operating conditions, downsize of engine displacement, and other means are adopted to enhance the energy efficiency and reduce emissions.

The hybrid electric powertrain can be classified into three categories in terms of structure—series, parallel, and series–parallel—based on the structural relationships among the drive system's mechanical structure, energy flow, and power flow. The

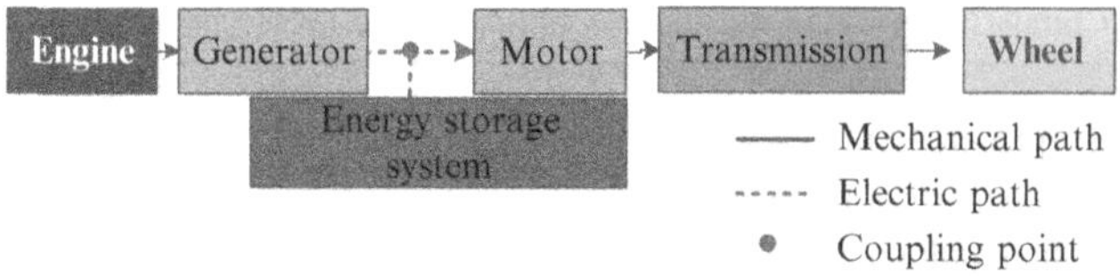

Fig. 4.18 Series HEV technology

hybrid electric powertrain with a series connection employs a type of electric coupling, that with a parallel connection uses mechanical coupling and that with a series–parallel connection combines electric and mechanical couplings.

4.3.1 Series HEV Technology

An example of series HEV technology is given in Fig. 4.18. Driven by the engine, the generator produces power, which, as electric energy, flows directly through the controller to the motor, where the electric energy is converted to driving torque to power the vehicle. In this system, the engine and generator are mechanically connected as an electricity generator; the motor, gearbox, and wheel are mechanically connected to transfer power; the generator and motor are linked by a cable, to which the battery, capacitor, or other energy-storing units positioned between them can be connected.

With series HEV technology, the energy-storage device (battery or super capacitor) functions by balancing the generator power and the input power of the motor. When the generator's power is greater than that required by the motor (for example, when the vehicle is slowing down, driving at low speed, or stops for a short time), the generator is operated by the control system, and it charges the energy-storage device. When the generator's power is lower than that required by the motor (such as when the vehicle starts, accelerates, and climbs), additional power is provided to the motor from the energy-storage device.

With a series hybrid powertrain, decoupling of the generator's power and the power required by the vehicle can be achieved. As a result, the engine can function stably in optimal conditions, therefore enhancing its performance, especially its economic performance in terms of gasoline consumption and emission characteristics.

4.3.1.1 Operating Modes

The typical operating modes for the series hybrid electric system appear in Fig. 4.19 and are described below.

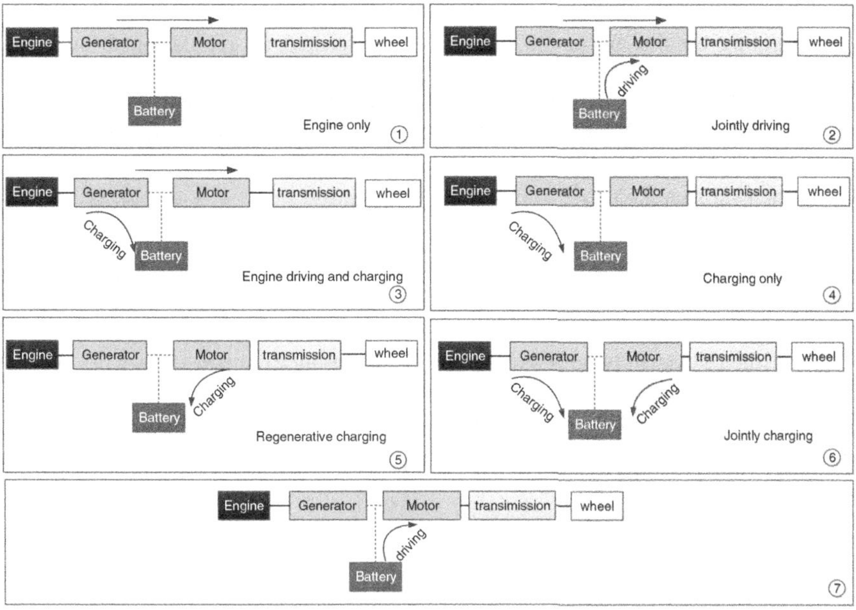

Fig. 4.19 Operating modes of series HEV powertrain

Normal Driving Mode

When a vehicle is in the normal driving mode, the engine is the sole power source. Usually, this occurs when vehicles are driving at constant speed. The engine is then working under optimal operating conditions, it is stable, and its economic performance and emissions are at their most favorable; no energy flows through the battery or capacitor, which avoids unnecessary energy losses in battery recharging or discharging process. Compared with conventional vehicles driving at constant speed, HEVs in the normal driving mode can be considered as having the additional function of stepless speed control. This is an advantage, but there are losses in each of the two energy transformations (transformation of mechanical energy to electric energy in the generator; transformation of electric energy to mechanical energy in the motor), so the total efficiency may be lower than with mechanical transmission.

Compound Driving Mode

The compound driving mode usually occurs when a vehicle accelerates or climbs. When this drive mode applies, the power need is greater than that supplied by the generator, so the battery in parallel connects with the engine output and, under the control system, provides complementary power to the motor. Even under these conditions, the engine still works at optimal status. When the load changes, the operating condition can be smoothly transformed by regulation through the control system; therefore, there is no sudden increase in emissions as occurs with conventional vehicles under the variation conditions.

Drive Charging Mode

Subsequent to some operation (e.g., the compound driving mode), the capacity (state of charge, SOC) of the battery falls to a lower level, so it must be recharged to a position for the next cycle demands. If the compound charging mode is also required, the battery may be at a low power level owing to self-discharge if a vehicle has not been used for a long time.

Compound Charging Mode

This mode generally occurs in extreme conditions (e.g., long-distance climbing), when the battery capacity is at its lowest status. Subsequently, when driving downhill, the motor may generate power through braking. This energy is collected and stored in the battery together with the power produced by the generator.

Generator Charging Mode

Powered by the engine, the generator produces electric energy, all of which is completely used for recharging the battery rather than in driving the vehicle. This mode usually operates when the battery is at low capacity or the vehicle stops for a short period, such as when stopping for a traffic light in a busy city.

Brake-Energy Recycle Charging Mode

This mode usually operates when a vehicle is braking and slowing down. At such times, the motor generates power to achieve electric braking. As a result, the energy produced in braking is stored in the battery and reused when necessary. This is a very important path to reduce energy consumption for the hybrid system.

Electric Driving Mode

This situation only occurs when the vehicle is powered by the electric energy stored in the battery. At such time, the battery is full, and so its energy must be used up to ensure that there is capacity for effective collection of braking energy during the next braking cycle. This mode also applies when a vehicle is driving on certain areas that have very strict requirements with regard to emissions and noise. In this mode, series HEVs can achieve zero emissions.

4.3.1.2 Benefits

With series HEV technology, energy conservation and reduction in emissions are achieved mainly through braking-energy recycling, optimization of the engine's working parts, and reduction in engine displacement.

Braking-Energy Recovery

When an HEV brakes, the wheel is connected by the gearbox to the motor, which is in the power-generation mode to transform the vehicle's kinetic energy to electric energy; this is commonly called "electric braking." Comparatively speaking, more energy is collected with a four-wheel-drive vehicle than with a single-axis vehicle; this is because the latter may apply electric braking to the driving wheels only, with the other wheels being used for mechanical braking. The amount of energy collected in the braking process depends on such factors as the road characteristics, the safety features of the vehicle's braking systems, the vehicle's braking power, and battery-recharging features.

Optimization of Engine's Operating Region

Transmission technology is employed in conventional vehicles to make an engine's working loads closer to the optimal area. Theoretically, the more gears a vehicle has, the better it is for the engine to work at the optimal point; thus, the engine's working loads may be shifted to the optimal area under different operating conditions by adjusting the rotating speed according to the stepless speed-change principle. However, because it is limited by the actual rotating speed-change section and torque working section, the motor does not often work at the optimal operation area. With series HEV technology, the motor may reach the optimal area for any power requirement at any rotating speed and torque through conversion by means of various advanced technologies (for example, voltage change). This is because the energy-flow process involves electric energy rather than conventional mechanical energy transmission.

Downsizing of Engine Displacement

To supply the reserves of power to meet certain dynamic requirements (such as acceleration and climbing), it is necessary for conventional vehicles to have a large engine displacement. However, a partially loaded large-displacement engine is not

Table 4.13 Fuel consumption and emissions of American series hybrid electric buses

Item	NO_X emission (g/mile)	Particle emission (g/mile)	CO_2 emission (g/mile)	Fuel consumption (mpg)
Diesel system	18.8	0.24	2.84	3.6
BAE hybrid system	9.1	0.02	1.95	4.5
Reduction ratio	51.6	91.7	31.3	25

efficient. With HEVs, however, a reduction in displacement means that the engine is often able to work with high efficiency and produce low emissions, and the power need for reserve use is supplied by the traction motor and battery.

Series HEVs are especially suitable for driving at low speeds in the city. When starting and driving at low speeds in crowded urban conditions, series HEVs may use the battery for power output only by cutting off the internal-combustion engine and thereby achieve zero emissions characteristics. However, nothing is perfect. The engine's output energy must be completely transformed to electric energy before it is transformed to mechanical energy; thus, the utilization rate of gasoline energy is low owing to the lower efficiency in energy transformation between mechanical and electric energy, and battery recharging and discharging.

4.3.1.3 Performances

Recently, the series hybrid electric system has been popular for commercial vehicles in the world. One representative overseas model is the Orion VII hybrid electric bus made by Daimler (the hybrid electric system is made by the American BAE System Company). In the United States, the series hybrid electric system is currently being used in over 3,000 vehicles, 1,700 of which are in New York (Wang 2011). Within China, commercial vehicles using series hybrid electric systems are the TEG6128SHEV city bus by Hunan CSR Times Electric Vehicle Co., Ltd. (CSR) and the LCK6112/6120/6110GHEV city bus by Zhongtong Bus. However, series hybrid electric systems are seldom used by companies that make passenger cars; an exception is the Volt made by General Motors.

Compared with conventional diesel bus, both fuel consumption and emissions with series HEVs may be lower to some extent. A comparison in terms of performance between the American Orion VII hybrid electric bus and a conventional diesel one appears in Table 4.13 (Beijing SinoHytec 2009). With regard to performance, series hybrid electric bus is lower in NO_X emissions, particle emissions, CO_2 emissions, and fuel consumption by 51.5, 91.7, 31.3, and 25 %, respectively, than conventional diesel bus. With conventional diesel bus, particles easily form during rapid acceleration, but the BAE hybrid electric system overcomes this problem, greatly reducing particle emission by over 90 %. In addition, by recycling energy in braking and optimizing engine operating condition, fuel consumption is reduced by 25 %.

Table 4.14 Fuel conservation of domestic series hybrid electric buses

Enterprise	Vehicle model	Basic model for comparison	Fuel saving (%)
Hunan CSR Times Electric Vehicle Co., Ltd.	TEG6128SHEV	TEG6128G	29.26
	TEG6129SHEV	TEG6129G	34.71
	TEG6119SHEV	TEG6119G	42.24

Table 4.14 shows test results for typical series hybrid electric bus in China. The results were selected from recommended vehicle models in terms of energy conservation and new-energy vehicles; the data were released by the China Vehicle Technology Service Center, which is recognized by the Ministry of Industry and Information Technology of the People's Republic of China. As seen in the table, the three urban series hybrid electric vehicles produced by CSR have a fuel saving of 29–42 % depending on the model.

4.3.2 *Parallel Hybrid Technology*

Parallel hybrid electricity technology mainly refers to the power of the generator and engine, which are mechanically combined to power the vehicle. Vehicles may be driven by the generator and engine together or separately. According to the position of the mechanical bonding points, parallel structures may be classified into five categories: type 1, before-clutch configuration; type 2, after-clutch configuration; type 3, after-gearbox configuration; type 4, dual-clutch configuration; and type 5, hybrid through the road (TTR). These types are shown in Fig. 4.20. The transmission device and hybrid electric device are required for the parallel structure; so compared with the structure in series hybrid technology, the transmission mechanism is to some extent more complicated with parallel hybrid technology.

BSG technology and ISG technology with intelligent start-stop functions are classified as type 1; such vehicles are also called micro-hybrid and mild-hybrid electric vehicles. Types 2, 3, 4, and 5 are for high-power motors, generally 30 % greater than driving power; such vehicles can be driven by electricity only at certain time and are called full-hybrid systems (Sidebar 4.12).

Sidebar 4.12: Series Hybrid Electric Systems

Micro-hybrid, mild-hybrid, and full-hybrid types are classified by the proportion of motor power among all power sources. Among the three kinds of hybrid power, the micro-hybrid electric system has the lowest motor power—generally suitable only for rapid engine cutoff and stopping. With the micro-hybrid system, the functions of motor-assistant force and accumulation

(continued)

(continued)
of braking energy are added to the mild-hybrid system. With the full-hybrid system, the motor power is sufficiently high to drive the vehicle without assistance from other sources, and so battery electric drive is possible for that system. The modes of the parallel system's engine include start-stop, pure electric mode, assistant power mode, and regenerative braking mode. Only systems where the clutch is between the engine and mixing point are suitable for the pure electric mode, as shown in Fig. 4.21. When a vehicle moves slowly forward or backward, the clutch in front of engine is released to stop the engine; under this condition, the driving force is provided by the battery. In the pure electric mode, the engine could stop completely, thereby reducing gasoline consumption.

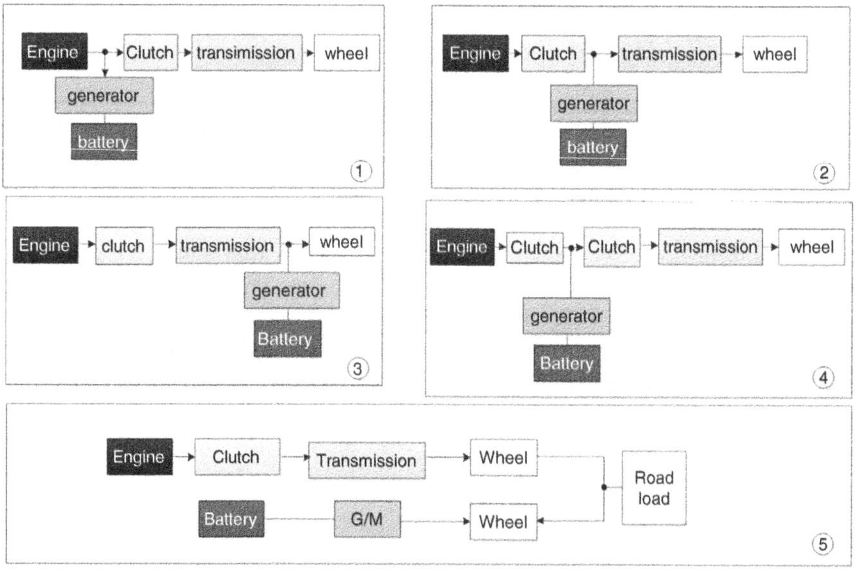

Fig. 4.20 Parallel hybrid electric system structure

At economical speeds, the parallel hybrid electric system is more cost-effective in terms of fuel consumption than the series hybrid electric system; this is because the parallel system is directly powered by the engine through mechanical transmission, which produces a higher energy utilization rate. However, because of structural limits, the parallel hybrid electric system may result in transient changes to the engine's operating condition. This changing status may affect fuel consumption efficiency and emissions. This system is therefore particularly suitable for vehicles

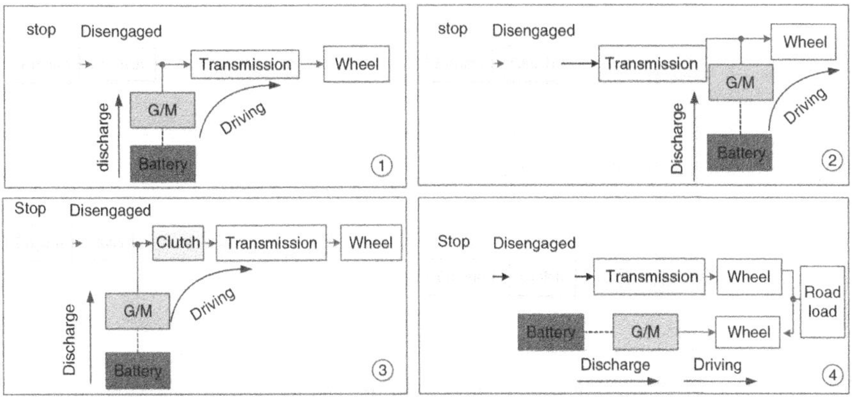

Fig. 4.21 Pure electric mode with parallel hybrid technology

that drive largely under stable operating conditions on urban roads and expressways, but it is not suitable where road conditions change, such as in crowded urban traffic.

At present, parallel hybrid electric vehicle technology is relatively successful and has undergone large-scale application. It is integrated as the starter generator (ISG, described above) in passenger cars, for example, Honda's Insight and Civic, the Zhixiang of China's Changan, and the Roewe 750 of SAIC Motor. The effect of parallel hybrid electric vehicle technology on fuel consumption is discussed in Sect. 4.2.2.2 (ISG Technology).

Examples of hybrid electric commercial vehicles are the Allison hybrid electric system of GM and the gasoline-power parallel hybrid electric system made by Eaton. In China, parallel power technology is widely adopted by companies such as the following: JAC Coaches; Dongfeng Yangtse Motor (Wuhan) Co., Ltd.; Dandong Huanghai Automotive Company Limited; Hunan CSR Times Electric Vehicle Co., Ltd.; Suzhou Kinglong Higer; Beiqi Foton Motor Co., Ltd.; Xiamen King Long United Automotive Industry Company, Ltd.; Youth Automobile Company; and Zhongtong Bus Holding Co., Ltd.

Since 2009, delivery vehicles of American shipping company UPS have used Eaton's parallel hybrid electric system in the city of Phoenix. As shown by the US Department of Energy's follow-up evaluation, under practical road driving conditions, a vehicle using the parallel hybrid electric system showed reduced fuel consumption by 28.9 %—from 17.96 to 23.06 L/100 km—compared with conventional diesel-powered vehicles. Under laboratory conditions and in test drives in the central business district of a city, a vehicle with a parallel hybrid electric was found to have reduced fuel consumption of 31 and 34 %, respectively (Lammerf 2009). After the Allison parallel hybrid electric system of GM was used for a bus system in King County, it was found that this system may increase the fuel efficiency of heavy buses (hinge type) by 27 %—from 3.15 to 21.5 mpg (Chandler and Walkowicz 2006).

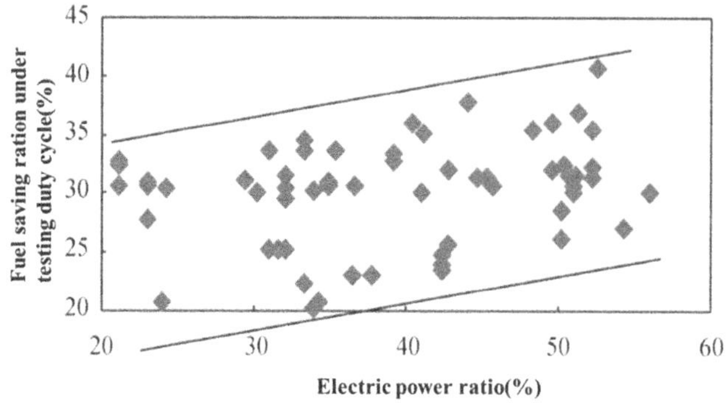

Fig. 4.22 Fuel-saving rate parallel hybrid electric bus in China (from published data)

In contrast to series HEV technology, parallel HEV technology is widely used in bus in China. From the model catalogues of various manufacturers, among 76 hybrid electric bus models, there were three series HEVs, 72 parallel HEVs, and one series–parallel vehicle. In Fig. 4.22, the relationship between the fuel-saving rate and electric power rate is presented. The fuel-saving rate of parallel hybrid electric bus in China ranges from 20 to 40 %, and there is an upward trend with both the electric power and fuel-saving rate.

4.3.3 Series–Parallel HEV Technology

Series–parallel HEV technology was developed through a combination of series HEV technology with parallel HEV technology. In series–parallel HEV technology, part of an engine's output power is mechanically transmitted to the vehicle's driving system; the rest is used for power generation by the generator. Power produced by the generator flows to the motor or battery; the motor then produces driving torque, which is transmitted to the driving system through the compound power device. With a series–parallel driving system, when a vehicle drives at low speed, the driving system works in series mode, but the parallel mode operates for high-speed driving.

The series–parallel hybrid electric driving system was developed by making full use of the advantages of series and parallel systems. With optimized matching of the engine, generator, motor, and other components, the vehicle always works under the most favorable status—even under complex physical operating conditions; it is thus easier to control emissions and gasoline consumption. Since the series–parallel system is more complex than the parallel system, it accordingly makes higher requirements of the compound power device. At present, the series–parallel structure uses a planet gear as the compound power device's basic structure. Generally, series–parallel technology is mainly used for passenger cars, including Toyota's Prius and Lexus (Sidebar 4.13); sometimes, it is used for commercial vehicles, including the SWB6127HE2 urban hybrid electric passenger cars of SAIC

Motor. Developed by FAW and using the series–parallel structure, the Benteng B70 HEV has a power assembly that comprises a 1.3-L engine (67 kW) and a hybrid electric assembly that is equipped with mechanical automatic manual transmission. Under urban operating conditions, the vehicle has a gasoline consumption of 6.9 L/100 km and general gasoline consumption of 6 L/100 km. This amounts to a reduction in gasoline consumption of 37 % compared with gasoline manual transmission.

Sidebar 4.13: THS System

The Toyota Hybrid System (THS) is made with global state-of-the-art HEV technology. The second-generation THS is equipped with a 1.5-L, in-line, four-cylinder gasoline engine (maximum power, 57 kW) and a 50-kW motor (output torque, 400 Nm). According to the newly released gasoline-consumption data by the US Environmental Protection Agency, gasoline consumption of the second-generation Prius is 4.6 L/100 km under urban conditions and 4.9 L/100 km under expressway condition. In 2009, the third-generation Prius, with a 1.8-L engine, was put on the market. The new generation shows improved power performance, and gasoline consumption is 3.9 L/100 km—8 % higher than the previous generation.

The Prius has excellent gasoline and emission performance because it has extensively assimilated a number of advanced energy-conservation technologies based on its hybrid electric system technology. For example, the vehicle uses Atkinson cycle engine technology, which reduces gasoline consumption by 8.5 %; if it is equipped with a cooled EGR, there is an additional 1.7 % saving in gasoline. In addition, the vehicle is equipped with intelligent-variable timing and an electric water pump.

Moreover, the THS is also used for the Lexus and Camry. The Camry hybrid electric automobile consumes 5.4 L/100 km gasoline at a constant driving speed of 90 km/h, and its overall gasoline consumption is 6.0 L/100 km for urban road driving—a saving of up to 41.75 % gasoline compared with the normal Camry. The Lexus RX450h, GS450H, and LS 600 h have gasoline consumption of 6.3 L/100 km, 7.6 L/100 km, and 9.3 L/100 km, respectively.

By 2011, the THS had been used in over three million vehicles. In 2010, 7.53 million Toyota cars were sold worldwide. Of those, 690,000 were HEVs, accounting for almost 10 % of the total.

4.4 Electric Driving Powertrain Technologies

Electric driving powertrain technologies include plug-in electric powertrain, extended-range electric powertrain, battery electric powertrain, and fuel-cell electric powertrain. The characteristics of these types are shown in Table 4.15.

Table 4.15 Characteristics of electric driving powertrain technologies

Type	Type of energy source	Main energy source
Plug-in hybrid electric system	Internal-combustion engine + traction battery	Engine
Extended-range electric driving system	Internal-combustion engine + traction battery	Traction battery/fuel cell
Electric powertrain	Traction battery	Traction battery
Fuel-cell electric powertrain	Fuel cell + traction battery	Fuel cell

Table 4.16 Performance of selected plug-in HEVs

Vehicle	Engine	Battery capacity (kW · h)	AER (60 km/h)	Fuel consumption (L/100 km)	Fuel saving (%)
Prius PHEV	60 kW	5.2	23.4	1.75 (PHEV cycle)	76
BYD F3DM	1.0 L, 50 kW	18	70	2.2	65.4
RS PHEV	1.0 L, 35 kW	10	50	2. 8	65
ROEWE 550PHEV	1.5 L, 50 kW	11.8	50	2.7	70
Benteng B50PHEV	1.6 L, 74 kW	10	45	5	44

4.4.1 Plug-in HEVs

Based on the normal hybrid electric system, plug-in hybrid electric system has greater battery capacity than a normal hybrid electric system to have certain all-electric range (AER), with recharging connections. As a result, there is no emission of pollutants when driving in urban commuting conditions (within the AER range), and it overcomes such problems as a lack of power (common to electric vehicles) and irregularity in the energy supply. There are two operating modes—charge-depleting mode and charge sustainable mode; the former mode can be subdivided into a pure electric mode and hybrid mode. In both modes, there is consumption of the battery's electric energy acquired externally so as to replace the fuel used for driving. The charge sustainable mode can be regarded as a normal hybrid electric operating mode, and the capacity of its power battery is maintained at around a relatively constant value. However, its working process is similar to that of a hybrid electric system, and it thus achieves fuel-saving effects by a further mixture of power. Plug-in HEVs achieve a reduction in fuel consumption of over 50 % compared with conventional vehicles. Currently, plug-in HEV technology is recognized as the most feasible transition technology toward pure electric power.

Around the world, vehicles that employ this technology mainly include Toyota's Prius, the BYD F3DM, the Benteng B50 PHEV and RS PHEV, and Changan's Zhixiang; the latter has successfully appeared in the Chinese market. Table 4.16 lists the performance indicators of several Chinese plug-in HEVs compared with foreign vehicles.

4.4.2 Extended-Range Electric Vehicles

Extended-range electric vehicles are battery electric vehicles that are primarily powered by a battery and they use a small engine for auxiliary power. The vehicles are driven by a motor, so the battery has a large capacity; however, the engine does not offer driving power to it directly. In addition, the engine functions only to recharge the battery when necessary. The powertrain of the vehicle comprises a battery system, transmission, vehicle control unit, and auxiliary power system. Among those systems, the vehicle control unit completes the control strategy, and the battery may be recharged by a ground charging point or vehicle-loaded charger. As required, the operating mode may be battery electric or vehicle-loaded power mode.

The engine of extended-range electric driving vehicles functions in power generation, and it works under optimal rotation speed under continuous operation once it has started. In addition, its output power and torque usually remain constant. Therefore, its efficiency, emission, and reliability are maintained at optimal status, and there is no need for it to engage in motor coupling. The battery of the system has a large capacity for a greater driving range so that it may effectively accumulate energy incurred in vehicle braking and driving downhill. In addition, the battery may offer sufficient power to the engine for start-up, acceleration, and climbing and act as an auxiliary motor.

When the electric vehicle market was first developed, the maximum driving range of the vehicles was limited owing to a lack of recharging facilities and an inadequate charging network. The charging of electric vehicles has now become much more convenient. However, this whole problem is overcome with extended-range electric driving vehicles, which employ a traditional internal combustion. Extended-range electric vehicles thus present a feasible technical solution in the transition period to battery electric vehicles. Extended-range electric vehicle technology is quite similar to electric vehicle technology, but the battery cost is much lower than with battery electric vehicles. Extended-range electric vehicles have remarkable technological features: they combine the advantages of battery electric vehicles with those of HEVs. This ensures long-distance electric driving and the transfer to gasoline mode for continuous driving when the vehicle-loaded battery system lacks sufficient power. The core of this technology is power system control, which maximizes energy use.

Typical vehicles around the world that use this technology are the Audi A1 e-tron, Audi A5 e-tron Quattro, and Volt of GM. GM has developed series structural and series–parallel extended-range electric driving vehicle systems. The Volt uses series structure and may drive for 64 km continuously in the pure electric mode. Beyond that range, the battery will function for further driving after it has been recharged by a small engine, which is installed below the trunk. The 254-mL engine has a peak rotation speed up to 5,000 rpm, which is capable of driving the 15-kW generator, though the total weight of the engine and generator is only 70 kg. The electric system

Table 4.17 Performance parameters of Chery's extended-range electric cars

	M3-REEV	M1-REEV	QQ3-REEV
Seating capacity	4	4	4
Weight (kg)	1,050	1,150	1,000
Maximum speed (km/h)	80	120	60
Maximum gradeability (%)	25	25	15
Time for acceleration from 0 to 50 km/h (s)	8	7	17
Driving range under electric mode (km)	100 (ECE)	104 (ECE)	90 (ECE)
Type of motor	Permanent magnet synchronous motor		
Power of motor (kW)	12/18	29/40	3
Type of traction battery	Lithium-ion battery		Lead–acid
Capacity of traction battery	60	60	150

may collect data relating to the journey, so at start-up the extended-range engine becomes automatically engaged as appropriate with respect to the destination and route conditions. In addition, the extended-range engine has a 12-L gasoline tank, and its range is 200 km. The Volt employs a series–parallel structure, and when it drives under 112 km/h, the pure electric mode is engaged. However, when driving at speeds over 112 km/h, the electric system is turned off and the vehicle is powered by the engine.

Based on the A1, the extended-range electric Al e-tron has an engine of 45 kW (peak power 77 kW) and peak torque of 150 Nm; it can achieve acceleration to 100 km/h in 10.2 s. The maximum speed is 130 km/h, and it is very suitable for urban driving. Under the electric driving mode, it can drive continuously for 50 km. Once the battery power is almost exhausted, the 0.245-L gasoline rotary engine may drive the engine at a constant 5,000 rpm, thereby extending the range to 200 km. The Al e-ton has general gasoline consumption of about 1.9 L/100 km and CO_2 emission of only 45 g/km.

Within China, extended-range electric technology is already used in buses and passenger cars. Chery Auto has developed several small-range extended electric cars (Table 4.17). The first extended-range vehicle developed by Chery was the Qilin M1-REEV, which appeared in 2010, and offers multiple options for driving mode. The M1 extended-range electric vehicle has a 15-kW range extender, allowing a maximum range of over 350 km. The M1-REEV adopts a rotary engine range extender, which is highly integrated with design and intelligent control technologies.

Extended-range electric technology is suitable for commercial vehicles, especially urban bus. Short-distance buses are also suitable for extended-range electric technology: if there is deficient energy, power generation can be supplied from the generator, which is driven by a small power engine at constant rotating speed; this may be recharged even when the vehicle is stationary. In addition, with extended-range electric vehicles, the problem of in-vehicle air conditioners may be effectively solved (Sidebar 4.14).

Sidebar 4.14: Air-Conditioning Energy Consumption

Power consumption for air-conditioning is a great problem for electric buses. If a vehicle drives on average at 16 km/h, it will take over 6 h to travel 100 km; under such conditions, it requires 60–70 kW · h, of which 10 kW · h is used for air-conditioning. If an electric vehicle's air conditioner is not being used, 60–70 kW · h is required to drive 100 km, but the power consumption increases to 150 kW · h if the air conditioner is being used in hot weather. That is to say, most of the electricity in an electric vehicle's battery is used for air-conditioning. However, using extended-range vehicles is an effective way to reduce power consumption by the battery since the engine may generate power for the air conditioner.

The Suzhou Kinglong Higer KLQ6129GQHEV1 extended-range electric bus adopts a small ICE, and it requires no complicated mechanical coupling. Only 30–40 % of the battery capacity of an electric vehicle is required by an extended-range electric bus, which greatly reduces costs. Moreover, the low-speed charging may be adopted for power recharging without the necessity for battery replacement; this reduces costs for constructing charging-station facilities and battery replacement. The Higer KLQ6129GQHEV1 may attain a driving range of 50 km in the electric mode; its diesel consumption is 30–50 % lower than that of conventional vehicles, and its emission and acceleration indicators are likewise superior.

4.4.3 Electric Vehicles

As motor-driven vehicles with power from in-vehicle energy-storage devices—such as in-vehicle batteries, super capacitors, and flywheel—electric vehicles are approved for driving on normal roads in China, subject to road traffic safety regulations. According to speed, electric vehicle may be classified into the following categories: low-speed vehicles (speed under 40 km/h), low- to middle-speed vehicles (40–70 km/h), mid- to high-speed vehicles (70–100 km/h), and high-speed vehicle (over 100 km/h). According to usage, electric vehicles may be classified into the following categories: electric cars, electric buses, electric trucks, and some special electric vehicles, such as mail vehicles, engineering vehicles, golf buggies, and tourist vehicles.

With no greenhouse gas emission during use, electric vehicles are completely powered by electricity. They thus represent diversification of primary energy and represent a break from the reliance on petroleum. At present, the mainly critical technical issues include the following. First, the battery has low specific energy, that is, energy produced by the battery, which results in high energy consumption and a short driving range, thus reducing the overall power performance of the vehicle.

Table 4.18 Performance of elected electric vehicles

Parameter	Leaf EV	RS EV	BYD e6
Seating capacity	5	5	5
Weight (kg)	1,521	1,200	2,295
Maximum speed (km/h)	144	95	140
Time to accelerate (s)	11.9 (0–100 km/h)	6 (0–50 km/h)	10 (0–100 km/h)
All electric range (km)	175 (NEDC)	100 (60 km/h)	300
Motor	AC synchronous	Permanent magnet DC brushless	Permanent magnet synchronous
Power of motor (kW)	107 hp	58	90
Type of traction battery	Lithium-ion	Lithium iron phosphate	Lithium iron phosphate
Battery (kWh)	24	15	48
Energy consumption (kWh/100 km)	21.1	13	21.5

Second, it takes a long time to recharge the battery, thereby lowering the vehicle's overall mobile performance. Third, the battery is very expensive, and this has proved to be a bottleneck for promoting this type of vehicle.

Battery technology is the core technology for the development of electric vehicles. Lithium-ion battery technology has achieved a great breakthrough in recent years, its safety and cost performance having greatly improved. Manufacturers of electric vehicles, largely represented by Japanese brands, have moved to the mass marketing of these vehicles. As of November 2011, over 20,000 Leaf electric vehicles had been sold over the world, and it is expected that up to 40,000 of these vehicles will be sold in 2012. However, Chinese electric vehicles have now reached the stage of demonstration, promotion, and technical verification. Currently, the JAC RS is the most used domestic electric vehicle in China; there are 585 of these vehicles. The performance parameters of several typical electric vehicles are given in Table 4.18.

4.4.3.1 Special Electric Commercial Vehicles

Compared with passenger cars, electric commercial vehicles have more apparent fuel-saving and emission-reduction effects. If a bus in China drives 220–280 km a day, the daily fuel consumption is about 90–120 L, which is equivalent to the fuel consumption of 30 passenger cars. The number of ordinary buses accounts for only 1.7 % of all motor vehicles in China, but their fuel consumption accounts for 25 % that of all vehicles, and their emissions account for 30 % that of all vehicles. Electric buses have zero emissions during use, which reduces carbon emissions by about 300 kg/day/bus. Electric commercial vehicles were emphasized among China's electric vehicles in the Twelfth Five-Year Plan in the public service area. Subsequent to large-scale use during the Beijing Olympic Games and Shanghai World Expo and the "Ten Cities, Thousand Vehicles" program (a large-scale promotion project

Table 4.19 Parameters of Chinese commercial electric vehicles

Parameter	Electric bus for olympic games	Yutong ZK6129EGQA	BYD K9	Wuzhoulong FDG6113EVG
Curb weight (kg)	13,700	11,500	18	14,200
Max speed (km/h)	80	80	7	75
Accelerate time (0 to 100 km/h) (s)	22	20	23	23
All electric range (km)	180	250 (urban operating mode)		150
Motor power (kW)	150	240	180 (90 × 2)	90
Battery type	Lithium iron phosphate			
Battery (kW · h)	140	169	2	100
Energy consumption (kWh/100 km)	94 (40 km/h)	102	13	100

for energy conservation and new-energy vehicles), commercial electric vehicles are on the way to major industrial production. The Beijing Olympic Games in 2008 used 50 electric buses; 120 electric vehicles were operated during the 2010 World Expo in Shanghai; in the 2010 Asian Games in Guangzhou, 20 electric buses were used. According to statistics from the Ministry of Science and Technology, 629 electric buses had been used for demonstration purposes in various cities by March 2011; that amounts to 7 % of the new-energy vehicles in public service use. Major commercial electric vehicle manufacturers have been established in China, including Wuzhoulong, Yutong, Ankai, and Kinglong. Details relating to commercial electric vehicles are given in Table 4.19.

4.4.4 Fuel-Cell Electric Vehicle Technology

As energy-saving, nonpolluting, environment-friendly vehicles, fuel-cell electric vehicles appear to be appropriate for future vehicle development. As indicated in the name, these vehicles use fuel cells as the power source, but the overall structure is almost the same as in battery electric vehicles. The only difference from battery electric vehicles is that most of the power energy for driving the vehicle comes from the in-vehicle fuel-cell stack. The fuel-cell stack works by transforming the chemical energy of hydrogen and oxygen into electric energy without burning. It is the reverse process of water electrolysis.

Fuel-cell electric vehicles can be divided into two categories with respect to driving mode—fuel-cell driving mode and hybrid driving mode. At present, fuel-cell electric vehicles mainly adopt the hybrid driving mode, with two common power structures—fuel cell + traction battery and fuel cell + battery + super capacitor.

In China, fuel-cell electric vehicles are made using a special electricity–electricity hybrid powertrain. The technical features of this platform include overall

Table 4.20 Performance of selected fuel-cell passenger cars

	SAICr	Dyke F-cell	Honda Clarity	Toyota FCHV adv	GM Provoq
Curb weight (kg)	1,833	17	1,625	1,880	1,978
Time to accelerate from 0 to 100 km/h (s)	15	1	11		8.5
Maximum speed (km/h)	150	17	160	155	160
Driving range (km)	300[1]	616[2]	570	830[3]	483
Hydrogen consumption (kg/100 km)	1.2[1]	1.195[3]	1.035[3]	0.91[2]	1.24[3]
Fuel-cell system (kW)	55	80	100	90	88
Hydrogen pressure (MPa)	35	70	70	70	70
Cold temperature (°C)		−25	−30	−30	−25
Motor power/torque (kW/Nm)	90/210	100/290	100/260	90/260	150/–

[1] Urban driving cycle in China
[2] New European driving cycle (NEDC driving condition)
[3] Test condition of Environmental Protection Agency (EPA operating condition)

vehicle adoption of the electricity–electricity platform, electricity–electricity hybrid energy power control, an in-vehicle high-pressure hydrogen-storage system, and the use of purified hydrogen (industrial byproduct). Jointly developed by Shanghai Volkswagen, Shanghai Fuel Cell Vehicle Power Vehicle Co., Ltd., and Tongji University, the Passat fuel-cell vehicle was initiated as a national vehicle product in April 2008.

Domestic fuel-cell electric passenger cars are almost the same as foreign vehicles in terms of basic performance indicators with respect to dynamics and driving range: the maximum speed is 150–170 km/h, and the time to accelerate to 100 km/h is 10–15 s. Thanks to the 70 MPa in-vehicle hydrogen-storage system adopted from an overseas maker, the driving range is greatly enhanced. However, with regard to basic configuration of the power system, the greatest difference is in the fuel-cell engine's output power and motor torque output. The power output of overseas fuel-cell engines is 80–100 kW, much higher than the 55 kW achieved with Chinese engines. In addition, foreign fuel-cell engines have high mass power density and volume power density. An overseas motor has a higher torque output capacity than domestic motors of the same power output capacity by 50–80 Nm—about 25–40 %. Details of Chinese and foreign fuel-cell electric vehicles are given in Table 4.20.

Researched and developed using the technical principles from different international authorities, China's urban fuel-cell electric bus is a breakthrough in terms of three systems (electric chassis, hybrid electric, and overall vehicle control) and three technologies (endurable fuel-cell power system, hydrogen-power safety, and efficiency of fuel); the project to build the bus came about after having been proposed in the Tenth Five-Year Plan. The work was carried out after comparing two different fuel-cell hybrid electric methods—energy hybrid and power hybrid. The bus complies with the national standard of hydrogen consumption for an urban

Table 4.21 Performance of domestic and overseas urban fuel-cell buses

	BAIC Group	Benz Citaro FC	Benz Citaro FC-Hybrid	AC Transit FC-Hybrid	Toyota FC-Hybrid
Curb weight (kg)	14,200	14,200	13,200	16,330	11,600
Time for acceleration from 0 to 50 km/h (s)	25	20			0–40 km/h, 9.3
Max speed (km/h)	80	80		113	80
H_2 consumption (kg/100 km)	8.5 (operating statistics)	20–24 (CUTE)	10–14 (CUTE)	8.29 (operation statistics)	9.06 (Aichi Expo)
Driving range (km)	300	>200	>250	400	250
Fuel-cell power (kW)	100	250	120	120	180
H_2 pressure (MPa)	35	35	35	35	35
Driving motor	Single motor		Dual-motor driving		
Total power (kW)	180	205	160	170	160

bus cycle $\leq$8.5 kg/100 km, and it amounts to one of the highest levels in the world. Though the bus provides similar performance and has similar configurations, its manufacturing cost is only 30 % that of similar overseas vehicles. With a maximum speed of around 80 km/h and acceleration of 0–50 km/h in 20 s, China's urban fuel-cell hybrid electric bus performance is almost equivalent to that of overseas vehicles as well as in other dynamic indicators; its driving range is 250–400 km. This vehicle has a good hydrogen consumption index, and the highest pressure of its hydrogen-storage cylinder is about 35 MPa. Details of this and other fuel-cell electric buses are given in Table 4.21.

4.5 Conclusions

1. The internal-combustion engine is mainly used for vehicles in China and accounts for over 99 % of its vehicle power technology. The percentage of gasoline engines has increased from 75 % in 2005 to 80 % in 2010, and this trend is continuing. Over 99 % of passenger cars are powered by gasoline engine, and among those, gasoline engines below 1.6 L account for up to 70 %. Diesel-powered passenger cars are mainly SUVs. Over 73 % of commercial vehicles are diesel-driven types. Over 99 % of large buses, medium-sized trucks, and heavy trucks are diesel-driven types. Commercial gasoline vehicles mainly consist of mini-trucks, light trucks, and light buses. The vehicle's power system is in the initial stage of electrification; but over 50 % of two-wheeled motor are electrified.
2. There is an upward tendency in gasoline and diesel consumption among Chinese vehicles, but the increase in gasoline consumption is lower than with diesel; thus, more diesel is consumed than gasoline. Of the total gasoline consumption,

gasoline consumed by Chinese vehicles accounts for over 85 %; among that, passenger cars accounted for 84 %, buses 6 %, and trucks 10 %. The proportion of diesel consumed by Chinese vehicles has increased to over 40 %; among diesel consumption, that by trucks accounts for 85 %, by buses 14 %, and by passenger cars 1 %. Along with the development of alternative-energy and new-energy vehicles, a certain amount of natural gas, LPG, ethanol, methanol, and power is consumed by Chinese vehicles for each year.

3. In China, the most efficient clean-vehicle technologies are undergoing research and development. Gasoline engine HCCI technology is still under development; GDI technology and supercharging technology are at a preliminary stage; VVT is technologically mature and used widely. Start-stop technology can make 5–12 % reductions in fuel consumption to meet regulations of phase III. In addition, ISG technology has been developed; it can lead to a fuel consumption reduction by over 20 % to make all passenger cars meet regulations on fuel consumption for phase III. There is a bottleneck in the development of diesel-powered passenger car, that is, the quality of diesel is lower than needed.
4. HEV technology is more widely used in commercial vehicles than in passenger cars in China. In China, parallel technology is more widely used than series technology for HEVs: of the 76 released commercial vehicle models, 72 use parallel technology, achieving savings in gasoline consumption of 20–40 %. Only a few vehicle models have adopted series–parallel power technology, and it does not appear to have a bright future.
5. PHEV vehicle and extended-range electric vehicle technology have been applied only in a few vehicle models, still at the stage of technical verification. In China, the typical models include BYD F3DM, FAW B50PHEV, JAC RS PHEV, and Chery M1-REEV. Battery electric vehicle technologies have been applied only in some special usage for commercial vehicles, for example, cars for sampling application, sanitation vehicles, and urban buses, but for a passenger car, especially for small size car, the battery electric car technology has been used on many models, and thousands of battery electric passenger car have operated in demonstration. Hydrogen fuel-cell technology has been technically verified for urban buses and other passenger cars; it has been used in a number of public demonstrations in cities and areas around China. In the future, the promotion and development of electric driving technology will rely on the development of energy-saving technology, such as batteries, capacitors, and hydrogen storage.

References

Beijing SinoHytec Co., Ltd. (2009) Sampling operation evaluation report on U.S. Urban Hybrid electric bus

Chandler K, Walkowicz K (2006) King County Metro Transit Hybrid articulated buses: final evaluation results. http://www.nrel.gov/vehiclesandfuels/fleettest/pdfs/40585.pdf. Accessed 17 Jan 2013

China Association of Automobile Manufacturers (2010) China automotive industry yearbook 2009. Beijing, China
China Association of Automobile Manufacturers (2011) China automotive industry yearbook 2010. Beijing
Lammerf M (2009) Twelve-month evaluation of UPS diesel hybrid electric delivery vans. http://www.nrel.gov/docs/fy10osti/44134.pdf. Accessed 17 Jan 2013
National Bureau of Statistics of China (2011a) National economy and social development statistic bulletin in 2010. China Statistics Press, Beijing
National Bureau of Statistics of China (2011b) China energy statistical yearbook 2010. China Statistics Press, Beijing
National Bureau of Statistics of China (2011c) China statistical yearbook 2010. China Statistics Press, Beijing
National Bureau of Statistics of China (2011d) Major data bulletin of the 6th nationwide census. Beijing, China
Sun Li (2010) Development of light electric vehicle in China. EVS25 conference paper collection, Shenzhen
Wang HW (2011) Commercialization of hybrid electric transit bus in North America and China: presentation at US-China clean vehicle cooperation conference, Beijing, 2011
Wang Jianxin, Wang Zhi (2010) Research progress of high efficient and clean combustion of automotive gasoline engines. J Automot Saf Energy 1(3):167–178
Zhang Guobao (2010) Chinese energy development report 2010. Economics Science Press, Beijing

Chapter 5
Petroleum-Derived Liquid Fuels

Li Zheng, Fu Feng, Ma Linwei, Liu Pei, Zhou Zhai, Zhang Jianbing, and Jiang Xiaolong

Abstract In this chapter, after a brief introduction, we examine the development status and historical trends of oil development in China. A Sankey diagram of the oil flow in China from crude oil supply to the final sectors of oil product consumption is mapped to indicate the physical patterns of oil supply and consumption. Following this, we review the current status and historical trends of oil reserves, oil imports, oil refining, oil demand, oil prices, and related policies to present the multidimensional status of oil development in China. Then, we review existing opinions on future oil demand, especially that by road vehicles, future oil production, and energy security issues, and we summarize the future challenges facing oil development in China. Based on a scenario analysis of Chinese oil consumption up to 2030, we discuss a coping strategy for energy security and emission reduction, and we conclude with several policy suggestions for the future development of petroleum-derived fuels for road vehicles.

Keywords Petroleum • Liquid fuel • Oil • Scenario

5.1 Introduction

Globally, petroleum is both the dominant energy source for transportation and the most important primary energy source. According to global energy statistics issued by the International Energy Agency for 2008 (IEA; IEA 2010a), petroleum-derived

Li Zheng • Fu Feng • Ma Linwei (✉) • Liu Pei • Zhou Zhai • Zhang Jianbing • Jiang Xiaolong
State Key Laboratory of Power Systems, Department of Thermal Engineering,
Tsinghua-BP Clean Energy Center, Tsinghua University, Beijing 100084, China
e-mail: malinwei@tsinghua.edu.cn

CAERC, Tsinghua University, *Sustainable Automotive Energy System in China*,
DOI 10.1007/978-3-642-36847-9_5, © Springer-Verlag Berlin Heidelberg 2013

fuels[1] accounted for 93.5 % of energy consumption in the transportation sector and that sector accounted for 61.4 % of the total petroleum consumption; further, petroleum accounted for 33.2 % of global primary energy consumption. However, the high dependency on oil (petroleum and petroleum-derived fuels) for transport and overall global energy consumption is creating serious problems involving energy security, greenhouse gas emissions, and pollution. Therefore, the major oil-consuming nations are highly concerned about conserving oil, oil security, making cleaner use of oil, and developing alternatives to oil.

At the beginning of the twenty-first century, China entered a stage of simultaneous industrialization, urbanization, and motorization, and the broad popularization of automobiles has become one of the distinctive characteristics of the current era. Propelled by various increasing demands, particularly the rising vehicle population, the total oil consumption has sharply increased. Owing to relatively low domestic petroleum reserves, China's oil imports are rapidly rising. By 2009, China's oil import dependency (OID) had surpassed 50 %. Meanwhile, volatility in the international price of oil in recent years has added to concerns over energy security. In some major cities in China, automobile exhausts have become the dominant source of NO_x and CO in the air (Lu et al. 2008). In addition, oil is second only to coal in terms of China's primary energy consumption, and oil accounts for roughly 18 % of China's primary energy[2]; thus, oil use has to be a focus of greenhouse gas (GHG) mitigation in China.

The above issues have raised concerns and produced differing views among experts and scholars over future perspectives regarding petroleum-derived fuel as the major source of automotive energy in China. Their views cover almost all possibilities: some believe we should completely disengage from our dependence on petroleum-derived fuels; others support the proper development of alternatives as a prudent supplement to petroleum-derived fuels; still others maintain that we should be solely dependent on petroleum-derived fuels. All these opinions are reasonable to a certain degree, but no single view has prevailed. It is therefore more important to discuss how measures might be adopted against existing problems before attempting to prove one or other of the above beliefs.

Thus, this chapter does not aim to make accurate projections regarding the future development of petroleum-derived fuels for automotive energy. It attempts to analyze the currently available information (predominantly for 2000–2009) by examining and balancing various views, discussing key issues on the future development of petroleum-derived fuels, and presenting possible scenarios that may result in addition to various internal mechanisms of such scenarios. Following this, the present chapter tries to make some preliminary strategic suggestions with respect to the future development of petroleum-derived fuels for automotive energy.

[1] Named "oil products" in the IEA publication.

[2] The actual proportion depends on the calculation method: the figure is 18.8 % when calculated in terms of calorific value calculation, 17.9 % in terms of the coal equivalent.

It is also worth noting that the discussions in this chapter focus on the future prospects for automotive energy. However, owing to the fact that these issues involve the whole energy-supply chain—from oil production, oil imports, and oil refining to the final consumption of oil products, which are not only used for road vehicles—the subject matter of this chapter is the entire oil-supply chain, rather than just that relating to the road transportation sector.

The main contents of this chapter are as follows. Section 5.1 is an introduction to the study background and subject matter. Section 5.2 presents the development status and historical trends, which includes a literature review and data analysis of the development status and historical trends of oil reserves, domestic crude oil production, strategic petroleum reserves (SPR), oil refining, oil demand, oil prices, and oil-related environmental policy. In this way, Sect. 5.2 provides data to support subsequent discussions. Section 5.3 examines future perspectives and corresponding measures. It first summarizes previous studies on future perspectives of China's oil development before conducting a scenario analysis and strategy analysis on key issues, including oil consumption, vehicle gasoline and diesel consumption, binding targets of China's energy security, the oil gap, pollution, and GHG mitigation. Section 5.4 offers conclusions and suggestions. It summarizes the major opinions regarding the future development of petroleum-derived fuels for automotive energy in China.

5.2 Development Status and Historical Trends

5.2.1 Oil Flow in China

To offer an overall understanding of China's oil supply and consumption on a purely physical basis, we mapped the entire oil flow in 2009 in the form of a Sankey diagram. For this, we used data from the 2009 Oil Balance Sheet in the *China Energy Statistical Yearbook 2010* (DESNBS and DGANEA 2011) as well as energy statistics from Wang (2010). The diagram appears in Fig. 5.1.

In this diagram, the entire oil flow in China is presented by means of a series of arrows according to sector and oil type. The width of the arrows denotes the scale, while the color distinguishes the different types of oil. The overall oil chain is divided into five subprocesses.

1. Crude oil: this includes crude oil production, oil imports, oil exports, oil stocks, own use of oil (consumption by the energy industry, including the consumption in energy exploitation, transformation, and allocation, predominantly in the petroleum industry), oil transportation loss, and industrial consumption of crude oil.
2. Oil refinery: here crude oil is refined into various oil products, though there is a significant amount of refining loss.

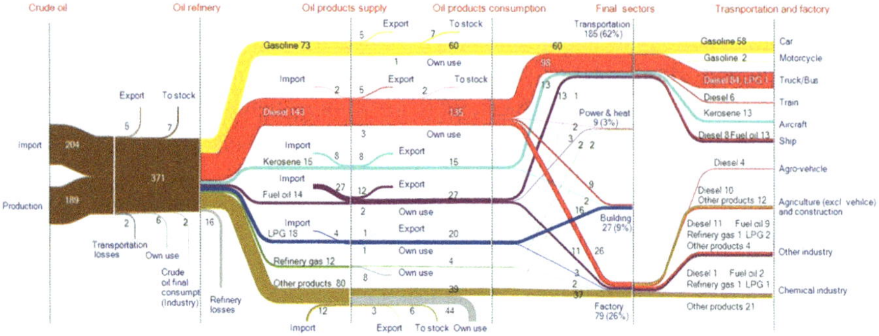

Fig. 5.1 Oil flow in China (2009; unit, Mt) (*Data source*: DESNBS and DGANEA 2011; Wang 2010)

3. Oil products supply: this includes the import, export, stock, and own use of various oil products.
4. Oil products consumption: this includes consumption by transportation, power and heat generation, buildings (including the service industry, residential consumption, and other consumption in the oil balance sheet), and other industries (manufacturing, agriculture, and construction).
5. Final sectors: this part further presents a detailed breakdown of consumption of oil products in transportation and by factories: the former includes cars and motorcycles, trucks and buses, trains, aircraft, and ships; the latter includes agricultural vehicles, agriculture, the construction industry, the chemical industry, and other industries (Sidebar 5.1).

Sidebar 5.1: Data Adjustments for the Oil Balance Sheet to Map the Oil Flow Diagram

Although the 2009 Oil Balance Sheet for China provides data on oil consumption by sector and oil type, certain adjustments are required owing to issues related to the statistical methods employed. The major adjustments are as follows:

1. Adjustments to transport oil consumption: oil consumption by transportation in the balance sheet applies only to transport-operating sectors; some oil is also consumed in other sectors by road vehicles. According to the method suggested by Wang (2010), 95 % of gasoline and 35 % of diesel consumed by light and heavy industry, the construction industry, and service industry is moved to "transportation" in the balance sheet, whereas all the gasoline and 95 % of diesel consumed in living is moved to "transportation."

(continued)

(continued)

2. Adjustments to industrial oil consumption: oil consumption by certain energy-related industries is included in "industrial consumption" in the balance sheet. This oil is in fact consumed by energy-supply industries; thus, it should not be considered the final consumption; it should be accounted as own use of oil by the energy-supply sector. Five departments should be excluded from industrial oil consumption: mining and washing of coal, extraction of petroleum and natural gas, processing of petroleum, coke, and nuclear fuel (which accounts for roughly 76 % of the total own use), production and distribution of electricity and heat, and production and distribution of gas.

From Fig. 5.1, the main characteristics of China's oil production and consumption of oil can be summarized as follows:

1. High import dependency on crude oil: in 2009, domestic crude oil production was 189 Mt, while imports amounted to 204 Mt.
2. Oil products are mainly supplied by domestic oil refineries: in 2009, domestically produced oil products amounted to 355 Mt, whereas imported oil products stood at 53 Mt.
3. Oil products are predominantly consumed by transportation, especially road transport: in 2009, China's total oil consumption (including losses and own use) was 385 Mt, and the final consumption of oil products amounted to 300 Mt. The transport sector consumed 185 Mt of oil products, which was 48 % of the total oil consumption and 62 % of the total final consumption of oil products. In particular, road transport (cars, motorcycles, trucks, and buses) accounted for 78 % of the total consumption of oil products in the transport sector.
4. Gasoline and diesel is mainly consumed by road transport, whereas diesel consumption by road transportation is greater than that of gasoline: almost all the gasoline is consumed by road transport. In contrast, diesel is used more diversely, and only 62 % is consumed by road transportation. In total, road transport consumes 84 Mt of diesel and 60 Mt of gasoline.
5. Light and heavy industry is the second-largest oil-consuming sector, and it mainly consumes "other products": in 2009, 79 Mt of oil products and 81 Mt of oil (plus crude oil) were consumed by factories, which accounted for 26 % of the final consumption of oil products and 21 % of the total oil consumption. Among factory-consumed oil products, 45 % were "other products," 60 % of which were used in the chemical industry. Diesel consumption by agriculture and the construction industry is relatively large in scale; in particular, the construction industry used almost one-third of the "other products" of factories. Among other industries, the materials industry accounted for the greatest consumption of oil products, mainly through consuming fuel oil and diesel.

6. There was large-scale own use, especially for processing petroleum: in 2009, own use of oil amounted to 66 Mt, which was 17 % of the total oil consumption. Of this own use, 83 % was consumed in processing petroleum, coking, and processing nuclear fuel.

5.2.2 Oil Reserves and Crude Oil Production in China

According to BP's statistics, proven oil reserves in the world have been continuously increasing over the past 10 years, amounting to 181.7 billion tons in 2009 (BP 2010). According to the magazine *Petroleum and Natural Gas*, the proven oil reserves are even higher—184.7 billion tons (IEA 2010b). According to IEA, the global recoverable reserves of conventional oil are 488 billion tons, and it suggests that only one-third of these reserves are actually proven (IEA 2008). Taking into account advances in oil exploration and extraction technology and the potential for using unconventional reserves, the global recoverable oil reserves may be as high as 887 billion tons (Sidebar 5.2).

Sidebar 5.2: Global Oil Reserves and Unconventional Resources

(a) *IEA Assessment of Global Oil Resources in WEO 2008 (Summary)*
The world is far from running out of oil. Remaining proven reserves of oil and natural gas liquids worldwide at the end of 2007 amounted to about 1.2 trillion to 1.3 trillion barrels (including about 0.2 trillion barrels of Canadian oil sands). These reserves have almost doubled since 1980. Though most of the increase has come from data revisions made in the 1980s by the Organization of Petroleum Export Countries (OPEC) rather than from new discoveries, modest increases have continued since 1990, despite rising consumption. The volume discovered has fallen well below the volume produced over the last two decades; the volume of oil found on average each year since 2000 has exceeded the rate in the 1990s owing to increased exploration activity (with higher oil prices) and improvements in technology.

Ultimately recoverable conventional resources—a category that includes initially proven and probable reserves from discovered fields, growth in reserves, and economically recoverable oil that has yet to be found—amounts to 3.5 trillion barrels. Only a third of this total has thus far been produced. Undiscovered resources account for about one-third of the remaining recoverable oil, the largest volumes of which are thought to lie in the Middle East, Russia, and the Caspian region. Unconventional oil resources are also large. Worldwide, oil sands and extra-heavy oil

(continued)

(continued)

resources amount to around six trillion barrels, of which between one trillion and two trillion barrels may ultimately be economically recoverable. These resources are largely concentrated in Canada (mainly in Alberta) and Venezuela (in the Orinoco Belt). There is additional potential from oil shales, but their production costs and the environmental impact of their commercialization are very uncertain.

(b) *Oil Shales, Heavy Oil, and Oil Sand* (*Chen et al.* 2009)

Oil shales are combustible mineral resources formed with minerals and organic matter. The composition of the minerals is very complicated: they largely include quartz, kaolinite, montmorillonite, and illite as well as external clay, sand, and salts. The organic matter is mainly composed of kerogen and bitumen, which are dispersed homogeneously in the mineral matrix.

Heavy oil is also referred to as thick oil, and it has high density (API < 22), high viscosity (>10 mPa · s), and a high bituminous content (nonhydrocarbon macromolecular compounds, which contain most of the sulfur and 90 % of the oil's heavy metals). Extra-heavy oil refers to heavy oil with an API lower than 10.

Bituminous oil sand, also referred to simply as oil sand, is a mixture of natural bituminous sand and soil. The natural bitumen extracted from oil sand has the characteristics of heavy oil, whereas the density and viscosity are greater than with heavy oil.

Although petroleum is still abundant globally, which suggests that the "oil peak" will not occur in the near future (according to the oil peak theory, when more than half of global resources have been exploited, crude oil production will plummet), the problem is the very imbalanced regional distribution of global oil resources. Almost 57 % of proven oil reserves are concentrated in the Middle East, whereas the intensively oil-consuming countries in North America, Europe, and the Asia-Pacific region are running short of oil reserves (BP 2010). Nearly all previous global oil crises and fluctuations in oil prices can be attributed to the high dependency on the global oil supply of oil-exporting countries in the Middle East. In addition, along with the depletion of conventional oil reserves, oil from enhanced oil recovery, deep-ocean resources, and unconventional resources is likely to account for an increasingly higher share in the global oil supply, which will drive up exploitation costs (IEA 2008) and raise oil prices in the future. Therefore, regional patterns of oil supply and demand and the cost of oil exploitation are more important than the total scale of oil reserves.

Regarding China, although proven oil reserves in 2009 were only two billion tons and the reserve/production ratio was as low as 10.7:1 (BP 2010), the country still has considerable potential for expanding its oil reserves. According to the results of

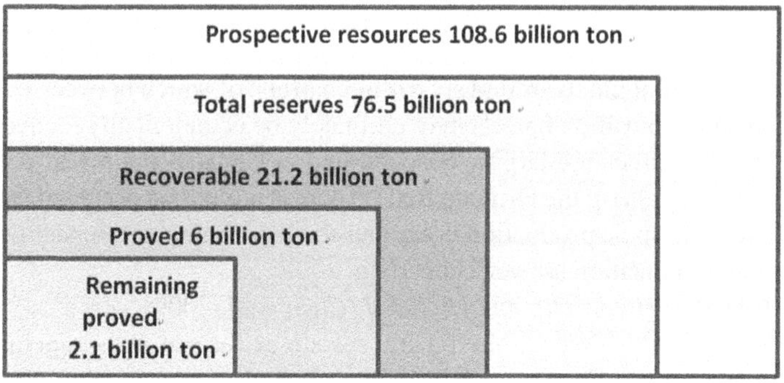

Fig. 5.2 Oil reserves of China—statistics from the Ministry of Land and Natural Resources

an assessment of national oil and gas reserves[3] conducted by the Ministry of Land and Natural Resources in 2007, China's recoverable conventional oil reserves stood at 21.2 billion tons (Fig. 5.2, Sidebar 5.3); only around 1/10 of those are proven. In addition, recoverable reserves of unconventional resources, such as oil shales and oil sands, are estimated to be 12 and 2.3 billion tons, respectively; these can be important supplements to conventional resources. Currently, the exploration and production of crude oil in China are making steady progress: crude oil production is expected to attain around 200 Mt by 2030. According to CAE (2011a), a crude oil production level of 180–200 Mt will be a secure and sustainable target by 2050.

Sidebar 5.3: Definitions of Oil Resources and Reserves in China

1. Prospective resources: the potential amount of oil or gas, that is, the probable maximum that can be exploited. In the assessment of resources conducted by the Ministry of Land and Natural Resources, prospective resources have a probability of 5 %.
2. Geological resources: the amount of oil or gas amount that can be proven using current techniques, including proven and unproven re-sources.
3. Recoverable resources: the amount of resources that can be exploited with future foreseeable technologies and economic conditions.
4. Proven recoverable reserves: under the specified economic, technological, and policy conditions, the developed and undeveloped oil or gas reserves.
5. Remaining proven recoverable reserves: for developed oil or gas reserves, the remaining proven recoverable reserves are equal to the recoverable reserves minus the amount of extracted oil or gas.

[3]Refer to the report by the Xinhua News Agency on national petroleum and gas assessment: http://news.xinhuanet.com/newscenter/2008-08/18/content_9480784.htm

Table 5.1 China's oil production, imports, exports, and consumption, 1995–2009

Year	Production (Mt)	Imports (Mt)	Exports (Mt)	Stock (Mt)	Consumption (Mt)	OID (%)
1995	150.1	36.7	24.5	1.5	160.7	6.6
1996	157.3	45.4	27.0	−0.7	176.4	10.8
1997	160.7	66.0	28.2	4.5	194.1	17.2
1998	161.0	57.4	23.3	−2.2	197.4	18.4
1999	160.0	64.8	16.4	−1.2	209.6	23.7
2000	163.0	97.5	21.7	12.4	226.3	28.0
2001	164.0	91.2	20.5	2.6	232.0	29.3
2002	167.0	102.7	21.4	−0.9	249.3	33.0
2003	169.6	131.9	25.4	0.7	275.4	38.4
2004	175.9	172.9	22.4	5.2	321.2	45.2
2005	181.4	171.6	28.9	−1.3	325.4	44.3
2006	184.8	194.5	26.3	3.7	349.3	47.1
2007	186.3	211.4	26.6	4.6	366.5	49.2
2008	190.4	230.2	29.5	18.0	373.2	49.0
2009	189.5	256.4	39.2	22.1	384.6	50.7
Total increase	39.4	219.7	14.7	20.6	223.9	44.1

Data source: Table 5.6 from China Energy Statistics 2010 (DESNBS and DGANEA 2011)

Some major oil fields in China, such as Daqing Field, have been experiencing a continuous decline in oil output owing to depletion of oil reserves (Tang et al. 2010). However, through the exploration of new oil fields, advances in oil exploitation technology, and the potential in unconventional oil resources, the oil peak that was previously forecast by scholars as occurring in 2020 will probably take place considerably later (see Sect. 5.3.1.3). This conclusion, though, is built on the precondition that crude oil production will stabilize at 180–200 Mt. If that is not the case, China's oil peak will come earlier. In 2010, the crude oil production of China surpassed the 200-Mt mark (NEA 2011).

5.2.3 Oil Imports and Strategic Petroleum Reserves of China

Globally, the regional imbalance in oil production and consumption has led to large-scale, ever-increasing oil trade. In 2009, the total amount of the global oil trade amounted to 2.61 billion tons, whereas global oil consumption was only 3.88 billion tons. Currently, the major oil-exporting regions are the Middle East, countries of the former Soviet Union, and Western Africa, whereas major oil-importing regions include the United States, Europe, Japan, and China.

China became a net oil-importing country in 1993, and the amount it imports has been increasing ever since. From 1995 to 2009, China's crude oil production showed a slight increase of only 40 Mt; by contrast, its oil imports increased by 220 Mt, resulting in its OID increasingly dramatically from 6.6 to 50.7 % (Table 5.1).

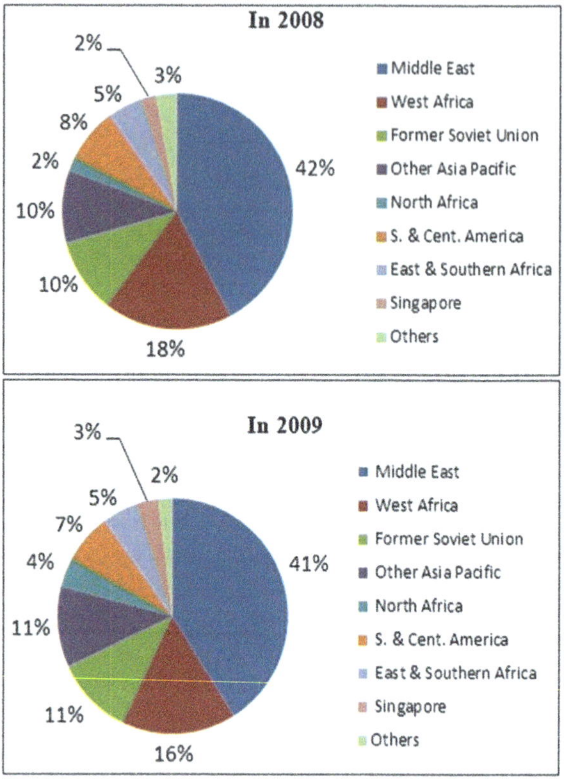

Fig. 5.3 Major regions supplying China's oil imports, 2008 and 2009 (*Data source*: global energy data review by BP 2009, 2010)

In 2009, China's oil imports accounted for 10 % of global oil trade—behind only Europe (26 %) and the United States (22 %) (BP 2010).

China is highly dependent on oil imports from the Middle East and West Africa, most of which is crude oil. However, compared with 2008, oil imports from the former Soviet Union, Asia-Pacific, and other regions increased slightly in 2009, as shown in Fig. 5.3. Meanwhile, the share of crude oil in oil imports decreased from 82 to 80 %, which suggests that China's oil imports are diversifying both in terms of category and channels to ensure energy security.

The SPR is one of the basic measures for ensuring energy security for oil-importing nations. In 2004, China proposed a three-stage plan to establish its national SPR. The first stage involved establishing oil reserve bases in four coastal cities—Dalian, Huangdao, Zhenhai, and Zhoushan; this stage has already been completed. The second stage has also been initiated. The target of the plan is to build a reserve of 85 Mt, which can sustain the country's oil consumption for 100 days, which is slightly higher than the standard set by the IEA. The whole plan is expected to be completed by 2020. As evident in Table 5.1, China's oil stock has been experiencing fast expansion since 2006. According to the National Energy Administration (NEA), national SPR in 2010 exceeded 23 Mt (NEA 2011).

In addition, the Revitalization Plan of the Chemical Industry issued by the State Council in 2009 proposed an increase in the national reserve of petroleum products. The China State Reserve Bureau has long been working on a plan to increase its reserve locations. Previously, its ten reserve locations were mostly in central and east China; the next candidates are Yunnan and Sichuan provinces.

5.2.4 Oil Refining in China

China's petroleum-refining industry is large in terms of overall scale, but it is small with regard to volume per unit. From 2000 to 2009, the scale of China's petroleum-refining industry increased from 269 to 418 Mt/a, which is equivalent to an annual increase of 18 Mt/a; in 2009 alone, there was an increase of 41 Mt/a (BP 2010). In 2009, China's oil-refining volume accounted for 9.5 % of the world total—second to that of the United States (19.5 %). Sinopec and CNPC rank third and ninth globally in terms of their total oil-refining capacity, and they have the greatest capacity in Asia.

Scaling-up has become a major tendency in the global oil-refining industry, and newly built refineries are on average over 10 Mt/a in capacity. In 2009, the average global per unit capacity was 6.57 Mt/a (True and Koottungal 2009). In recent years, the per unit capacity of Sinopec and CNPC has also continuously risen, amounting to 5.7 and 5.4 Mt/a, respectively, in 2009, though this is still far behind the world average. Apart from Sinopec, CNPC, and ten large-scale local refineries, China still has many technically backward small-scale oil refineries (CPN 2010). In 2009, the largest oil refinery in China was Sinopec Zhenhai refinery, which has a capacity of 20 Mt/a, ranking it only 19th in the world that year, behind some large-scale refineries in India and South Korea.

Compared with the world capacity share,[4] the share of catalytic cracking[5] in China's oil-refining technology is high—similar to that of the United States. The share of hydrotreating[6] is also higher than the world average, though the shares

[4]Capacity share refers to the proportion of the processing capacity in the atmospheric distillation capacity.

[5]FCC is the process of cracking heavy oil into gas, gasoline, diesel oil, and other products under high temperatures and with a catalyst. This is one of the main processes used in heavier oil refineries.

[6]Hydrocracking is the process of cracking heavy oil into gas, gasoline, jet fuel, diesel oil, and other products under heating, high hydrogen pressure, and catalytic action. Its feedstock is usually heavy distilled oil. Its main characteristic is high flexibility in controlling the output rate.

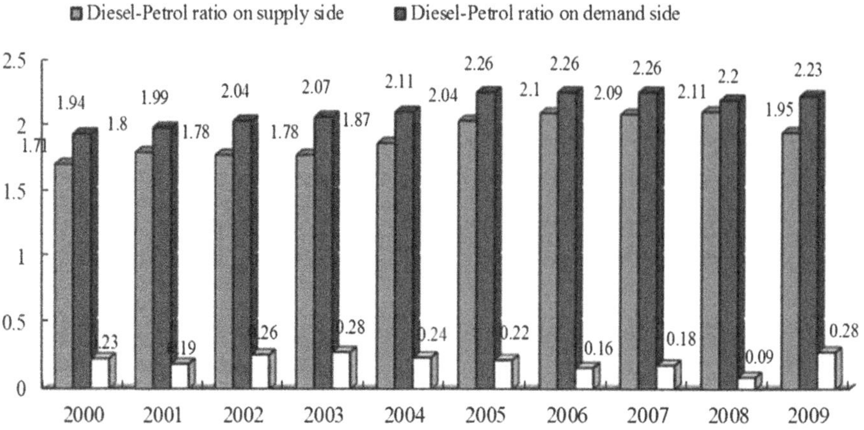

Fig. 5.4 Comparison of supply-side and demand-side diesel-to-gasoline ratios in China (2000–2009) (*Data source*: DESNBS and DGANEA 2010, 2011)

of catalytic reforming[7] and hydrofinishing[8] are substantially lower. This situation can be traced back to the history of China's oil-refining industry. Traditionally, China's oil refineries were mainly configured for domestic, low-sulfur, and heavy crude oil input to produce a variety of output products that lacked an emphasis on low-emission transportation fuels (e.g., low-sulfur diesel and gasoline). Therefore, the share of catalytic cracking is high while that of catalytic reforming is low. To accommodate rising imports of high-sulfur crude oil and booming demands for clean transportation liquid fuels, China has in recent years introduced advanced technology for catalytic cracking, hydrotreating, and coking during the expansion of its oil-refining industry. However, hydrotreating and hydrofinishing, which are needed to improve oil quality and remove impurities, are making slow headway in the industry. To adapt to changes in feedstock and meet the demand for clean transportation fuels, China's oil refineries should continue improving their technology and place greater emphasis on light, clean liquid fuel and lower petroleum consumption per unit output (Walls 2010).

A key challenge is presented by the changing diesel-gasoline ratio on the demand side: owing to flourishing demand for diesel, the diesel-gasoline ratio on the demand side has been increasing; it reached 2.2:1 in 2005 (Fig. 5.4). One major solution to deal with this trend involves increasing the diesel-gasoline ratio in the oil-refining

[7]Catalytic reforming is the process of transforming light-distillation gasoline fractions (or naphtha) into high-octane-value gasoline under heating, high hydrogen pressure, and catalytic action. Catalytic reforming is a major method for improving the quality of gasoline, and it is commonly employed in modern refineries and petrochemical joint enterprises.

[8]This is the most important method for finishing petroleum products. It refers to removing harmful impurities, such as sulfur, oxygen, and nitrogen, by converting them into the corresponding hydrogen sulfide, water, and ammonia in order to improve the oil products' quality.

Table 5.2 Import and export of China's gasoline and diesel, 2000–2009 (unit, Mt)

	2000	2001	2002	2003	2004	2005	2006	2007	2008	2009
Diesel imports	0.52	0.55	0.79	1.12	3.04	0.61	0.81	1.74	6.33	1.93
Diesel exports	0.78	0.47	1.45	2.44	0.87	1.71	1.03	0.93	0.89	4.79
Net import of diesel	−0.26	0.08	−0.66	−1.33	2.17	−1.10	−0.22	0.80	5.44	−2.86
Gasoline imports	0.00	0.00	0.00	0.00	0.00	0.00	0.06	0.23	1.99	0.04
Gasoline exports	4.68	5.86	6.30	7.54	5.41	5.60	3.51	4.64	2.03	4.92
Net export of gasoline	4.68	5.86	6.30	7.54	5.41	5.60	3.44	4.42	0.05	4.88

Data source: DESNBS and DGANEA (2010, 2011)

output, but this is hampered by a number of barriers. Since oil-refining units were originally configured to produce certain categories of products and meet a particular diesel-gasoline demand ratio, it is difficult to adjust their output structure.

Both hydrotreating and FCC have certain flexibility regarding adjustment of the output diesel-gasoline ratio. But from an international viewpoint, hydrotreating is more suitable to the output diesel-gasoline ratio, whereas FCC is better for improving the output share of gasoline. In Europe, hydrotreating is mainly carried out to increase diesel output; most US oil refineries adopt low- and medium-pressure hydrotreating, which can simultaneously increase gasoline output and improve diesel quality. In China, the dominant technology is FCC, which is used to further treat heavier fractions after distillation. Following advances with catalysts and related technology, China has managed to achieve a fairly high diesel-to-gasoline ratio, which is higher even than that of Europe. However, owing to the low hydrotreating capacity and already-installed large-scale atmospheric-vacuum catalytic cracking, a diesel-to-gasoline ratio of 2.1:1 is already near the upper technological limit with the facilities currently in service (Sun et al. 2009).

As a result, the imbalance between the supply and demand diesel-to-gasoline ratio is becoming increasingly pronounced (Fig. 5.4), and this has led to oversupply and even export of gasoline (Table 5.2). At the same time, Chinese oil refineries have substantially reduced fuel oil output (Table 5.3); this has resulted in large-scale imports of fuel oil, while refining losses and output of other products have increased accordingly. In 2009, the output ratios of gasoline/diesel and transportation liquid fuels (gasoline, diesel, kerosene) were 58.2 and 62.2 %, respectively, compared with 59.9 and 66.5 % for Europe and 67.8 and 75.3 % for North America (IEA 2011).

5.2.5 *Oil Demand in China*

China is the second-largest oil consumer in the world, second only to the United States. Over the period of 2000–2009, China's oil consumption increased from 226 to 385 Mt, with an annual growth rate of 6.1 %. By 2009, China's oil consumption

Table 5.3 Input and output of China's oil-refining industry, 2000–2009

	Input of crude oil (Mt)	Loss[a] (%)	Gasoline (%)	Kerosene (%)	Diesel (%)	Fuel oil (%)	LPG (%)	Refinery gas (%)	Other products (%)
2000	203.1	3.6	20.4	4.3	34.9	10.1	4.5	3.4	18.9
2001	204.1	3.0	20.4	3.9	36.7	9.1	4.7	3.3	19.0
2002	215.8	4.1	20.0	3.8	35.7	8.6	4.8	3.2	19.7
2003	238.4	4.1	20.1	3.6	35.8	8.4	5.1	3.0	20.0
2004	277.4	4.3	19.0	3.5	35.5	7.3	5.1	3.0	22.3
2005	290.4	4.0	18.7	3.5	38.2	6.1	4.9	3.1	21.5
2006	310.5	4.1	18.0	3.1	37.9	5.7	5.6	3.2	22.3
2007	328.3	3.9	18.0	3.5	37.6	6.0	5.9	3.1	21.9
2008	341.0	4.0	18.6	3.4	39.3	5.1	5.6	3.2	20.7
2009	371.1	4.4	19.7	4.0	38.5	3.6	4.9	3.2	21.6

Data source: DESNBS and DGANEA (2010, 2011)
[a]Including transportation loss and refining loss

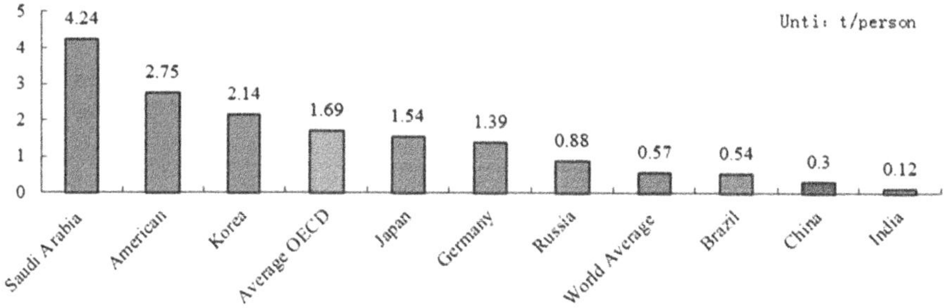

Fig. 5.5 International comparison of oil consumption per capita (2009) (*Data source*: oil data from BP 2010; population data from Population Reference Bureau 2010: 2009 World Population Data Sheet; OECD countries account for 18 % of the global population, according to the OECD factbook 2009)

accounted for 10.4 %[9] of global consumption (21.7 % for the United States). However, the oil consumption per capita per year was only 0.3 tons, which is half the global level and lower than that of most major oil-consuming nations, whose total oil consumption is over 100 Mt (Fig. 5.5).

The issue of oil demand is quite complicated in China because it involves the demand by various sectors for different oil types. Using the same data-processing method (Sidebar 5.1) as that adopted for Fig. 5.1, figures for by-sector oil consumption in 2000, 2003, 2005, 2007, and 2009 can be obtained, as presented in Table 5.4. The main features of China's oil consumption and structure can be summarized as follows:

[9]According to BP world energy statistics, 2009 Chinese national oil consumption was 404.6 Mt, which was slightly higher than the figure of 384.6 Mt provided by *China Energy Statistics*.

Table 5.4 China's oil consumption by sector (2000, 2003, 2005, 2007, 2009)

	2000		2003		2005		2007		2009		Total increase in 2000–2009	Contribution to total oil demand growth in 2000–2009
Sector	Mt	%	Mt	%	Mt	%	Mt	%	Mt	%	Mt	%[f]
Losses[a]	9.1	4	11.3	4	13.2	4	14.8	4	18.1	5	8.9	6
Own use	42.5	19	51.9	19	58.2	18	63.4	17	66.5	17	24.0	15
Power	16.0	7	19.1	7	20.1	6	13.1	4	8.8	2	−7.2	−5
Agriculture	7.0	3	9.4	3	12.9	4	12.3	3	11.4	3	4.4	3
Agricultural vehicles	–	–	11.0[b]	–	10.0[c]	78[d]	9.1	74	4.4	58	–	
Industry	36.8	16	41.8	15	44.3	14	50.5	14	52.7	14	15.8	10
Chemical	15.1	41	20	48	22.2	50	27.4	54	26.8	51	11.7	74
Mineral	7.8	21	6.8	16	7.8	18	8.4	17	7.3	14	−0.5	−3
Construction	6.5	3	9.7	4	12.0	4	14.9	4	15.6	4	9.1	6
Transportation[e]	89.3	40	109	40	142	44	171	47	185	48	95.3	60
Cars/motorcycles	33.4	37	39.2	36	47.4	33	53.9	32	60.6	33	27.2	28.5
Trucks/buses	34.6	39	43.4	40	60.0	42	72.1	42	83.8	45	49.2	51.7
Trains	2.7	3	3.3	3	4.6	3	5.6	3	6.1	3	3.4	3.5
Planes	5.4	6	7.4	7	9.5	7	11.3	7	13.1	7	7.8	8.2
Ships	13.3	15	15.1	14	20.6	14	27.2	16	21.0	12	7.7	8.1

(continued)

Table 5.4 (continued)

Sector	2000		2003		2005		2007		2009		Total increase in 2000–2009	Contribution to total oil demand growth in 2000–2009
	Mt	%	Mt	%	Mt	%	Mt	%	Mt	%	Mt	%[f]
Commercial	1.5	1	1.6	1	2.1	1	2.5	1	2.3	1	0.8	0
Residential	9.4	4	11.6	4	13.7	4	16.8	5	15.5	4	6.1	4
Urban	7.8	83	9.8	85	10.8	79	13.1	78	11.7	76	3.9	64
Rural	1.6	17	1.8	15	2.9	21	3.7	22	3.8	24	2.2	36
Others	6.2	3	6.0	2	6.6	2	7.4	2	8.5	2	2.3	1
Total[f]	224	100	271	100	325	100	366	100	384	100	159	100

Data source: DESNBS and DGANEA (2010, 2011); Wang (2006, 2007, 2008, 2009, 2010)

[a]Losses include both transportation losses and refinery losses

[b]This data is questionable since the total value for agricultural vehicles exceeds the total diesel consumption in the agricultural sector

[c]Estimated as 10 Mt according to the available data for 2003 (11 Mt) and 2007 (9 Mt)

[d]The percentages in italics are the portion of each subsector within the total oil consumption of the broader sector

[e]The data for the subsectors in transportation before 2007 were estimated assuming that the structure of diesel use was the same as in 2007, when data were available. The gasoline use by motorcycles in 2009 was 1.8 Mt, but no data are available for other years

[f]The proportion of each sector's contribution to the total increase of oil consumption in 2000–2009

1. Transportation is the main propelling factor behind the booming oil consumption. In 2000–2009, the share of transport in total oil consumption increased from 40 to 48 %; increased oil consumption by transportation accounted for 60 % of the total increased oil consumption during this period.
2. Road vehicles, especially diesel vehicles, are the main factor behind the growth in oil demand in the transport sector. Increased diesel consumption by road vehicles accounted for 51.7 % of the total increased oil consumption by transportation; this was followed by gasoline consumption by road vehicles, which accounted for 28.5 %. Kerosene consumption by planes and diesel consumption by trains accounted for 3.5 and 8.2 %, respectively. Diesel and fuel oil consumption by ships accounted for 8.1 %; this consumption began to decrease after 2007.
3. Own use and losses have shown a continual rise. In 2000–2009, own use and losses accounted for a steady annual 22 % of the total oil consumption, and the increase in these areas amounted to 21 % of the total increased oil consumption, second only to that in the transportation sector. This was due to increased losses and transportation losses together with expansion of oil refining and distribution of petroleum products. Over the same period, China's total energy consumption grew at almost 10 % per year, which also drove up own-use oil.
4. Industrial oil consumption has grown steadily and ranks third among the sectors that have contributed to the total increased oil consumption. Consumption in this sector is mainly made up of other products. In 2000–2009, industrial oil consumption accounted for 14–16 % of the total oil consumption, and its increase accounted for 10 % of the total increased oil consumption. Among all sectors, the chemical industry accounts for around half of the total industrial oil consumption. In other oil-intensive industries, such as the mineral industry, oil consumption is generally declining. Overall, it is relatively easy to find alternatives for oil in other industrial sectors; for example, coal and gas can be used as substitutes for oil in providing heat and steam.
5. Oil consumption by the construction industry has experienced a significant increase, and this is notable as a factor behind oil consumption. Although oil consumption by this sector accounted for only 3–4 % of the total oil consumption in 2000–2009, its increase accounted for 6 % of the total increased oil consumption during that period. Construction mainly consumes other products and also some diesel.
6. Oil consumption by the agricultural, commercial, residential, and other sectors has also grown. These sectors accounted for around 10 % of the total oil consumption in 2000–2009, and their increase accounted for around 8 % in the total increased oil consumption over the same period. These sectors mainly consume liquefied petroleum gas (LPG) and diesel. Although rural LPG consumption was less than in urban areas during this period, it grew at a higher rate. In addition, diesel consumption by agricultural vehicles showed a continuous decline, while that by agricultural machines showed a constant rise.
7. Power and heat generation was the only sector that experienced a decline in oil consumption from 2000 to 2009; this was due to restrictive policies on using oil for power and heat generation and also due to its inadequate technology.

Table 5.5 China's oil consumption by oil type, 1996–2009 (unit, Mt)[a]

	Crude oil	Diesel	Gasoline	Kerosene	LPG	Fuel oil	Refinery gas	Other	Total
2000	6.4	65.8	35.0	8.7	13.9	27.4	5.6	36.7	199.5
2001	6.5	69.2	36.0	8.9	14.1	26.9	5.6	36.9	204.1
2002	6.8	74.4	37.5	9.2	16.2	26.8	5.7	43.3	219.9
2003	8.1	81.4	40.7	9.2	17.9	28.8	5.9	48.6	240.6
2004	8.4	95.6	47.0	10.6	20.1	31.3	6.7	61.0	280.6
2005	8.7	106.1	48.5	10.8	20.4	29.5	7.9	60.0	291.9
2006	9.8	115.2	52.4	11.2	22.0	32.7	8.2	64.6	316.1
2007	9.8	122.7	55.2	12.4	23.2	34.6	8.7	72.0	338.6
2008	11.9	133.5	61.5	12.9	21.1	27.6	8.8	69.7	347.0
2009	8.3	136.0	61.7	14.4	21.5	25.3	9.3	80.3	356.9
Total increase	1.9	70.3	26.7	5.7	7.6	−2.1	3.7	43.6	157.4
Contribution rate	1.2 %	44.6 %	17.0 %	3.6 %	4.9 %	−1.3 %	2.4 %	27.7 %	100.0 %

Data source: DESNBS and DGANEA 2010, 2011; Wang 2006, 2007, 2008, 2009, 2010
[a]Own use is not listed in this table; therefore, final consumption may be slightly higher than that indicated in Table 5.4, especially for refinery and other petroleum products

Other power-generation technologies, such as coal, wind, and hydropower, are all superior to oil power in terms of economic and environmental performance. However, oil-based power and heat could not be completely replaced owing to peaking and emergency-power generation needs (Leung 2010).

Seen in terms of final oil consumption by oil type (Table 5.5), in 2000–2009, there was a significant increase in diesel consumption, which accounted for 44.6 % of the total final oil consumption. In recent years, China's diesel consumption has rapidly increased, mainly because of its wide applications, and there has been a simultaneous increase in diesel consumption in a number of areas. A detailed explanation is summarized below and in Table 5.5.

1. Accelerated industrialization and the huge scale of building infrastructure have led to a dramatic increase in demands for freight traffic. Therefore, road, railway, and ship transportation have all exhibited rapid growth in diesel consumption. This is the dominant reason.
2. Accelerated urbanization and motorization have led to an increase in the transportation volume by rail and public transport systems, which mostly consisted of passenger traffic.
3. There has been expansion of energy-intensive industries. Although this segment of diesel consumption may be gradually replaced, oil is still an important energy source for these industries.
4. There has been an increase in military diesel consumption. Although data are lacking in this field, it is estimated that the military sector contributes a great deal to the increase in diesel consumption.

5. With regard to mechanical power, diesel consumption by the construction industry has also risen.
6. Demand for distributed power and heat generation and emergency-power generation has also grown in the residential and commerce sectors.

5.2.6 Oil Prices

International oil transactions are usually priced according to benchmark oil prices in specific regions. There are five major oil spot markets in the world: the United States, Singapore, the Caribbean, the Mediterranean, and northwest Europe. The major oil futures markets are the New York Mercantile Exchange, the London International Oil Exchange, and, emerging in recent years, the Tokyo Commodity Exchange (Wei and Lin 2007).

International oil prices are influenced by many factors, and it is almost impossible to predict future trends. Overall, the international oil prices mainly depend on long-term supply and demand. But at the same time, they are also influenced by such uncertain factors as cyclical fluctuations in the oil industry, world military and political events, and financial speculation. Regarding the international oil price fluctuation that occurred in 2003–2008, studies have shown that this was due to the combined influence of growth in oil demand, lack of investment, shortages in oil-refining capacity, geopolitical uncertainty, the weak dollar, and many other factors (Kesicki 2010). Because of the fall in oil demand caused by large-scale economic recession in developed countries, the 2009 international price of oil (Brent crude price, 2009 dollar exchange rate) fell from US$96.9/barrel in 2009 to $61.7/barrel (BP 2010). However, in 2010 and 2011, the international price of oil rose again, indicating that this price is still volatile.

Although China's oil prices are increasingly in line with world prices, they are still highly dependent on government policy. The most recent oil-pricing policy was the Oil Price Management Measures (trial), which was issued in May 2009 by the NDRC. The main areas of the policy include the following. (1) Crude oil prices should be set independently by enterprises in China with reference to international market prices. (2) Domestic prices of petroleum products should be based on international market prices of crude oil in addition to average domestic manufacturing costs, taxes, and reasonable and appropriate circulation costs and profit margins. To avoid the impact of sharp fluctuations in international market prices on domestic oil prices, when the moving average price in international market changes more than 4 % for 22 consecutive weekdays, domestic gasoline and diesel prices can be adjusted. (3) When the international market price of crude oil falls below $80/barrel, normal processing profit margins should be added to petroleum product prices. If the international market price rises above $80/barrel, processing profit margins should be properly reduced to stabilize oil prices until the zero

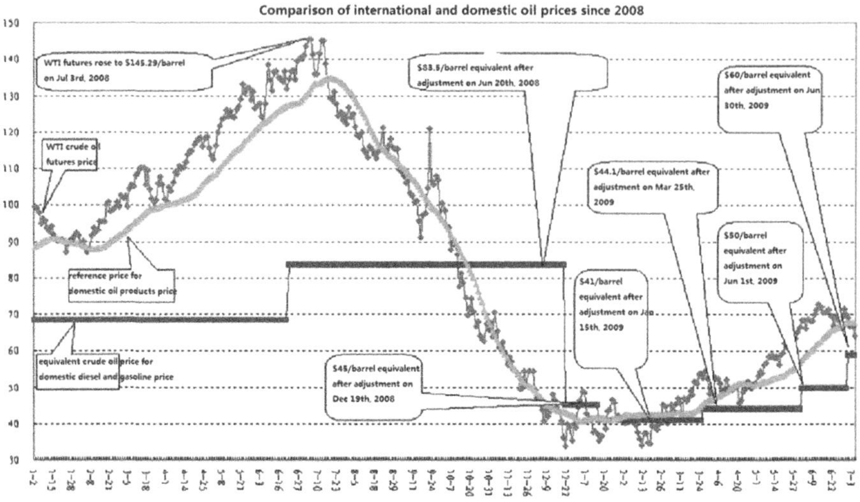

Fig. 5.6 Comparison of domestic and international oil prices, 2008–2009 (*Data source*: http://www.sdpc.gov.cn/xwfb/t20090715_290975.htm)

processing profit margin is reached. If the international market price rises above $130/barrel, taking into account producers' and consumers' interests and the need to maintain a stable national economy, appropriate fiscal policies should be undertaken to guarantee smooth production and supply of petroleum products.

With the intervention of national policy, the impact of international oil price fluctuations on domestic petroleum product prices was smaller in the period of 2008–2009. From the international and domestic oil price comparison in Fig. 5.6, it is clear that China's oil price has been characterized by lagging fluctuation and smaller amplitudes.

Although fuel tax has not been directly imposed in China, it has been indirectly implemented through oil consumption tax. This was achieved through the Notice on Implementing Petroleum Product Prices and Tax Reform promulgated by the State Council in December 2008 (hereinafter, the Notice). The Notice stipulated that gasoline consumption tax would be increased by RMB 0.9/L and diesel by RMB 0.7/L. Together with the original consumption tax, consumption tax on gasoline, naphtha, solvent oil, and lubricating oil would be RMB 1/L, while consumption tax on diesel oil, fuel oil, and kerosene would be RMB 0.8/L.

Compared with other countries, the current price of gasoline in China including tax is higher than in the United States, but lower than in Europe. The reason for this is that the European fuel tax rate is relatively high while that of the United States is low. And China's gasoline price excluding tax is on the same level as in the United States and Europe, the difference being mainly due to the different costs involved in purchasing crude oil (NDRC 2009).

5.2.7 Oil-Related Environmental Policy

Oil-related environmental policies mainly refer to those that control motor vehicle exhaust emissions and vehicle fuel economy policies to improve the efficiency of energy utilization. Fuel quality standards in developed countries have generally become stricter in recent years as a result of global climate changes, and these countries have begun to focus more on controlling GHG emissions. According to an analysis by the IEA (IEA 2010b), to contain temperature rises and keep the global atmosphere CO_2 concentration at 450 parts per million (PPM)[10] or so, the global oil consumption should peak before 2020 and then begin to decline.

According to an analysis by the U.S. National Energy Technology Laboratory (NETL), GHG emissions from crude oil mining and oil refineries account for only a small part of the total emissions. By contrast, emissions in final consumption (mobile sources, such as cars) form the major part, accounting for 80–84 % of the total GHG emissions (NETL 2008). Therefore, for oil GHG mitigation, primary measures should be made to improve the motor vehicle fuel efficiency and emission control standards. The European Union has proposed to set the 2012 GHG emissions for passenger cars at 120 g CO_2/km; in April 2009, it approved a law controlling passenger vehicle CO_2 emissions and fuel consumption. In May 2009, the United States also put forward a policy goal of improving car fuel economy to 36 gal/km (about 7 L/km) by 2016 (Wang 2010).

In China, the problem of conventional pollutants by motor vehicles is still serious. Although in recent years, China has been continuously improving motor vehicle exhaust emission standards, it still has a long way to go compared with developed countries, especially in terms of sulfur content. In June 2008, China began implementing China III emission standards, which are equivalent to Euro III (2000). More stringent standards are implemented in China's big cities, where vehicle exhaust pollution is more severe. For example, Beijing began implementing China IV standards, which are equivalent to Euro IV (2005), in March 2008, and it intends to execute China V standards, which are equivalent to Euro V (2008), by 2012.

According to the plan, China began implementing a gasoline sulfur control standard of 150 PPM in January 2010, and it introduced a diesel sulfur control standard of 350 PPM in July 2011. By 2009, most of the sulfur content of vehicle fuel met only China II standards (equivalent to Euro II), that is, 350 PPM in gasoline and 2,000 PPM in diesel. A survey in 2008 in north China by Zhang et al. showed that 88 % of gasoline had a sulfur content below 350 PPM and 41 % of gasoline had a sulfur content below 150 PPM; only 20.5 % of diesel had a sulfur content below 2,000 PPM and only 17 % of diesel had a sulfur content below 350 PPM (Zhang et al. 2010a). In the United States in 2010, the sulfur content of vehicle diesel was limited to 15 PPM. The European Union reduced the sulfur content in vehicle diesel to 10 PPM in 2009.

---

[10]1 ppm means 1 part per million by volume.

In terms of fuel economy, China in September 2004 began conducting the first and second phases of passenger car fuel economy standards. A recently published study on China's third-stage fuel economy standards proposed that the fuel economy of new passenger cars should be improved to 7 L/km by 2015 and 5 L/km by 2020 (Wang et al. 2010).

China has no mandatory obligation to cut GHG emissions. However, at the 2009 United Nations climate change conference held in Copenhagen, China promised to decrease its CO_2 emissions intensity by 40–45 % from 2005 to 2020, and it has begun to focus on executing this. This policy will strongly promote the efficient use of oil and the development of low-carbon fuels as alternatives to fuel oil.

5.3 Future Challenges and Coping Strategies

5.3.1 Literature Review

Before conducting our analysis, we reviewed related studies on China's future development of oil that had been conducted since 2000. These studies mainly focused on the demand for oil and liquid fuels, the oil demand of road vehicles, domestic oil production, and energy security; however, few discussed an integrated development strategy for oil supply, demand, and security. Therefore, though these studies are an important source of information, they do not provide a complete view of the future development of oil in China owing to the lack of an integrated approach.

5.3.1.1 Future Demand for Oil and Liquid Fuels

Because oil (petroleum and petroleum-derived fuels) may be replaced by liquid fuels from other primary energy sources, some studies have focused on the demand for liquid fuels other than oil. The forecast results of future demand for oil and liquid fuels are normally expressed as tons of oil or tons of oil equivalent. Some international organizations, such as the IEA, OPEC, and U.S. Energy Information Administration (EIA), made predictions for the future demand for oil and liquid fuels in China, as listed in Table 5.6. On the whole, the forecast result by OPEC was higher, that of IEA was lower, and that of EIA was intermediate between the other two.

Compared with the reports of 2009, the forecasts of OPEC and EIA in 2010 predicted higher demand. However, the result of IEA was more conservative in light of China's policy for reducing GHG emissions. In the EIA report, uncertainties with respect to the international price of oil were a main influencing factor for future oil demand in China.

Table 5.6 Forecasts of demand for liquid fuels in China by international organizations (unit, Mt)

Data source	2010	2015	2020	2025	2030	2035
OPEC (2009)	413	518	612	702	792	–
IEA (2009)	–	490	557	646	758	–
EIA (2009)	423	498	602	687	762	–
Average	–	502	590	687	758	–
OPEC (2010)	433	543	652	747	832	–
IEA (2010b)	–	528	583	647	712	762
EIA (2010)		515	598	699	793	873
Average	–	529	610	700	780	–

Note: The results in this table are all from reference scenarios

However, some Western scholars believe that these results are still too conservative. For example, Nel and Cooper (2008) plotted curves for China's smallest increase in oil demand per capita in the light of international experience, and they made predictions based on GDP, population data, and the data presented in IEA's "World Energy Outlook 2006." The result showed that, according to the smallest-increase curve, China's oil demand may reach 1.43 billion tons in 2030.

Because of considerations of energy security, domestic oil demand forecasts are normally conservative. A recent study conducted by the Chinese Academy of Engineering determined that China's oil demand would be 500–600 million tons in 2020, 600–700 million tons in 2030, and 700–800 million tons in 2050 (CAE 2011a).

In addition, foreign studies have all recognized that the rapid increase in oil demand in transportation (especially road transport) would be the main propelling force for demand in liquid fuels and that the future demand for diesel would be greater than that for gasoline. Taking IEA and EIA (IEA 2009; EIA 2010) as examples, results showed that 80 % of the increased demand for liquid fuels in China from 2007 to 2030 would be caused by rising demand from the transport sector. The IEA's study about China showed that the rapid increase in the number of passenger and freight vehicles would be the leading cause of swift growth in demand for liquid fuels (IEA 2007). Cambridge Energy Research Associates also conducted a study on China in 2008. The results showed that the future speed of growth in diesel demand would be much greater than that for gasoline; it was also determined that the diesel-gasoline ratio would be 3.9:1 in 2030.

5.3.1.2 Future Oil Demand by Road Vehicles

Research results regarding future oil demand by road vehicles in China differ quite significantly. The main reason for this is that different studies make different assumptions about the future vehicle population, use, fuel economy, and development of alternative fuels. He et al. (2005) made a scenario analysis of oil consumption and carbon emission of Chinese road vehicles up to 2030. Their results showed that oil consumption by such vehicles would rapidly increase and that vehicle fuel economy improvement was absolutely critical to saving oil and reducing carbon emissions. By improving fuel economy, oil consumption by road vehicles could be reduced from 360 to 280 Mt. Wang et al. (2010) estimated that 39.2 Mt of oil could be saved if the third-phase fuel economy standard could be passed and implemented.

A study by Zhang et al. (2010b) determined that the development of biofuels could considerably lower oil consumption by road vehicles. Under the business-as-usual (BAU) scenario, oil consumption by transport will amount to 992 Mt by 2030. Among that, 392 Mt could be saved by promoting bioethanol gasoline, 135 Mt by improving fuel economy, and 204 Mt by popularizing biodiesel. Ou et al. (2010) designed six scenarios of oil consumption by Chinese road vehicles up to 2050; the scenarios added such new technologies as coal-derived liquid fuels, electrical vehicles, and carbon capture and storage. As a result, under the BAU scenario, oil consumption will be 412 Mt in 2050, but that figure could be reduced to 165 Mt if all possible oil-saving and oil-substitution technologies are employed.

5.3.1.3 Future Production of Crude Oil

Based on oil peak theory, many Chinese and international scholars have proposed oil peak prediction models, which were used to forecast Chinese crude oil production and reserves. Among them, three have been widely used: the Hubbert model, Weng model, and HCZ model (Feng et al. 2007). Comparing the results of these models, the forecast results are mostly not positive: the forecast oil peak would be reached at the latest by 2020. In a recent study, Feng et al. (2008) predicted that the Chinese oil peak would occur in 2011 in the absence of substantial technological development.

However, those studies did not properly consider improvements in exploration and extraction technology regarding conventional and unconventional resources. The Chinese Academy of Engineering (CAE) proposed in 2003 that the future trend of annual crude oil production in China would be as follows: oil peak production would become stable at 180–200 Mt; China has already entered the peak period, and this peak period may last until after 2035 (CAE 2003). In 2011, CAE suggested that crude oil production in China should and could be maintained at 180–200 Mt up to 2050 (CAE 2011a).

Recently, international studies have been increasingly influenced by the results of studies conducted in China. The IEA predicted in 2007 that the Chinese oil peak would be 194 Mt before 2015; thereafter, crude production would decline quickly to 130 Mt by 2030 (IEA 2007). However, in 2008, the IEA adjusted that prediction according to changes in circumstances in China, whereby China would maintain oil production at 200 Mt up to 2030 (IEA 2008).

5.3.1.4 Energy Security Issues

In a previous study by the present authors (Ma et al. 2011), it was pointed out that national energy security is more complex than the measures currently

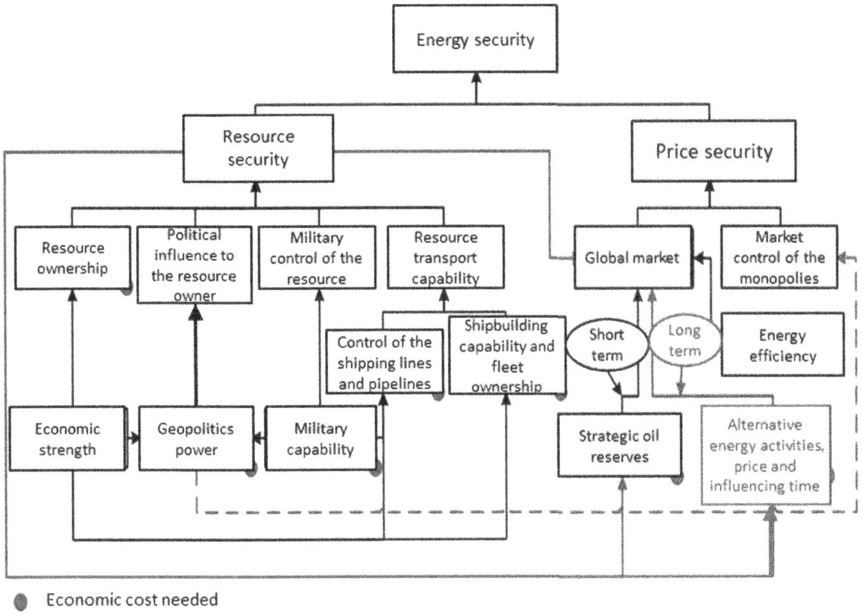

Fig. 5.7 Proposed framework for an energy security system for China

implemented by China would tend to suggest. Those measures include strategies to encourage domestic oil companies to become more deeply involved in global markets, setting up SPR, diversifying oil imports, and developing alternative fuels. The national energy security system is a huge system that involves many factors and players, including the military, politics, the economy, diplomacy, and technology. Implementing some of those factors demands high costs. That study proposed a national energy security system as illustrated in Fig. 5.7, where the arrows represent the decisive sources and influencing issues, and the red arrows and red points signify that costs are required.

In that figure, oil security includes resource security and price security, which are closely correlated with each other. For a country, resource security mainly depends on direct ownership of domestic and foreign resources as well as the political influence, military control, and resource transport capabilities of the resources owner. Price security is influenced by the supply–demand balance of international markets and control of the oil price by international oil monopolies. With a certain demand, the global supply–demand market is influenced by resource security, SPR amount, energy efficiency, and alternative-energy amount. The international oil monopolies pursue maximum economic profit, but their attempts and behavior to control market prices are under the constraints of geopolitics; overly high prices may lead to sluggish demand and long-term economic interest losses.

Some scholars have investigated the measures that could be adopted for China's energy security. Feng and Mu (2010) suggested that a proactive strategy in Africa faced economic and diplomatic problems in addition to cultural challenges. Zhang et al. (2009) conducted research into the optimal scale of SPR in China and concluded that the amount should be 44 Mt in 2017 under the BAU scenario. Zhang et al. also determined that if the oil supply becomes at high risk, the reserve should be increased to 75 Mt. Wu et al. (2010) proposed that the risk of importing product oil is smaller than that of importing crude oil.

Recently, research into energy security has involved the interactions among Chinese oil demand, international oil prices, and the US dollar exchange rate. Skeer and Wang (2007) examined the influence between Chinese oil demand and oil prices. Their results showed that the demand for Chinese oil-based liquid fuel would sharply grow and push up the oil price, thereby promoting the development of energy security measures, such as oil exploitation, oil diversification, and developing alternative fuels. Du et al. (2010) pointed out that although Chinese economic development was more closely linked with oil prices than ever before, China still did not have sufficient power to influence the international oil price. Benassy-Quere et al. (2007) showed that the rise of China in the world oil and international exchange markets would heavily influence the relationship between the world oil price and the US dollar exchange rate. Though the fluctuation in international oil prices has not retarded the economic development of China in the past, international experience suggests that that could happen in the future.

5.3.2 *Scenario Analysis of Chinese Oil Consumption*

Depending on China's oil development, the demand for liquid fuels will continue to show rapid growth if current trends continue. Because it will be difficult to bring about an increase in domestic crude oil production, oil imports will continue to grow and energy security risks will thereby increase. China may therefore control its total oil consumption in the future to keep it within 600–700 Mt per year to ensure the safety of its energy supply (CAE 2011a). To realize such a target, China faces various choices in its oil-conservation strategies, and these will have important influences on vehicle diesel and gasoline consumption.

The following analysis is based on different oil-consumption control targets and oil-conservation methods under three possible scenarios up to 2030:

1. Linear Extrapolation Scenario: a continuation of the historical linear increase trend of 2000–2009.
2. Total Consumption Control Scenario A: oil consumption is kept to under 700 Mt, and there is an emphasis on oil conservation in the nontransport sector.
3. Total Consumption Control Scenario B: based on the second scenario, oil conservation is further enhanced in the transport sector, and alternative fuels are developed to keep oil consumption under 600 Mt.

Table 5.7 Oil consumption in China in 2020 by sector and by oil type under the Linear Extrapolation Scenario (unit, Mt)

	Crude	Gasoline	Kerosene	Diesel	Fuel oil	Liquid gas	Refinery gas	Other oil production	Gross oil product
Loss	29.1								29.1
Own use	7.5			1.9		0.5	13.5	74.2	97.5
Power and heat				0.6			3.8	3.7	8.1
Agriculture				16.7		0.1			16.8
Factories	3.4	0.3		18.9	6.4	4.9	0.4	37.8	72.0
Buildings		0.2	0.2	4.4	0.3			21.6	26.7
Transportation		93.8	22.7	166.4	17.4	1.0			301.3
Commerce		0.1	0.5	1.9	0.0	0.7			3.2
Residential				0.6		22.7			23.2
Other				11.3		0.8		0.3	12.4
Gross	40.0	94.4	23.3	222.5	24.1	30.8	17.7	137.6	590.5

Table 5.8 Oil consumption in China in 2030 by sector and by oil type under the Linear Extrapolation Scenario (unit, Mt)

	Crude	Gasoline	Kerosene	Diesel	Fuel oil	Liquid gas	Refinery gas	Other oil production	Gross oil product
Loss	39.1								39.1
Own use	8.6			0.8			18.0	101.4	128.8
Power and heat							5.1	5.3	10.4
Agriculture				21.5		0.1			21.7
Factories	4.6	0.4		24.8	3.1	6.4		50.7	89.9
Buildings		0.3	0.3	5.9	0.4			30.0	36.8
Transportation		124.0	31.3	228.8	21.9	1.5			407.4
Commerce		0.2	0.6	2.5		0.8			4.1
Residential				0.9		29.9			30.8
Other				14.8		1.1		0.5	16.4
Gross	52.4	124.8	32.2	300.0	25.3	39.7	23.1	187.8	785.2

5.3.2.1 Linear Extrapolation Scenario

By assuming that oil consumption by sector and by oil type will continue the same trend as existed from 2000 to 2009, we can derive the oil consumption for 2030 and 2050 by linear extrapolation, as presented in Tables 5.7 and 5.8.

Under this scenario, Chinese oil consumption will amount to 591 Mt in 2020 and 785 Mt in 2030. With the new increased demand for oil, transport will account for 56 %, own use for 16 %, industry for 9 %, and loss for 5 %. In addition, because of the increase in oil demand in the nontransport sector, growth in demand for diesel demand will be larger than that for gasoline (Table 5.9), and the diesel-gasoline ratio will further increase to 2.36:1 in 2020 and 2.4:1 in 2030. Meanwhile, the rapid growth of other oil products may lead to tension in the supply of chemical materials.

Table 5.9 Structure of oil product demand up to 2030 under the Linear Extrapolation Scenario

	Crude (%)	Gasoline (%)	Kerosene (%)	Diesel (%)	Fuel oil (%)	Liquid gas (%)	Refinery gas (%)	Other oil production (%)
2009	17.3	4.0	38.5	7.9	6.0	3.3	23.0	100
2020	17.2	4.2	40.4	4.4	5.6	3.2	25.0	100
2030	17.0	4.4	40.9	3.5	5.4	3.2	25.6	100

Using linear extrapolation, we also can obtain the vehicle fuel consumption of diesel and gasoline under this scenario. In 2020, vehicle diesel and gasoline consumption will be 144 and 94 Mt, respectively, which will account for 40 % of the gross oil consumption and 46 % of the new increased oil consumption. In 2030, diesel and gasoline consumption will amount to 199 and 124 Mt, respectively, and the proportion of gross oil and increased oil will be 41 and 45 %.

5.3.2.2 Total Consumption Control Scenario A

Simple linear extrapolation using historical trends has its limitations. First, China is currently in a period of dynamic development: there are many uncertainties regarding the future, and so it is difficult to make forecasts using linear extrapolation. Second, the influence of policy needs to be considered. Policies with respect to energy conservation and energy security will be increasingly strict. For example, according to NEA (2011), Chinese oil consumption in 2010 showed a 12.3 % increase above the 2009 level; this was quite large given that the average increase rate from 2000 to 2009 was 6.1 %. As another example, from 2007 to 2009, oil demand in many nontransport areas, such as agricultural, commercial, residential, and energy-intensive industrial sectors, showed a decline; this was the reverse of the above trend and was caused by the influence of energy saving and development of alternative fuels.

Under Total Consumption Control Scenario A, we assume that future policies will be adjusted as appropriate and achievements will be made in some foreseeable problems, including the following: (1) restricting total oil consumption to no more than 700 Mt in 2030; (2) promoting oil conservation and developing alternative fuels in nontransport sectors, especially diesel conservation and diesel alternatives; and (3) preferential supply to the transport sector, especially liquid fuels to meet the demands of passenger transport (e.g., cars and planes); this is because the demand for liquid fuels by this sector will also be limited by restrictions on total oil consumption. In addition, changes in oil demand will have to conform with China's macroeconomic trends in the future. It is expected that the development of industrialization will accelerate up until 2020, that the trend of large-scale construction of the infrastructure will continue, and that thereafter all areas will show stable development. Up to 2030, urbanization and motorization will continue to show fast growth.

Regarding total oil consumption, the assumption is that up to 2020 consumption will maintain its rapid growth to 600 Mt, that is, intermediate between the level in the linear extrapolation scenario (590 Mt) and the average prediction made by international organizations (610 Mt), as indicated in Table 5.6. If energy saving and energy security policies continue to be successful up to 2030, the total amount of oil consumed will be effectively maintained under 700 Mt. On this basis and following the historical trends from 2000 to 2009, we can make assumptions about the changing proportion of oil consumption in China by sector and oil type up to 2030:

1. Oil loss: up to 2020 China will be busily expanding refinery construction, therefore, it may be difficult to achieve a significant drop in oil loss. Subsequently, because expansion of refineries will slow down, we assume that oil loss will be restricted to 4 % after 2020 and fall to 3 % by 2030.
2. Own use: the average annual own-use rate fell 1.11 % from 2000 to 2009. In light of this trend, the own-use rate will fall to 15 % by 2020 and 14 % by 2030 (though the demand for own-use oil for chemical raw materials will increase, natural gas chemicals, coal chemicals, and even the import of chemical materials can replace part of the demand). With regard to oil consumption by oil type, during 2000–2009, the proportions of other products and refinery gas rose significantly; the remaining types all showed a decline; and crude oil maintained a high proportion. Thus, we assume that in 2020, the composition of oil products will be as follows: other products, 78 %; refinery gas, 18 %; and crude oil, 4 %. In 2030, we assume that other products will account for 80 % and refinery gas for 20 %.
3. Power and heat generation: we assume that this proportion will fall to 1 % in 2020, and thereafter the level will remain unchanged since oil power cannot completely disappear. We assume that the composition will maintain the recent diverse proportional trend: 25 % diesel, 25 % fuel oil, 25 % refinery gas, and 25 % other products.
4. Agriculture: agricultural diesel consumption declined after 2005. We assume that this proportion will fall to 2 % in 2020 and 1 % in 2030; we further assume that all consumption in this sector will be diesel.
5. Industry: because alternative fuels have great potential, we assume that oil consumption by industry will gradually decline in the same manner as the trend from 2000 to 2009: the fall will be to 11 % in 2020 and 9 % in 2030. The general trend among industrial oil consumption from 2000 to 2009 was as follows: consumption of other petroleum products rose significantly; consumption of fuel oil was clearly reduced; and LPG and diesel consumption increased a little. Therefore, we assume that the composition of industry oil products will be as follows: in 2020—60 % other petroleum products, 25 % diesel, 10 % fuel oil, and 5 % LPG; in 2030—70 % other petroleum products, 15 % diesel, 10 % fuel oil, and 5 % LPG.
6. Construction: considering that large-scale infrastructure building will be mostly completed close to 2020, we assume that oil consumption by the construction

industry will be maintained at 4 % up to that year and then fall to 3 % by 2030. The composition of that consumption will remain as at present: 80 % other products and 20 % diesel.

7. Commerce and other: we assume that the proportion will remain constant until 2030: 1 % commerce and 2 % other consumption. The composition of oil consumption by commerce is assumed to be 75 % diesel and 25 % LPG up to 2020. In 2030, all consumption by commerce will be diesel. Consumption of oil by the other sector will be diesel.
8. Residential: after 2007, the consumption of LPG by the residential sector presented a downward trend, especially in urban areas; in rural areas, it showed almost no change. The main reason for this was that natural gas gradually replaced LPG. Therefore, we assume that the proportion of oil consumption by the residential sector will fall to 3 % in 2020 and 2 % in 2030; we assume that all that consumption will be LPG.
9. Transportation: the remaining oil consumption is all for transportation, and that fraction will be 57 % in 2020 and 64 % in 2030. With respect to oil type, the diesel-gasoline ratio showed an ongoing rise from 2003 to 2009, the LPG ratio evidenced no significant change, the proportion of gasoline initially declined but then rose, and the proportion of fuel oil first increased before dropping. We assume that the liquid fuel demand by road passenger transportation and aviation will persist, the rising trend of road-freight diesel demand could slow down after 2020, railway diesel consumption will diminish with improvements in electrification, and LPG and fuel oil demand will not increase. Considering all these factors, the composition of oil product consumption (except LPG) by the transport sector is assumed to be as follows: in 2020—55 % diesel, 35 % gasoline, 8 % kerosene, and 2 % fuel oil; in 2030—50 % diesel, 40 % gasoline, and 10 % kerosene.

From the above rough analysis, we can obtain figures for the total oil consumption in 2020 and 2030 according to sector and oil type, as presented in Tables 5.10 and 5.11. The transport sector will account for 83 % of oil-demand growth in 2009–2030, and the rest will be accounted for by own use (10 %), industry (4 %), construction (2 %), other (2 %), commerce (1.5 %), and loss (1 %); the power and heat (–1 %), agriculture (–2 %), and residential (–0.5 %) sectors will show a decrease.

Under Total Consumption Control Scenario A, the changes in oil consumption according to oil type in 2009–2030 appear in Fig. 5.8. The demand for diesel, gasoline, kerosene, and other transportation fuels all exhibit a rapid rise; the demand for refinery gas and other products likewise shows a significant growth, whereas demand for fuel oil and LPG decreases. After 2020, the growth in demand for diesel and other products will decelerate.

Table 5.12 presents the consumption of oil products in China up to 2030 excluding oil loss and own use. It is evident that the problem of the imbalance in the diesel-gasoline ratio will be largely solved during that period. Owing to reduced diesel consumption by industry and in energy supply, the diesel-gasoline ratio will

Table 5.10 Oil consumption in China in 2020 according to sector and oil type under Total Consumption Control Scenario A (unit, Mt)

	Crude	Gasoline	Kerosene	Diesel	Fuel oil	Liquid gas	Refinery gas	Other oil production	Gross oil product
Loss	24								24
Own use	3.6						16.2	70.2	90
Power and heat		1.5			1.5		1.5	1.5	6
Agriculture		12.0							12
Factories		16.5			6.6	3.3		39.6	66
Buildings		4.8						19.2	24
Transportation		188.1	119.7	27.4	6.8				342
Commerce		4.5				1.5			6
Residential						18			18
Other		12							12
Gross	27.6	239.4	119.7	27.4	14.9	22.8	17.7	130.5	600

Table 5.11 Oil consumption in China in 2030 according to sector and oil type under Total Consumption Control Scenario A (unit, Mt)

	Crude	Gasoline	Kerosene	Diesel	Fuel oil	Liquid gas	Refinery gas	Other oil production	Gross oil product
Loss	21								21
Own use							19.6	78.4	98
Power and heat		1.75			1.75		1.75	1.75	7
Agriculture		7							7
Factories		9.45			6.3	3.15		44.1	63
Buildings		4.20						16.8	21
Transportation		224	179.2	44.8					448
Commerce		7							7
Residential						14			14
Other		14							14
Gross	21	267.4	179.2	44.80	8.05	17.15	21.35	141.05	700

drop to 2:1 on the demand side by 2020; thereafter, because industrialization will tend to stabilize and the demand for freight transportation will slowly increase, the diesel-gasoline ratio will be further reduced to 1.49:1 by 2030.

Considering the growth trends in railway electrification and the reduction in oil consumption by ships, most of the future diesel consumption will be accounted for by road transportation. We assume that in 2020, 90 % of transportation diesel will be used for road vehicles and that that will rise to 95 % by 2030. We can estimate the supply of vehicle diesel and gasoline under Total Consumption Control Scenario A: in 2020, vehicle fuel will amount to 120 Mt of gasoline and 169 Mt of diesel, which will account for 48 % of the total oil consumption and 67 % of the consumption growth 2009–2020; in 2030, vehicle fuel will amount to 179 Mt of gasoline and 213 Mt of diesel, which will account for 56 % of the total oil consumption and 78 % of the consumption growth during 2009 to 2030. All these forecasts are greater than

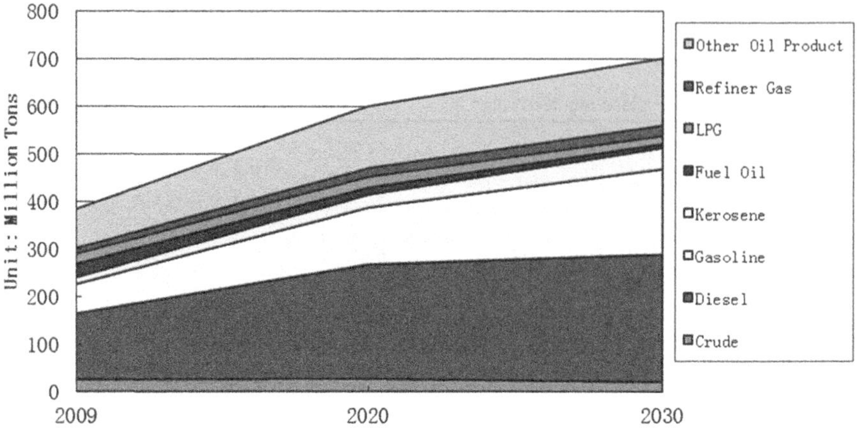

Fig. 5.8 Oil consumption in China according to by oil type under Total Consumption Control Scenario A (2009–2030)

Table 5.12 Consumption of oil products in China up to 2030 under Total Consumption Control Scenario A

Year	Gasoline (%)	Kerosene (%)	Diesel (%)	Fuel oil (%)	LPG (%)	Refinery gas (%)	Other oil products (%)	Total (%)
2009	17.3	4.0	38.5	7.9	6.0	3.3	23.0	100
2020	20.9	4.8	41.8	2.6	4.0	3.1	22.8	100
2030	26.4	6.6	39.4	1.2	2.5	3.1	20.8	100

those predicted under the Linear Extrapolation Scenario. This means that despite the reduction in the total oil supply, oil savings in the nontransport sectors could result in more gasoline and diesel for road vehicles.

5.3.2.3 Total Consumption Control Scenario B

Under Total Consumption Control Scenario B, oil demand will be confined to less than 600 Mt by 2030; this demands a further reduction by 100 Mt in the oil consumption used as the basis for Total Consumption Control Scenario A. Since Total Consumption Control Scenario A has fully accounted for the oil-saving potential in the nontransport sector, the 100-Mt reduction in oil demand will largely have to derive from transportation and mainly through energy saving and alternative fuels for road vehicles.

Energy saving and alternative fuels for road vehicles will be examined in detail in other chapters of this book. This chapter will therefore just briefly discuss the methods used in creating Total Consumption Control Scenario B; instead of carrying out quantitative analysis, views and opinions will be presented here.

Total oil demand by transportation can be expressed using the following formula:

$$O = M \cdot \sum_{\text{modes}} \sum_{\text{fleets}} A_m S_{m,\mathrm{f}} I_{m,\mathrm{f}} F_{m,\mathrm{f}} \tag{5.1}$$

where O signifies the oil demand—unit, tons of oil equivalent (toe); M signifies the transport demand—the unit for freight transport is tons · km and that for passenger transport is person · km; A_m signifies the share of a certain transport mode m (such as road, railway, water, and aviation); $S_{m,\mathrm{f}}$ signifies the share of a certain kind of transport fleet in the mth kind of transport mode (such as in road traffic, vehicles can be classed by weight or application); $I_{m,\mathrm{f}}$ signifies the average fuel efficiency of a certain kind of transport fleet in the mth transport mode—the unit is toe/(person · km) or toe/(t · km); and $F_{m,\mathrm{f}}$ signifies the supply efficiency from crude oil to the final transportation of oil products.

From this, we can examine ways of achieving traffic oil savings.

1. *Guiding Traffic Demand*
   The reduction in total traffic demand M can fundamentally reduce the oil demand by the transport sector. With freight transport, traffic demand can be reduced by optimization of the industrial structure and logistics operations. For example, coal transportation currently accounts for 40 % of rail transport capacity and a significant proportion of road transport capacity. Instead of transporting raw coal over a long distance, if coal conversion were carried out near coal mines (such as in coal power generation and production of coal chemicals), considerable reductions in transport demand for electricity and chemicals would be achieved. The traffic demand by urban residents could be reduced by optimizing urban land function, balancing the spatial distribution of working and living environments, and using information transmission instead of physical travel (Lu et al. 2008).

2. *Optimizing a Multilevel Transport Structure*
   The unit transport energy consumption by various kinds of transport modes is very different. For example, according to 2005 statistics, the unit transportation energy consumption of air transport was eight times greater than that of road transport; the average unit road transport energy consumption was 18 times that of rail transport and 22 times that of water transport (CAE 2011b). In urban passenger transport, the unit rail transport energy consumption is far lower than that of road traffic. Therefore, improving the proportion of rail and waterway traffic and developing urban rail traffic will help reduce the consumption of oil by transport.

   Among specific transport modes, there are great differences in the unit transportation energy consumption of different kinds of vehicles. For example, the unit transport energy consumption of buses is far lower than that of cars. The weight and emissions of cars can influence unit transport energy consumption. Therefore, promoting the use of public transport and encouraging the use of small cars will benefit energy saving by road transport.

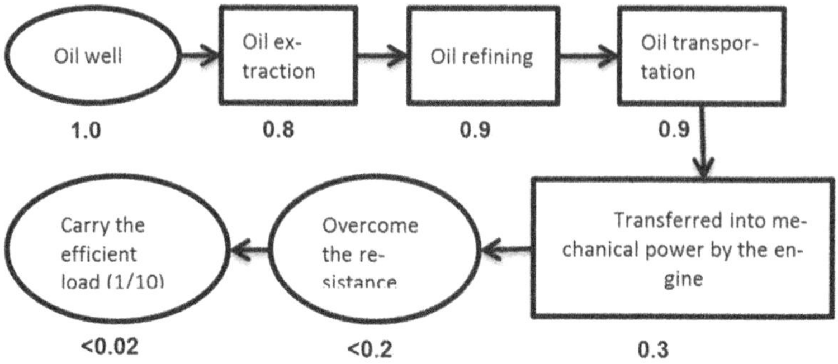

Fig. 5.9 Magnification effect of end-use energy saving by automobiles. *Note*: The rectangles represent the efficiency of a single sector; the ovals represent accumulated efficiency

3. *Improving the Efficiency of Vehicles*
   There is a huge potential for making cars more efficient in terms of energy. According to rough estimates, only 30 % of the energy of a gasoline car is transformed into mechanical energy and transferred to the wheels. The remaining 70 % is lost during the transmission system: lack of engine efficiency, energy consumption by environmental protection facilities, and loss of mechanical transmission efficiency during torque transmission to the wheels (Lovins et al. 2004). Since two-thirds to three-fourths of the energy loss is related to the car weight during application, the energy-saving potential of lightweight materials is enormous. Another effective measure is developing more hybrid cars.
   In addition, the energy efficiency of cars is also related to such factors as road conditions, load rate, and driver habits. Measures toward improving transport management and optimizing transport scheduling, encouraging car pooling, avoiding idle load and flameout, and improving traffic management have great potential for energy saving.
4. *Reduce Energy Loss in the Fuel Supply*
   Improving energy efficiency in the areas of crude oil extraction, transportation, refining, and distribution will help reduce the oil demand by the transport sector. However, there is greater potential for energy saving by the end user, such as the measures cited above: reducing traffic demand, optimizing the transport structure, and improving the energy efficiency of vehicles.
   According to preliminary estimates, in the well-to-wheel supply chain from oil exploitation to consumption by the vehicle, after oil extraction, refining, and distribution, only about two-thirds of the energy is effectively delivered to the motor vehicle fuel tanks. Ultimately, only one-fifth of the energy of the oil resource is effectively utilized; 90 % of this one-fifth energy is used to carry the vehicle's own weight and less than 10 % is used to carry the payload (passengers). Therefore, after various rounds of conversions and losses, only about 1/50 of the energy of the petroleum is effectively used, as shown in Fig. 5.9 (Ni et al. 2009). Therefore, for motor vehicles, the scale effect (magnification

effect) of final energy saving can be several times and even dozens of times the energy saving that can be achieved on the supply side. For energy saving by transportation, end-use energy saving should be prioritized. At the same time, terminal energy saving often faces the problem that at this stage, change is more difficult to implement.

In addition to transportation energy saving, another effective way to reduce traffic oil consumption is developing different kinds of alternative fuels, especially those for vehicles. However, it should be noted that the various kinds of alternative fuels for vehicles are still at an early stage of development. They all face problems regarding benefits and costs: they provide the positive benefit of reducing oil demand, though they also incur other negative costs or risks. For example, coal-derived fuel has low efficiency and produces additional carbon emissions, biofuels demand the collection of raw materials and the use of land and water resources, and the electric car technology faces innovation risks. Therefore, the development of alternative fuels for vehicles should not in general take priority over traffic energy saving. The development scale of alternative fuels for vehicles has to be carefully analyzed in terms of costs and benefits instead of simply attempting to introduce the fuels as quickly and on as great a scale as possible.

Considering medium- to long-term technological developments and economic feasibility toward replacing almost 100 Mt of oil with alternative fuels, coal-derived fuels and natural gas, which belong to the realm of fossil energy, may be more realistic choices. At the same time, technical breakthroughs and innovations with electric cars and the second generation of biofuels should be accelerated. After all, they represent the development direction of the future (Ma et al. 2009).

In addition, current automotive alternative fuels are mainly aimed at providing substitutes for gasoline; there are fewer choices for diesel alternative technology. Therefore, the large-scale development of alternative fuels for vehicles may address the imbalance in the diesel-gasoline ratio, which was dealt with in Total Consumption Control Scenario A. This question demands further study and examination.

5.3.3 Energy Security Constraints and Oil Gap

With China's present energy security policy, there are no clear constraint indicators, such as lowering the energy intensity (energy consumption per unit of GDP) and carbon emission intensity (CO_2 emissions per unit of GDP) in energy-conservation and emission-reduction policies (Xinhua Net 2011). Different energy security constraints will have a great effect on China's total oil supply and will further affect the supply of vehicle gasoline and diesel. Under strict energy security constraints, the future oil supply of China may be lower than the demand and create a gap in the oil demand—an oil gap—that requires additional alternative fuels if it is to be filled.

Table 5.13 Six scenarios for China's energy security constraints and the oil gap

	Scenario A	Scenario B	Scenario C	Scenario D	Scenario E	Scenario F
Energy security constraint	None	SOT 15 %	OID 70 %	SOT 12 %	OID 65 %	OID 60 %
Crude oil production (Mt)	200	200	200	200	200	200
Maximum oil import (Mt)	No limit	493	467	394	371	300
Oil gap in Scenario I (Mt)	No	92	118	191	214	285
Oil gap in Scenario II (Mt)	No	7	33	106	129	200
Oil gap in Scenario III (Mt)	No	No	No	6	29	100

*In Scenario A, the SOT of Scenarios I, II, and III will be about 18, 15, and 12 %, respectively, and the OID of Scenarios I, II, and III will be about 75, 71, and 67 %

*Once one energy constraint has been decided, the others can be calculated based on the basic setting. For example, the OID of Scenario B and Scenario D is about 71 and 66 %, respectively, and the SOT of Scenario C, Scenario E, and Scenario F is about 14, 11, and 9 %

*SOT signifies share of oil trade; OID signifies oil import dependency

To quantitatively examine the relationship between different energy security constraints and the oil gap, we designed six possible oil supply-and-demand scenarios for 2030, as indicated in Table 5.13. The design steps and the main considerations of these scenarios are detailed below.

5.3.3.1 Basic Boundary Conditions

For domestic oil production, stable domestic oil production of 200 Mt is assumed for 2030. For domestic oil demand, reference is made to the three 2030 scenarios presented in Sect. 5.3.2. For international oil trade, for consistency with the rest of this chapter, we use the predictions of OPEC (2010): in 2030, the global oil trade will be 3.287 billion tons.

5.3.3.2 No Energy Security Constraint Scenario (Scenario A)

Under the situation of no energy security constraints, the gap between domestic oil demand and production will be completely met by oil imports: Total Consumption Control Scenario B requires 400 Mt, Total Consumption Control Scenario A requires 500 Mt, and the Linear Extrapolation Scenario requires 585 Mt.

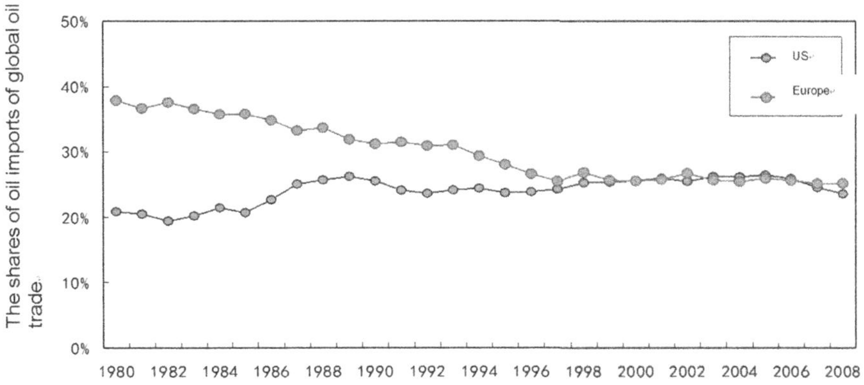

Fig. 5.10 Share of oil imports in the global oil trade of the United States and Europe, 1980–2008. *Note*: The figures for Europe before 1993 do not include those for central Europe (*Data source*: BP 2009)

5.3.3.3 Share of Oil Trade Constraint (Scenarios B and D)

China's oil imports are limited by global oil trade and the actions of other oil-importing countries: China cannot obtain as much oil as it wishes. With the increase in China's oil imports, its share in the global oil trade will be higher. To maintain a high share in the global oil trade, great costs in terms of energy security and comprehensive national strength are required.

The share of oil trade (SOT) of the United States and Europe since 1980 is presented in Fig. 5.10. The SOT of the United States has remained at 20–25 %; though the SOT of Europe has undergone reduction in recent years, it has also been around 25 %. Considering that these two strong regions can maintain just 20–25 % of the SOT, it will be quite difficult for China to attain the same level. However, in Scenario A, the SOT of China reaches 18 %.

If we set the SOT of 15 % as the energy security constraint (Scenario B), China's oil imports will be 493 Mt (OID 71 %) and its domestic oil production 200 Mt, which will be sufficient to meet the oil demand in Total Consumption Control Scenario B. However, there will be a 7-Mt oil gap in Total Consumption Control Scenario A and 92-Mt oil gap in the Linear Extrapolation Scenario.

If we set the SOT of 12 % as the energy security constraint (Scenario D), China's oil imports will be 394 Mt (OID 71 %): there will be a 6-Mt oil gap in Total Consumption Control Scenario B, a 106-Mt oil gap in Total Consumption Control Scenario A, and a 191-Mt oil gap in the Linear Extrapolation Scenario.

5.3.3.4 OID Constraint (Scenarios C, E, and F)

The rise of OID means that the total cost of importing oil will increase, and the impact of disruption in the oil supply and fluctuations in the international oil price on economic development will be greater. In domestic discussions about energy security, OID is a commonly used index. However, there are drawbacks in using only OID to evaluate energy security risks. To take a positive example, after 1980 the OID of the United States displayed a continuous increase, but in 2005 it reached a peak of 65–66 %. This is the main basis for some domestic experts advocating that China should maintain the OID below 65 % or 70 %. Another negative example is that of Japan, whose OID has been close to 100 % because it is almost totally reliant on imported oil supplies.

Considering that China will be similar to the United States in its future oil consumption and imports, there is some sense in using the OID to assess energy security constraints. If we set the OID of 70 % as the energy security constraint, China's oil imports will be 467 Mt (SOT 14 %) and domestic oil production 200 Mt, which is sufficient to meet the oil demand under Total Consumption Control Scenario B. However, there will be a 33-Mt oil gap under Total Consumption Control Scenario A and a 118-Mt oil gap in the Linear Extrapolation Scenario.

If we set the OID of 65 % as the energy security constraint, China's oil imports will be 371 Mt (SOT 11 %), and there will be a 29-Mt oil gap under Total Consumption Control Scenario B, a 129-Mt oil gap under Total Consumption Control Scenario A, and a 214-Mt oil gap under the Linear Extrapolation Scenario.

If we set the OID of 60 % as the energy security constraint, China's oil imports will be 300 Mt (SOT 9 %): there will be a 100-Mt oil gap under Total Consumption Control Scenario B, a 200-Mt oil gap under Total Consumption Control Scenario A, and a 285-Mt oil gap under the Linear Extrapolation Scenario.

The above scenario analysis results show that China's oil supply and oil gap are very sensitive to energy security constraints. This means that China needs to choose its energy security strategy with great care, and this will of course be closely related to the future situation for oil demand.

Under the Linear Extrapolation Scenario, the SOT and OID are up to 18 and 75 %, respectively, and the energy security risk is high. If we take the SOT of 15 % and OID of 70 % as the energy security constraints, the oil gap will amount to 100 Mt, and there still is hope to close the gap by means of large-scale development of oil alternatives. According to the estimation made by CAE (2011a), Chinese alternatives to oil in 2030 are expected to amount to 100 Mt. Under stricter energy security constraints, replacing oil could amount to as much as 200–300 Mt, which would be difficult to achieve.

Under Total Consumption Control Scenario A, only 7–30 Mt of oil needs to be replaced (mainly alternative fuels for vehicles); this can satisfy the SOT of 15 % and OID of 70 %. If we can make a further increase to 100 Mt of alternative oil, we still can achieve 12 % SOT and 65 % OID. However, it would be difficult to satisfy OID of 60 %, because the oil-replacement amount would be 200 Mt.

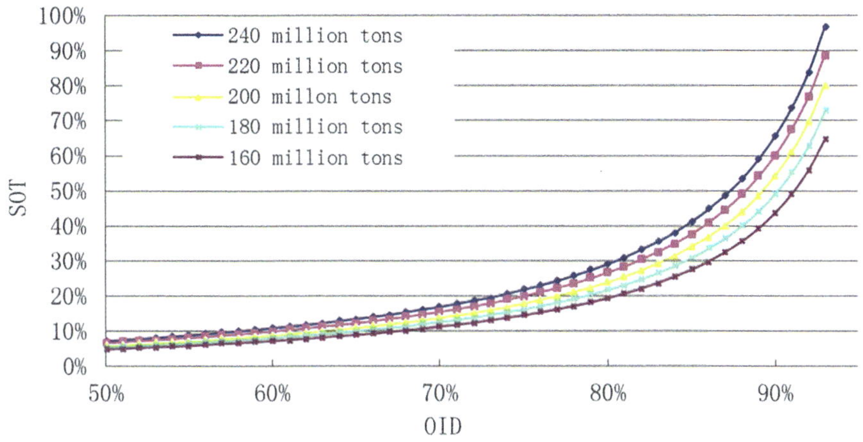

Fig. 5.11 Sensitivity-analysis curves among SOT, OID, and domestic crude oil production (for 2030)

Under Total Consumption Control Scenario B, only 6–29 Mt of oil needs to be replaced (mainly alternative fuels for vehicles), which can satisfy the SOT of 12 % and OID of 65 %. However, to meet the 60 % OID, additional oil replacement of 100 Mt is required. Because under Total Consumption Control Scenario B, we have already considered the potential of transport energy saving and vehicle alternative fuels, it is difficult to provide a further 100-Mt increase using alternative fuels.

In summary, China's energy security policy must be closely combined with a policy controlling the total amount of oil consumption. Based on the above scenario, analyses of oil demand, SOT of 15 % and OID of 70 %, are the least energy security constraint that China should aim for. Having 12 % SOT and 65 % OID would make energy security much safer and could be set as the basic policy. However, it would be difficult to control OID below 60 %.

It is should be pointed out that the above analyses have considered only limited future uncertainties. For example, there may be fluctuations in the 200-Mt domestic oil production before 2030. Figure 5.11 presents sensitivity-analysis curves among SOT, OID, and domestic crude oil production. It shows that the decline in domestic crude oil production will cause the curve to move to the lower right, which means that for the same SOT condition (when the global trade volume is fixed, the SOT is proportional to oil imports), OID will be higher. For example, if the SOT were 15 %, domestic oil production would fall from 200 to 160 Mt, OID would increase from 71 to 76 %, and the energy security risk would increase. However, if domestic oil production rose from 200 to 240 Mt, OID would drop from 71 to 67 %, and the energy security risk would be reduced.

However, if domestic oil production was excessively raised before 2030, although the energy security risk during that period would be reduced, it might result in an earlier oil peak and make long-term energy security risk much greater.

Therefore, China's oil exploration and development should be mainly devoted to improving and ensuring long-term stable oil production and avoiding volatility in the production.

In addition, there is uncertainty regarding the future global oil trade and the supply and demand of oil. An increase in the global oil trade is conducive to China's oil security to a certain extent. Rather than the OID, the proportion of imported oil costs in GDP and the impact of oil disruptions and oil price shocks on GDP may be more practical indicators for measuring energy security risk. However, since the relationships among China's macroeconomic development, oil demand, international oil prices, and the dollar exchange rate are very complex, there is considerable uncertainty over international oil prices. Therefore, these issues demand greater research and discussion.

5.3.4 Conventional and GHG Emissions

A major solution to oil-related pollution lies in a gradual improvement in the emission standards of motor vehicle exhausts with respect to international standards. However, this solution is still hindered by poor oil product quality, especially the sulfur content. The International Clean Transportation Committee points out that the high sulfur content of fuel is a significant barrier in applying reduction technology to mitigate other pollutants (such as particles).

China's fuel quality is dependent on the technological configuration of already-installed refining units (Zhang 2005). At present, the proportion of domestic FCC gasoline is as high as 75 %; this is because the feedstock of most FCC gasoline is mixed with vacuum residual fractions, and therefore the olefin and sulfur content of the FCC gasoline is quite high. This is the major reason for the poor quality of Chinese gasoline. Similarly, the proportion of catalytic hydrogenation diesel is almost 50 %, and part of the high-sulfur straight-run diesel is directly used as diesel harmonic components, which results in the high sulfur content and low hexadecane content in diesel products.

Hydrotreating and hydrofinishing can remove sulfur and impurities from oil, and these are major processes in oil refineries to improve product quality. Although in recent years China's hydrotreating and hydrofinishing capacity has continued to improve, the installation rate is still clearly lower than in developed regions. There are no technological difficulties here, but the high investment costs and inadequate market supervision in addition to other reasons deter some refineries from installing hydrotreating processing equipment. Another factor that threatens to deteriorate fuel quality is the high sulfur content in imported crude oil and the rising proportion of heavy oil; this is a common problem in world oil markets (IEA 2008; OPEC 2009). This issue will pose new barriers in improving the fuel quality of China's refineries.

In terms of reducing GHG emissions, a major potential lies downstream in the oil-supply chain, namely, traffic energy saving. However, there is also some potential in upstream areas, including oil fields and refineries. For example, in the atmospheric distillation at oil refineries, fluidized catalytic cracking and hydrocracking devices are the main source of CO_2 emissions. Fluidized catalytic cracking is in general more CO_2 intensive than hydrocracking; therefore, more hydrocracking will increase flexibility in the diesel-to-gasoline ratio and also be beneficial in reducing CO_2 emissions. In addition, carbon capture, utilization, and sequestration in oil refining are urgent issues that need to be studied.

Overall, however, to increase transport of liquid fuel production and improve fuel quality, additional technical processes need to be added to oil refineries, thereby driving up costs and energy consumption by oil refining. To lessen GHG emissions and other pollution resulting from oil consumption, substantial investment is required in addition to promoting end-use energy saving and developing oil alternatives.

5.4 Conclusions and Suggestions

Although this chapter has analyzed and discussed the whole oil-supply chain and the development of the entire transport sector, its starting point is examining petroleum-derived automotive fuels. Therefore, in this section, the focus will be on presenting proposals and insights relating to the future development of petroleum-derived automotive fuels.

5.4.1 Main Conclusions

5.4.1.1 Road Vehicles Account for the Greatest Oil Consumption

In 2009, China's road transportation (excluding motorcycles) consumed 97 % of the total gasoline, 62 % of diesel, and 5 % of LPG; this amounts to 48 % of the total consumption of petroleum products and 37 % of the total oil consumption. From 2000 to 2009, increased oil consumption by road transport accounted for 47 % of China's total increased oil consumption and was thus the main propelling force behind the growth in oil demand.

Both previous studies and the analysis in this chapter indicate that road vehicles will continue to be the main force behind future growth in oil demand. From a linear extrapolation of historical trends, oil consumption by road vehicles in 2009–2030 will account for 45 % of the total increased oil consumption; in 2030, oil consumption by road vehicles will amount to 41 % of the total oil consumption. Under Total Consumption Control Scenario A, which considers both the macroeconomic situation and oil-saving by nontransport sectors, oil consumption by road vehicles will account for up to 81 % of the total increased oil consumption, and in 2030 oil consumption by road vehicles will amount to 57 % of the total oil consumption.

5.4.1.2 Domestic Crude Oil Production Will Stabilize at Around 180–200 Mt

China's oil reserves and production have entered a stable growth stage. The country is generally not rich in terms of its oil resources, but because its reserves have huge growth potential, crude oil production up to 2050 is expected to stabilize at around 180–200 Mt. Already-proven reserves of crude oil may peak in 2020; however, considering additional reserves, technological development in exploration and extraction techniques, and unconventional resources, China's oil peak may be significantly delayed. However, all the above forecasts are based on China adopting a sustainable, stable crude oil-production strategy—as opposed to raising oil output in the near future and running the risk of future crude oil production volatility and depletion.

5.4.1.3 Oil Import Must Reflect Energy Security Constraints

Because domestic oil production is likely to remain stable, increased oil consumption will be met mainly through oil imports, which will in turn lead to increasing energy security risks. From a comprehensive consideration of the domestic oil production, the SOT available to China, and OID, an import level of 400–500 Mt would ensure energy security to a certain extent. This entails that China should avoid the high oil-consuming development path pursued by developed countries; China should greatly promote oil saving and enhance the development of alternative fuels to solve the short supply of petroleum-derived automotive fuels.

5.4.1.4 Optimizing Technical Configuration in Oil Refining

Faced with a series of serious problems, such as an imbalanced diesel-to-gasoline ratio and poor fuel quality, the overall technological configuration of China's refining industry requires urgent optimization urgently. The goals should be as follows. (1) Enhancing feedstock adaptability: with the rapidly rising proportion of oil imports, the refinery feedstock will be heavier and have a higher sulfur content. In addition, the refining scale of unconventional oil resources will also gradually increase. (2) Increasing clean transport fuel output and enhancing flexibility in the diesel-to-gasoline ratio: since oil will be increasingly used in transport, particularly road transport, newly built refineries should be technically configured to increase their output of gasoline, diesel, and kerosene. However, there may be large fluctuations in the future demand-side diesel-to-gasoline ratio, so the technical configuration of refineries must ensure great flexibility in this ratio. (3) Improving fuel quality: the key to improving fuel quality is to reduce the sulfur content so as to achieve adaptability in meeting constantly improving fuel quality standards. (4) Reducing energy losses and GHG emissions: the focus here is to reduce oil consumption per unit of output, and reducing GHG emissions demands urgent study.

5.4.2 Policy Suggestions

5.4.2.1 Increase Upstream Investment to Ensure Supplies of Vehicle Oil

Transportation energy will remain dominated by oil in the long term and such issues as energy security and control of conventional pollutions. Therefore, China should maintain long-term stable investment in crude oil exploration and exploitation, crude oil refining, and other related areas to promote technological progress. The priorities should include the following: (1) continuously promoting the exploration and exploitation of conventional and unconventional oil resources to ensure long-term stable oil supply and build strategic oil reserves; (2) optimizing refinery technical configuration (e.g., increasing the output ratio of transport fuels, such as gasoline and diesel, and reducing their sulfur content, improving the installation of hydrogenation processing and hydrotreating equipment, and enhancing flexibility in the diesel-to-gasoline ratio) to adapt to the ever-growing demand for clean transportation liquid fuel; and (3) improving carbon capture, utilization, and sequestration to handle the increasing pressure for reducing GHG emissions.

5.4.2.2 Control Total Oil Consumption and Build a National Energy Safety System

China needs to be fully aware of the complexity of energy security issues and the huge investment needed to resolve them. On the one hand, it has been proposed that a timely restriction on total oil consumption, such as upper limits of SOT of 15 % and OID of 65 %, should be carried out to restrict that consumption to the range of 600–700 Mt, thereby avoiding uncontrollable energy security risks produced by huge oil imports. On the other hand, facing the inevitable boom in oil imports in the short and medium term, China needs to do more than diversify its import channels. The country needs to implement a more proactive strategy and establish state strategic petroleum reserves. It should also resort to more unconventional measures, such as making efforts in energy security-related technology, the economy, military affairs, politics, and diplomatic relations, to further improve its energy security system and enhance its presence with respect to international oil prices and international oil resources.

5.4.2.3 Promote Oil Saving, Especially for Road Transport

To contain the growth of total oil consumption, China has to promote oil conservation at every part of the oil-supply chain and in every oil-related sector. First, oil saving should be enhanced in nontransport sectors by replacing oil with alternatives, such as coal, gas, electricity, or renewable sources. Second, it is necessary to improve oil saving in the transport sector through transport-demand management, transport-structure optimization, and improving transport energy efficiency. Since the main end use of oil is in automotive transport and because of the magnification effect in end-use oil saving, cars should be the priority in oil saving and also for reducing GHG emissions and other pollution. Fortunately, there is still significant potential for vehicle energy saving. Therefore, China should hasten the promulgation of stricter fuel economy standards and also the introduction of CO_2-mitigation standards.

References

Benassy-Quere A, Mignon A, Penot A (2007) China and the relationship between the oil price and the dollar. Energy Policy 35:5795–5805

BP (2009) BP statistical review of world energy, June 2009. BP p.l.c., London

BP (2010) BP statistical review of world energy, June 2010. BP p.l.c., London

CAE-Chinese Academy of Engineering (2003) China's sustainable development strategy research: oil and gas resources. Chinese Academy of Engineering, Beijing

CAE-Chinese Academy of Engineering (2011a) Research on the mid- and long-term (2030/2050) energy development strategy of China: volume of electricity, oil and gas, nuclear energy, and environment. Science Press, Beijing (in Chinese)

CAE-Chinese Academy of Engineering (2011b) Research on the mid- and long-term (2030/2050) energy development strategy of China: volume of energy saving and coal. Science Press, Beijing (in Chinese)

Chen JW, Chen XS, Li CN (2009) Review of petroleum alternatives. China Petrochemical Press, Beijing (in Chinese)

CPN-China Petrochemical News (2010) China's crude oil processing capacity becomes the world's second largest. http://enews.sinopecnews.com.cn/shb/html/2010-05/11/content_107741.htm. Accessed 22 Jan 2013

DESNBS-Department of Energy Statistics of National Bureau of Statistics of China, DGANEA-Department of General Affairs of National Energy Administration of China (2010) China energy statistical yearbook 2009. China Statistics Press, Beijing

DESNBS-Department of Energy Statistics of National Bureau of Statistics of China, DGANEA-Department of General Affairs of National Energy Administration of China (2011) China energy statistical yearbook 2010. China Statistics Press, Beijing

Du LM, He YN, Wei C (2010) The relationship between oil price shocks and China's macro-economy: an empirical analysis. Energy Policy 38:4142–4151

EIA (2009) International energy outlook 2009. U.S. Energy Information Administration, Washington, DC

EIA (2010) International energy outlook 2010. U.S. Energy Information Administration, Washington, DC

Feng G, Mu XZ (2010) Cultural challenges to Chinese oil companies in Africa and their strategies. Energy Policy 38:7250–7256

Feng LY, Tang X, Zhao L (2007) Peak forecast model based on Chinese oil production reasonable planning. Oil Explor Dev 4:497–501 (in Chinese)

Feng LY, Li JC, Pang XQ (2008) China's oil reserve forecast and analysis based on peak oil models. Energy Policy 36:4149–4153

He KB, Huo H, Zhang Q, He DQ, An F, Wang M, Walsh MP (2005) Oil consumption and CO_2 emissions in China's road transport: current status, future trends, and policy implications. Energy Policy 33:1499–1507

IEA (2007) World energy outlook 2007. IEA, Paris

IEA (2008) World energy outlook 2008. IEA, Paris
IEA (2009) World energy outlook 2009. IEA, Paris
IEA (2010a) Key world energy statistics 2010. OECD/IEA, Paris
IEA (2010b) World energy outlook 2010. IEA, Paris
IEA (2011) Monthly oil survey January 2011. http://www.iea.org/stats/surveys/archives.asp. Accessed 22 Jan 2013
Kesicki F (2010) The third oil price surge – what's different this time? Energy Policy 38:1596–1606
Leung GCK (2010) China's oil use, 1990–2008. Energy Policy 38:932–944
Lovins AB, Datta EK, Bustnes OE et al (2004) Winning the oil endgame: innovation for profits, jobs, and security. Rocky Mountain Institute, Snowmass
Lu HP, Mao QZ, Li Z (2008) Sustainable urban mobility in a rapid urbanization: theory and practice of China. China Railway Publishing House, Beijing
Ma LW, Li Z, Fu F et al (2009) Alternative energy strategies for China towards 2030 verso. Front Energy Power Eng China 3(1):2–10
Ma LW, Liu P, Fu F, Li Z, Ni WD (2011) Integrated energy strategy for the sustainable development of China. Energy 36:1143–1154
NDRC-National Development and Reform Commission of China (2009) Instructions on the hot issue of oil products' price from the National Development and Reform Commission. http://www.sdpc.gov.cn/xwfb/t20090715_290975.htm. Accessed 22 Jan 2013 (in Chinese)
NEA–National Energy Administration of China (2011) Energy and economic situation in 2010 and outlook of 2011. NEA, Beijing. http://nyj.ndrc.gov.cn/ggtz/t20110128_393339.htm
Nel WP, Cooper CJ (2008) A critical review of IEA's oil demand forecast for China. Energy Policy 36:1096–1106
NETL (2008) Development of baseline data and analysis of life cycle greenhouse gas emissions of petroleum-based fuels. http://www.netl.doe.gov/energy-analyses/pubs/NETL%20LCA%20Petroleum-Based%20Fuels%20Nov%202008.pdf. Accessed 22 Jan 2013
Ni WD, Chen Z, Ma LW, Fu F, Li Z (2009) Reflections on generalized energy saving. Sino-Glob Energy 14(2):1–8 (in Chinese)
OPEC (2009) World oil outlook 2009. OPEC, Vienna
OPEC (2010) World oil outlook 2010. OPEC, Vienna
Ou XM, Zhang XL, Chang SY (2010) Scenario analysis on alternative fuel/vehicle for China's future road transport: life-cycle energy demand and GHG emissions. Energy Policy 38:3943–3956
Skeer J, Wang YJ (2007) China on the move: oil price explosion? Energy Policy 35:678–691
Sun RJ, Qiu K, Chan LG et al (2009) Chinese oil refining development in the chemical industry. Sino-Glob Energy 14(10):1–5 (in Chinese)
Tang X, Zhang BS, Hook M, Feng LY (2010) Forecast of oil reserves and production in Daqing oil field of China. Energy 35:3097–3102
True WR, Koottungal L (2009) Special report: global refining capacity advances; US industry faces uncertain future. Oil gas. http://www.ogj.com/articles/print/volume-107/issue-47/processing/special-report-global.html. Accessed 22 Jan 2013
Walls WD (2010) Petroleum refining industry in China. Energy Policy 38:2110–2115
Wang QY (2007) Reference source of China Sustainable Energy Program: 2007 energy statistics. China Sustainable Energy Program funded by David and Lucille Pike Foundation, William and Flora Hewlett Foundation, and US Energy Foundation, Beijing
Wang QY (2008) Reference source of China Sustainable Energy Program: 2008 energy statistics. China Sustainable Energy Program funded by David and Lucille Pike Foundation, William and Flora Hewlett Foundation, and US Energy Foundation, Beijing
Wang QY (2009) Reference source of China Sustainable Energy Program: 2009 energy statistics. China Sustainable Energy Program funded by David and Lucille Pike Foundation, William and Flora Hewlett Foundation, and US Energy Foundation, Beijing

Wang QY (2010) Reference source of China Sustainable Energy Program: 2010 energy statistics. China Sustainable Energy Program funded by David and Lucille Pike Foundation, William and Flora Hewlett Foundation, and US Energy Foundation, Beijing

Wang Z, Jin YF, Wang M, Wu W (2010) New fuel consumption standards for Chinese passenger vehicles and their effects on reductions of oil use and CO_2 emissions of the Chinese passenger vehicle fleet. Energy Policy 38:5242–5250

Wei WX, Lin BJ (2007) The volatility of oil prices domestic and abroad and the interaction. Econ Res 12:130–141

Wu G, Liu LC, Wei YM (2010) Comparison of China's oil import risk: results based on portfolio theory and a diversification index approach. Energy Policy 37:3557–3565

Xinhua Net (2011) The outline of the 12th-five-year plan of socioeconomic development of China. http://news.xinhuanet.com/politics/2011-03/16/c_121193916.htm

Zhang DY (2005) The world oil refining industry structure adjustment and revelation to China. Petrochem Technol Econ 3:1–7

Zhang XB, Fan Y, Wei YM (2009) A model based on stochastic dynamic programming for determining China's optimal strategic petroleum reserve policy. Energy Policy 37:4397–4406

Zhang KS, Hu JN, Gao SZ, Liu YG, Huang XN, Bao XF (2010a) Sulfur content of gasoline and diesel fuels in Northern China. Energy Policy 38:2110–2115

Zhang QY, Tian WL, Zheng YY, Zhang LL (2010b) Fuel consumption from vehicles of China until 2030 in energy scenarios. Energy Policy 38:6860–6867

Chapter 6
Natural Gas

Ma Linwei, Gao Dan, Li Weiqi, and Li Zheng

Abstract In this chapter, after a brief introduction, we discuss the potential of natural gas as a form of automotive energy in China. The exploration, production, consumption, import, storage, transportation, and distribution of natural gas in China are comprehensively reviewed, and the development of coal-based synthetic natural gas is also discussed. Following this, we provide our perspectives on the supply and demand of natural gas and the potential for its use as a source of automotive energy in China. Then, we propose policy suggestions for developing the main technological pathways, including compressed natural gas (CNG) vehicles, liquefied natural gas (LNG) vehicles, and gas to liquids, based on an analysis of their technical features, development status in China and abroad, technological performance, and supporting conditions.

Keywords Natural gas • CNG • LNG • Vehicles

6.1 Introduction

In studies of natural gas as an energy form for automotive transportation, a distinction has to be made between two similar concepts: utilization of natural gas in vehicles and natural gas vehicles. The main distinction is that the former is defined on the basis of primary energy, while the latter is defined on the basis of the energy carrier. In addition to directly providing vehicle fuels in the form of CNG and LNG, natural gas can also provide energy for vehicle transport in the form of electricity, gas to liquids (GTL), methanol, and hydrogen. The term "natural gas vehicles" mainly refers to CNG or LNG vehicles. By contrast, the use of natural

Ma Linwei (✉) • Gao Dan • Li Weiqi • Li Zheng
State Key Laboratory of Power Systems, Department of Thermal Engineering,
Tsinghua-BP Clean Energy Center, Tsinghua University, Beijing 100084, China
e-mail: malinwei@tsinghua.edu.cn

DOI 10.1007/978-3-642-36847-9_6,

gas for vehicles can also be achieved by first converting the natural gas to other forms of energy. Examples of this are a mixture of hydrogen in vehicle fuels and electricity and liquid fuels derived from natural gas. Use of natural gas in vehicles may even refer to the use of synthetic natural gas (SNG), coke oven gas, and biogas as alternative fuels; these are also forms of "natural gas" in a broad sense. In some instances, even liquefied petroleum gas (LPG) vehicles are classified as natural gas vehicles.

This chapter examines natural gas as a form of automotive energy from the perspective of energy supply, and it focuses on the whole supply chain from exploitation of natural gas to end use. The two major issues involved are as follows: (1) the potential of natural gas as a form of primary energy for vehicle fuels and (2) developing trends in natural gas vehicles as end-energy use. These two questions involve multiple influencing factors; however, this chapter focuses on the following two issues. (1) The future supply and demand for natural gas: China's natural gas is still at an early stage of development; thus, there are considerable uncertainties concerning its supply and consumption. The authors maintain that judgments about natural gas as an automotive energy can be made only after the overall picture of its supply and demand are clear. (2) The technological pathways for using natural gas for vehicles: this chapter focuses on three pathways—CNG vehicles, LNG vehicles, and GTL. Since this is a preliminary study, the content in this chapter does not attempt to make a far-reaching judgment about the above two issues; rather, basic information is collected and analyzed, major related issues are discussed, and preliminary viewpoints and suggestions are provided.

In this chapter, Sect. 6.1 is an introduction to the background and content of the study. Section 6.2 presents a preliminary analysis of the potential of natural gas for automotive power in China. This section considers natural gas exploration, production, consumption, import, storage, transportation, and distribution as well as coal-based SNG in addition to the future supply and demand of natural gas and its potential for utilization in vehicles. In Sects. 6.3, 6.4 and 6.5, the development of technological applications, including CNG vehicles, LNG vehicles, and GTL, is analyzed. A preliminary analysis and discussion of technical issues and development in China and abroad are made; technical performance and supporting conditions are evaluated, and suggestions are made regarding future development. Conclusions and suggestions appear in Sect. 6.6.

6.2 Potential of Natural Gas as an Automotive Energy Source

6.2.1 Resource Exploration of Natural Gas in China

On the whole, China is not rich in conventional natural gas resources. However, natural gas has been less explored in China than other energy sources, and there is still a large potential for exploring unconventional natural gas resources; thus, the

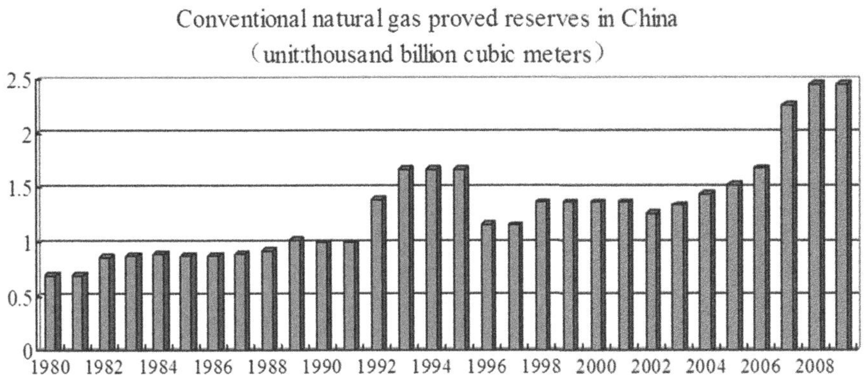

Fig. 6.1 Proven natural gas reserves in China (1980–2009) (*Source*: BP 2010)

country's natural gas reserves could increase in future. In 2009, China's proven conventional natural gas reserves stood at only 2.46 trillion m^3; the reserve/production ratio was only 28:1, which is less than half the world average of 64:1 (BP 2010). However, the proven reserves of conventional natural gas have continuously risen, as shown in Fig. 6.1. According to the latest domestic assessment of natural gas reserves, the prospective natural gas reserves onshore and in 115 offshore basins stand at 56 trillion m^3, the geological reserve amounts to 35 trillion m^3, and the recoverable reserve is 22 trillion m^3, which is nearly nine times the proven reserves in 2009. According to some domestic studies, the exploration of natural gas is 30 years behind that of petroleum; natural gas exploration is thus still at an early stage.

In addition, China possesses 11 trillion m^3 of coal bed methane reserves. China has also recently made rapid progress in exploration of shale gas and natural gas hydrate resources. These unconventional resources represent a powerful guarantee for the natural gas supply in the future (Sidebars 6.1 and 6.2).

Sidebar 6.1: Definition and Classification of Natural Gas Reserves

From the perspective of exploiting energy resources, natural gas refers to the mixture of hydrocarbon and non-hydrocarbon gases contained in the strata; it is mainly found in oil field gas, gas field gas, coal bed methane, mud volcano gas, and biogas. Natural gas primarily consists of methane in addition to such other gases as ethane, propane, and butane.

Processed natural gas that contains more than 90 % methane is called "dry gas"; processed natural gas that contains under 90 % methane but over 10 % of such alkanes as ethane and propane is called "wet gas." In the energy industry, natural gas is classified as follows.

(continued)

(continued)

1. Conventional Natural Gas
   Conventional natural gas offers the best exploration possibilities from a technical and economic viewpoint. It is further classified as follows: oil field gas, which is associated with petroleum production; natural gas liquids, which are associated with condensate gas fields; and gas field gas. Oil field gas is also called associated gas; the latter two are collectively known as non-associated gas. Natural gas from gas fields can be sold as natural gas only after having undergone certain treatments. For example, associated gas is usually subjected to dehydration, purification, and light hydrocarbon recovery; liquefied gas and light oil are then extracted. Natural gas liquids are separated into condensates and wet gas for LNG; use as pipeline gas first demands the separation of ethane and propane (Hua 2007).
2. Unconventional Natural Gas
   Unconventional natural gas is technically and economically more difficult to exploit. It is further classified as sandstone gas, coal bed gas, shale gas, and natural gas hydrate (also known as "combustible ice"). Since there are problems associated with the supply of natural gas and its price is rising, unconventional natural gas resources are receiving increasing attention. Among types of unconventional natural gas, coal bed gas has been exploited, and recently there have been commercial operations for shale gas in the United States and Canada. In addition, there are huge reserves of natural gas hydrate. According to rough estimates, the natural gas contained in natural gas hydrate is equivalent to more than 100 times the amount of conventional natural gas (Chen et al. 2009).
3. Synthetic Natural Gas
   This is not primary energy, but it is still included as a source of natural gas in this chapter. Since methane can be synthesized in various ways, such as from coke oven gas, coal-based SNG, and biogas, synthesized gas that has methane as its main content should also be included in the natural gas supply, and these gases may be collectively termed SNG. For example, before natural gas was developed on a large scale, China's city gas was mainly coal-based and oil-based gas. In terms of coal-based gas, the state of North Dakota in the United States set up a synthetic fuel plant called Great Plains in 1984; this plant produces 1.53 billion m^3 SNG annually using lignite as feedstock.

Sidebar 6.2: China Finds Onshore Natural Gas Hydrate with Prospective Reserves of over 35 Billion Tons Oil Equivalent

On November 25, 2009, the Ministry of Land and Resources announced that a natural gas hydrate sample had been successfully drilled from the permafrost in the southern part of the Qilian Mountains, Qinghai Province. The sample was tested and a series of raw data was obtained. This was a major breakthrough following the finding of natural gas hydrate in the northern part of the South China Sea in July 2007. The latter finding proves that China's permafrost regions are rich in natural gas hydrate; it was significant in determining the distribution patterns of natural gas hydrate toward exploring new natural gas resources.

China is the world's third-largest country in terms of permafrost. It has 2.15 million km^2 of permafrost, whose conditions are ideal for the formation and storage of natural gas hydrate. According to preliminary estimates, the prospective reserves amount to at least 35 billion tons oil equivalent (toe). China has long laid emphasis on surveying and researching natural gas hydrate resources in onshore permafrost regions.

This discovery of onshore natural gas hydrate in China makes China the first country in the world to find natural gas hydrate in a mid- to low-latitude permafrost region. It also makes China the third country in the world to obtain a natural gas hydrate sample onshore through drilling, following the finding of natural gas hydrate in Mackenzie Delta, Canada, in 1992, and the United States finding of natural gas hydrate in northern Alaska in 2007.

Source: Xinhua Net (2009a), Beijing, September 25, 2009, 17:54 (Liu)

6.2.2 Production, Consumption, and Import of Natural Gas in China

6.2.2.1 Production

After many years' efforts, the development of natural gas production in China has begun to accelerate. Since 1996, the startup of long-haul gas pipelines, such as those of Shaanxi-Beijing, Seninglan, the West-to-East Gas Pipeline, Zhongwu, and Shaanxi-Beijing II, has boosted the production of existing gas fields and put a number of new gas fields into use. By the end of 2007, China had established 12 gas fields, each with an annual output greater than one billion cubic m^3. Among these gas fields, Kela No. 2 gas field produces 10.7 billion m^3 each year. China has three large-scale natural gas production bases—Sichuan, Tarim, and Changqing; each of these has an annual output greater than 10 billion m^3.

China's natural gas production increased from 20.11 billion m^3 to 85.17 billion m^3 per annum from 1996 to 2008, with an average annual growth rate of 11.7 % (BP 2010). With rapidly growing reserves, demand for natural gas has continuously increased, and exploitation technology has undergone constant improvement. In addition to the construction of the Sichuan-to-East Gas Pipeline, No. 2 West-to-East Gas Pipeline, and Central Asia–China Gas Pipeline, China will establish a series of gas fields, which will substantially increase its natural gas output. It has been estimated that the annual output will amount to 200 billion m^3 by 2020.

Domestic studies based on the mining rate method, R/P control method, Weng method, and HCZ model have indicated that China's peak annual output of conventional natural gas could reach 240–280 billion m^3 (Lu 2009). This optimistic forecast does, however, have to be modified by various factors, such as low resource endowment, the harsh geographic environment, the great distance from consumer markets, and uncertainty over the exploitable reserve.

In addition to conventional natural gas, after 30 years of exploration, evaluation, and experimental work, coal bed methane production in China has been commercialized and is being rapidly developed. Although China is still conducting research into shale gas exploitation, estimates based on the experience in the United States suggest that China has substantial potential in this area.

Although there is some uncertainty with regard to natural gas hydrate reserves and their exploitability, large amounts of gas hydrate have been tentatively identified in the South China Sea, onshore permafrost regions, and the East China Sea. According to projections, onshore and offshore gas hydrate can be commercially exploited by 2030–2050. As an essential supplement to and substitute for conventional natural gas, unconventional natural gas can guarantee stable increasing production over the long term (Sidebar 6.3).

Sidebar 6.3: China Officially Initiates Exploitation of Shale Gas as a New Form of Energy

The Ministry of Land and Resources launched China's first shale gas exploration project recently in Qi County, Chongqing. This makes China the third country in the world to undertake shale gas exploration and exploitation after the United States and Canada.

It has been reported that the Ministry of Land and Resources is paying particular attention to the exploitation and utilization of shale gas in terms of the national energy strategy. In 2004, the Strategic Studies Center of Oil and Gas (under the Ministry of Land and Resources) at China University of Geosciences (Beijing) began studies into shale gas. From a comparison of metallogenic conditions of eight provinces, including Hunan and Sichuan, the research team concluded that the southern and southeastern regions of Chongqing were rich in lower Cambrian, lower Silurian, and Permian

(continued)

(continued)
deposits, which are likely to contain large amounts of shale gas. Qijiang of Chongqing, Wansheng, Nanchuan, Wulong, Pengshui, Youyang, Xiushan, and Wuxi were listed as the most favorable metallogenic belts of shale gas; they were thus established as the first areas for field exploration.

Shale gas is a natural gas; it is absorbed or occurs freely in dull mud or high-carbon mud shales. Shale gas is currently being exploited in the United States and Canada. In 2007, US shale gas output amounted to 45 billion m^3. China's shale gas reserve occurs in major basins and regions and is around 15–30 trillion m^3; it is therefore comparable with the 28.3 trillion m^3 of the United States.

Source: Want, L., October 5, 2009, Beijing, Xinhua Net (2009b)

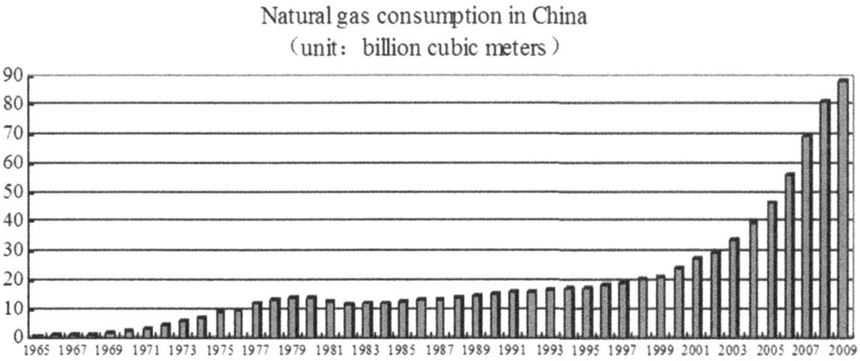

Fig. 6.2 Annual natural gas consumption in China from 1965 to 2009

6.2.2.2 Consumption

In recent years, China's demand for natural gas has maintained a rapid growth, especially after 2000, as shown in Fig. 6.2. In 2009, China's annual natural gas consumption amounted to 88.7 billion m^3. From 2000 to 2009, annual natural gas consumption in China displayed an almost 3.6-fold increase, with an annual growth rate of 15.4 %; this was much higher than the increase in energy consumption (9.4 %) over the same period (BP 2010).

However, in terms of overall primary energy consumption, natural gas accounts only for a fairly small share. In 2009, natural gas accounted for just 4.1 % of the total primary energy production in China, or 3.9 % of the total primary energy consumption based on the coal equivalent method (Fig. 6.3). Natural gas accounts for 23.8 % of the world's total primary energy consumption (BP 2010).

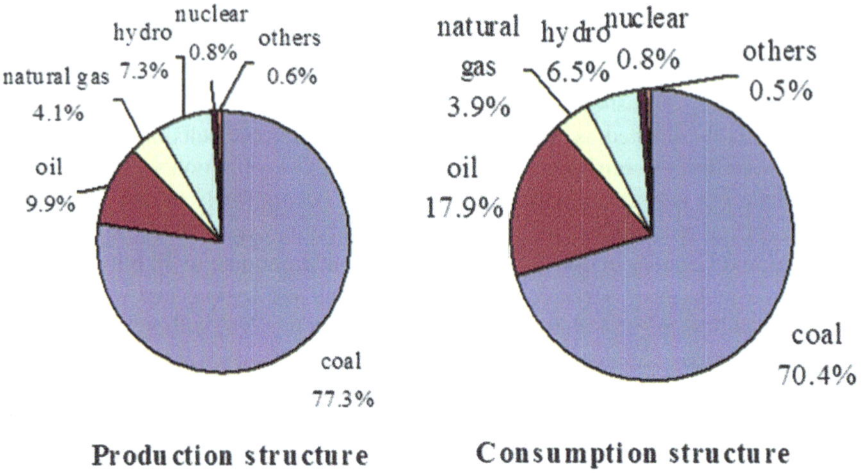

Fig. 6.3 Primary energy production and consumption in China in 2009 (*Source*: NBS and NEA (2011) (based on the coal equivalent method))

Table 6.1 Natural gas consumption and structure in China (2000, 2005, 2008)

	2000		2005		2008	
	Billion m^3	%	Billion m^3	%	Billion m^3	%
Chemical industry	8.23	33.6	14.14			24.6
Power generation	0.64	2.6	1.88	4.0	7.39	9.1
Industrial fuel	11.03	45.0	16.86	36.1	25.77	31.7
Transportation[a]	0.58	2.4	1.97	4.2	3.27	4.0
Urban gas[b]	4.02			25.5	24.86	30.6
Total	24.50	100.0	46.76	100.0	81.29	100.0

Source: Wang (2010, Table 78)
[a]Gas consumption figures from the transport sector
[b]Total of commerce, city gas, and other consumption

China's natural gas is mainly consumed by industry, including feedstock of the chemical industry and industrial fuel. City gas and power generation account for a smaller share and the amount of natural gas used by vehicles is smaller still, as indicated in Table 6.1. In recent years, the natural gas consumed in each sector has grown rapidly. The structure of natural gas consumption has changed: the proportion accounted for by the chemical industry and electricity generation has decreased, while that of city gas and power generation has risen rapidly; natural gas consumed by vehicles accounted for 4 %.

China's natural gas is a prime energy source that is clean and of high quality; however, since at present it is scarce, it cannot meet huge demands from various sectors. Thus, the imbalance in China's natural gas supply and demand cannot be resolved over the short term. Comprehensive use of natural gas in the future is still at the research and planning stage (Sidebar 6.4).

Sidebar 6.4: Policies on Natural Gas Consumption by National Development and Reform Commission (NDRC) (NDRC 2007)

Considering the social, environmental, and economic benefits of natural gas utilization and also according to the characteristics of different users, the use of natural gas can be classified into four areas: priority, permitted, restricted, and prohibited.

1. Priority
   (a) Urban gas for cooking, domestic hot water, and so on, especially in large- or medium-sized cities
   (b) Natural gas for public facilities, such as airports, government offices, staff canteens, nursery schools, colleges, hotels, restaurants, supermarkets, office buildings
   (c) Natural gas vehicles, especially dual-fuel vehicles
   (d) Distributed heat–electricity cogeneration system, heat–electricity–cooling coproduction system
2. Permitted
   (a) Concentrated heating gas, in central regions
   (b) Household heating gas
   (c) Central air-conditioning system
   (d) Natural gas used as a substitute for oil and LPG in such sectors as building materials, the mechanical and electrical industry, textile industry, petrochemical industry, and metallurgical industry
   (e) Natural gas used as a substitute for coal in projects that provide good environmental and economical benefits, in the building material industry, mechanical and electrical industry, textile industry, petrochemical industry, and metallurgical industry
   (f) Users whose gas feedstock permit interrupting in the building material industry, mechanical and electrical industry, textile industry, petrochemical industry, and metallurgical industry
   (g) Natural gas power-generation usage in power load centers with rich gas resources
   (h) Natural gas-based synthetic hydrogen usage that provides good economic benefits and where the gas consumption is not great
   (i) Natural gas-based synthetic nitrogenous fertilizer projects using gas that is not suitable for transport toward the outside or items a. and b. above
3. Restricted
   (a) Natural gas power-generation projects in unimportant power load centers

(continued)

(continued)

(b) Extension projects to existing synthetic ammonia plants using natural gas, fuel-switch (from coal to natural gas) projects for synthetic ammonia plants
(c) Chemical projects for one-C products, such as acetylene and chloromethane with methane as a feedstock
(d) Natural gas-based synthetic ammonia projects apart from item i. in area 2 above

4. Prohibited

(a) Natural gas power-generation projects as basic load built in 13 large-scale coal bases
(b) Natural gas-based synthetic methane projects that have been newly built or expanded
(c) Fuel-switch synthetic methane projects, from coal to natural gas

However, with the future development of natural gas, these policies need to be adjusted to meet new circumstances.

6.2.2.3 Imports

Natural gas can be imported in two ways—onshore pipelines and LNG ocean shipping. Similar to the situation with petroleum, there is a serious regional imbalance in the distribution of natural gas around the world, and this has produced a rapid increase in the international natural gas trade. In 2009, the total natural gas trade volume by pipeline was 633.77 billion m^3 per annum, while that of LNG was 242.77 billion m^3 per annum; LNG accounted for about 30 % of total gas consumption in 2009 (BP 2010). According to research by the International Energy Agency (IEA; IEA 2010), the natural gas trade will continue to rise rapidly, and it will be increasingly transported by sea as LNG. According to the IEA's predictions, the total volume of natural gas traded in the world will increase by 79 % from 2008 to 2035, and the share of LNG will increase from 31 to 42 %.

China is close or adjacent to countries of the Middle East, the former Soviet Union, and the Asia-Pacific region (Australia, Indonesia, and Thailand) that are rich in natural gas. China is thus in a favorable geographic position for large-scale imports of natural gas. In 2009, China imported around 7.63 billion m^3 of LNG (BP 2010), which is equivalent to 5.6 million tons of LNG, mainly through LNG import projects in Guangdong and Fujian provinces. Four agreements for LNG import projects have already been signed; these can ensure a continuous supply of natural gas over the next 25 years, and the total annual supply will be about 13 million tons (CAE 2011). China plans to build 10 LNG receiving terminals, as shown in Table 6.2. On completion of these projects, the total LNG import volume will amount to 48 million tons, which is equivalent to 66 billion m^3 of LNG.

Table 6.2 Ten LNG projects in China, planned or under construction

Project	Company	Planned scale
Guangdong LNG project	China National Offshore Oil Corporation (CNOOC)	5 Mt/year
Fujian LNG project	CNOOC	5 Mt/year
Shanghai LNG project	CNOOC	6 Mt/year
Hainan LNG project	CNOOC	3 Mt/year
Zhejiang LNG project	CNOOC	3 Mt/year
Qinhuangdao LNG project	CNOOC	3 Mt/year
Shandong LNG project	Sinopec	5 Mt/year
Jiasu LNG project	CNPC	3.5 Mt/year
Guangxi LNG project	CNPC	–
Liaoning LNG project	–	–

Sources: Status quo and solution of ocean shipping of imported natural gas in China "China Ocean Transport Announcement" 2005

Onshore import pipeline projects will be completed at a later time in China, though the domestic natural gas pipeline network has already been completed. Import pipeline projects are also being planned or are under construction; they include the Central Asia–China Gas Pipeline, Russia–China pipeline, and Myanmar–China pipeline. The Central Asia–China Gas Pipeline project has already reached completion and is in use, as detailed in Sidebar 6.5.

Sidebar 6.5: Central Asia–China Gas Pipeline Construction

The second facility at the No. 1 compression station in Kazakhstan started up at 22:24 on April 19, 2011. Plans call for the annual gas supply to amount to 17.7 billion m^3.

Imported natural gas from Central Asia passes through 16 provinces, municipalities, and autonomous regions via the Second West–East Gas Pipeline, which is connected to the Central Asia–China Gas Pipeline. In the first quarter of last year, 3.1 billion m^3 of gas was piped to the domestic market via the Central Asia–China Gas Pipeline. As of April 24, 2011, it has been operating safely and stably for 509 days, and the total amount of natural gas supplied has been 8.5 billion m^3. The Central Asia–China Gas Pipeline provides an effective guarantee for the rapid growth of the domestic natural gas market.

According to construction plans, five new compression stations will be built over the next 2 years. The total annual capacity of this transborder gas pipeline will amount to 30 billion m^3 when all eight compression stations have been completed. In addition, a coordination mechanism relating to safety operations along this pipeline among Turkmenistan, Uzbekistan, Kazakhstan, and China will be set up this year. Including the south line of the Kazakhstan gas pipeline project, which is already being developed, and line C of the

(continued)

(continued)
Central Asia–China Gas Pipeline project, which is still at the blueprint stage, when the natural gas transmission network across the Asian hinterland has been completed and put into use, the domestic population benefited by it will exceed 400 million.

Source: CNPC News Center (2011)
April 27, 2011
Website: http://news.cnpc.com.cn/system/2011/04/27/001331940.shtml

Table 6.3 Comparison of natural gas prices in China and abroad

	LNG	West–east natural gas transmission project	Russia to Europe
RMB/m^3	2.5–2.8	2.05	3.18
Notes	Average price in Jiangsu and Zhejiang provinces	Price of civil natural gas Beijing, in 2009	EU price in 2008 (2009 exchange rate)

Source: *China Natural Gas Bimonthly*

However, China's imports of natural gas are restricted by high cost and security issues. Imported natural gas is more expensive than that produced domestically, as evident in Table 6.3 (which takes city gas as an example). With an increasing trading volume in natural gas and increasing international interdependence in trading natural gas, global natural gas supply security issues become increasingly apparent. The two cases when Russia suspended natural gas exports to Europe via Ukraine are typical of security issues with natural gas supply (Sidebar 6.6).

Sidebar 6.6: Russia–Ukraine Natural Gas Disputes

1. First suspension. In December 2005, Russia demanded a price hike from \$50/1,000 to \$230/1,000 m^3. Russia suspended its natural gas supply to Ukraine at 9:40 a.m., January 1, 2006, for 3 days.
2. Second suspension. In December 2008, Russia and Ukraine had another dispute on natural gas price and loans. Russia suspended its natural gas supply to Ukraine on January 1, 2009. Russia announced suspension of natural gas exports to Europe via Ukraine. Natural gas exports were resumed on January 12, 2009, following negotiations.
3. Implications. There were soaring prices, which had an impact on everyday life (winter heating), natural gas-related industries, and caused an oil price hike in New York. In this way, people recognized the significance of natural gas supply security. Thus, a diversified, multichannel supply system needs to be established (Yin and Huang 2009).

6.2.3 *Storage, Transportation, and Distribution of China's Natural Gas*

6.2.3.1 Transportation

Most of China's inland natural gas transport is made by pipeline; a small proportion is carried in tankers or ships in the form of LNG. China has made substantial progress in its inland natural gas pipeline transportation in recent years. A number of long-haul natural gas pipelines have been constructed, such as the Shanxi-Beijing I/II, the first stage of the West–East Natural Gas Transmission Project, Zhongwu, and Seninglan, which form a preliminary natural gas pipeline network. For example, the pipeline in the first stage of the West–East Natural Gas Transmission Project extends west as far as the Tarim Basin of Xinjiang and east as far as Baihe in Qingpu, Shanghai; the designed annual transmission volume of the pipeline is 12 billion m^3, and it is being built over a period of 30 years.

Future deployment of onshore pipeline construction will take both domestic and overseas natural gas distribution into consideration. For example, to address the seasonal imbalance of consumption in such large cities as Shanghai and Guangzhou, China will launch the Russia to China Natural Gas Project (the Russia–China pipeline mentioned above) to follow up on the West–East Natural Gas Transmission Project. In addition, the second pipeline of the West–East Natural Gas Transmission Project, which is at the planning stage, will start from Xinjiang Province and end at Guangzhou. Natural gas in this pipeline will be transported from such Central Asian countries as Turkmenistan and Kazakhstan to southern provinces in China that lack natural gas. The designed transport volume of this project is about 30 billion m^3/year, with an actual transport volume of 26 billion m^3/year.

Because China has a large number of scattered small-scale gas fields that are unsuitable for pipeline transport, transport by tankers and ships in the form of LNG constitutes an essential complement to inland pipeline transport. China boasts the largest inland LNG industry in the world, and it has more than 100 LNG vaporizing stations in operation.

For imported LNG, shipping LNG by sea is also an important complement to pipeline supply. After LNG receiving terminals vaporize LNG received through the pipeline, the LNG can be directly transported to target markets, where LNG can be vaporized and distributed (Hua 2007).

6.2.3.2 Storage and Distribution

As more and more natural gas is consumed by urban residents and industry, the issue of consumption volatility has become increasingly prominent, especially in the production cycles of large-scale industry users. To address seasonal fluctuations in city gas consumption and those owing to changes in industrial production intensity, it is necessary to examine storage and peak sharing.

Interconnection projects among major natural gas pipelines have considerably boosted the systematization of China's natural gas pipeline network. To better use natural gas as a prime energy and increase the overall efficiency, it is necessary to examine and implement some major transformations in the storage, transport, and distribution of natural gas:

1. Speed up construction of underground natural gas reservoirs
2. Cogeneration modules of power, heating, and cooling with load flexibility and using combinations of different heat pumps
3. Take full advantage of the pressure energy in pipelines, vaporization latent heat in exhausts, and low temperatures in LNG
4. More appropriate deployment of natural gas pipelines
5. Optimizing the urban natural gas system through full use of information network technology, for example, the integrated dispatching and control of urban heat, power, and gas networks

6.2.4 Coal-Based SNG

In the 1970s and 1980s, the United States and Europe launched research into producing SNG from coal-based synthetic gas through the methanation reaction to guarantee long-term supply security of natural gas. Construction began in 1981 of Great Plains, a facility operated by a synthetic fuel company. Located in North Dakota, United States, the plant began producing SNG in 1984, and it is the only large-scale SNG commercial plant in the world (Chandel and Williams 2009). Great Plains uses 14 Lurgi fluidized bed gasifiers (two standby), consumes six million tons of lignite annually, and produces 1.53 billion m^3 of SNG. It also produces 1,200 tons of anhydrous ammonia and 5,000 tons of high-purity CO_2 per day as by-products. This CO_2 is later sold to oil fields in Canada to enhance the oil-recovery ratio (Doug Huxley 2006).

Owing to limits in coal gasification technology and the subsequent discovery of natural gas reserves, coal-based SNG has not been developed elsewhere. However, because of increasing natural gas prices, the growing issue of supply security, and the wide application of large-scale spouted fluidized bed technology, SNG is starting to gain attention again in various countries. China is rich in coal but not in gas; thus, with the ever-increasing demand for natural gas, China is very interested in coal-based SNG.

From the perspective of overall energy strategy, coal-based SNG offers promise as a new way of making clean use of coal and natural gas. Thus, the technological issues in this area need to be addressed. Since the security issue with the natural gas supply has become increasingly prominent, a certain amount of coal-based SNG can also help guarantee supply security of natural gas. However, considering that coal-based SNG may cause additional CO_2 emissions and lower efficiency

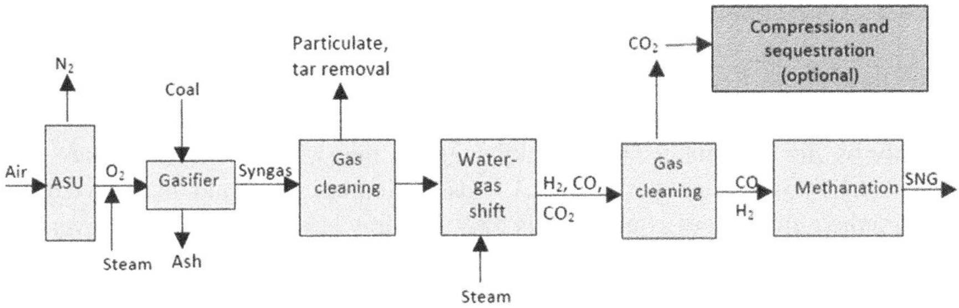

Fig. 6.4 Pure oxygen route of coal-based SNG

in the service time, coal-based SNG projects should be appropriately distributed, moderately developed, and effectively utilized; in addition, the scale should not be too large.

6.2.4.1 Technical Background of Coal-Based SNG

The basic technology of coal-based SNG is coal gasification and the synthesis to methane. The main difference in the various technical routes employed lies in gasification: pure oxygen route, hydrogasification route, and steam gasification. Among these, the pure oxygen route is the most mature and is the technology to have been proven through commercial demonstration.

As shown in Fig. 6.4, the pure oxygen route gasifies coal under high temperature and pressure with oxygen as the gasification agent; the main components of the synthesis gas are CO and H_2. After cooling and removing such components as dust, asphalt, and tar, the hydrogen content is increased through the water–gas shift reaction: $CO + H_2O = CO_2 + H_2$. Then, CO_2 and H_2S are removed through Rectisol; CO and H_2 are converted to CH_4 in a fixed-bed reactor by the catalytic methanation reaction: $3H_2 + CO = CH_4 + H_2O$.

In the hydrogasification route, the main reaction is a direct reaction of hydrogen and methane to obtain methane; the hydrogen can be added externally or obtained by reforming a proportion of the CH_4 generated. In 1970s, there was a demonstration of hydrogasification in Germany.

In the steam gasification route, gasification and methanation occur under catalytic action within the same reactor; thus, the heat released from methanation is fully utilized. After methane is extracted from the synthesis gas, the remaining CO and H_2 circulate afresh into the gasifier.

Compared with the pure oxygen route, the hydrogasification and steam gasification routes do not require oxygen being separated from the air; in addition, the gasification and methanation temperatures are relatively low with these methods and they thus can achieve higher efficiency. However, these two technologies are still at the research and development stage and have not been commercially demonstrated.

6.2.4.2 Conversion Efficiency of Coal-Based SNG

The Great Plains SNG plant adopts pure oxygen technology and achieves a load factor as high as 90–92 % and energy efficiency of 59–61 % (NETL 2007). However, a study by the University of Kentucky suggests that without carbon capture, the SNG conversion efficiency is only 60.1 % using Kentucky coal and that with carbon capture the efficiency of conversion is 58.9 % (Gray et al. 2007). SNG conversion efficiency may be higher with the other two technical routes, but this has not been validated by commercial demonstration. For example, the theoretical efficiency of hydrogasification is 79.6 % and that of steam gasification is 72.7 %.

6.2.4.3 Economics of Coal-Based SNG

In coal-based SNG projects that adopt pure oxygen technology, the gasifier, at 21 %, accounts for the largest share of total investment. This is followed by purifying equipment, at 15.8 %; the air separation and compression segment account for 11.3 %, and the methanation reactor accounts for 11 % (Gray et al. 2007).

Without CO_2 capture and sequestration (CCS), the production costs of SNG from Illinois coal and Rio Grande river coal are RMB 2.03/m^3 and RMB 2.3/m^3,[1] where coal prices are RMB 220/tons and RMB 156/tons, respectively, and 1 ton of coal can be converted into 254.9 m^3 natural gas. Increases in the coal price have minimal impact on SNG production costs: the SNG production cost increases by just 12.5–18.8 % when the coal price doubles (Chandel and Williams 2009). When integrated with polygeneration, the economics of SNG can be further improved.

6.2.4.4 Carbon Emissions of Coal-Based SNG

The hydrogen/carbon ratio required by methanization is around 3:1 or even higher; thus, a large amount of CO_2 will be emitted during production. This has long been a problem for the international development of SNG. In the case of pure oxygen technology, about two-thirds of the total carbon is converted into CO_2; the rest is converted into SNG. The CO_2 emissions of specific plants are determined by the type of coal and technical routes. According to a study by the US Department of Energy, when using Illinois and Rio Grande river coal, the CO_2 emission per ton of coal in SNG production is 1.9 and 2.3 tons, respectively. Since higher conversion efficiency reduces CO_2 emission, this emission may be dramatically reduced when using hydrogasification or steam gasification.

Moreover, owing to high concentrations of CO_2 emitted during SNG production, the capture cost of CO_2 is fairly low. Thus, with CCS, CO_2 emissions can be reduced

[1] In the original reference, the unit is dollars/ton (2007), which has been converted into renminbi (RMB) at the 2009 exchange rate.

Table 6.4 Prediction of annual natural gas production in China up to 2050 (unit, billion cubic meters)

Year	2020	2030	2050
Conventional natural gas	200	250	250
Coal bed methane	20	40	50
Shell gas and others	–	10	–
Total	220	300	Over 300

Source: CAE (2011)

Table 6.5 Prediction of annual natural gas import in China up to 2050 (unit, billion cubic meters)

Year	2020	2030	2040	2050
Russia–China pipeline	–	30	60	80
Central Asia–China Pipeline	30	30	30	30
Myanmar–China pipeline	–	10	10	10
Imported LNG	50	80	100	130
Total	80	150	200	250

Source: CAE (2011)

dramatically at a reasonable cost. The CO_2 sequestration technology of choice should be enhanced oil recovery, which is the most economical; this is followed by enhanced coal bed methane extraction and enhanced gas production. Another possibility is sequestration in saline aquifers, which offers the greatest sequestration potential but is also the most expensive method.

6.2.5 *Natural Gas Supply and Demand and Utilization by Vehicles*

6.2.5.1 Trends in Natural Gas Supply and Demand up to 2050

On the whole, the natural gas supply in China will maintain fast growth in the future. The reasons are that the reserves of conventional natural gas and their production are increasing and that unconventional natural gas (mainly coal bed methane and shell gas) also has great exploitation potential. With improvement in the global natural gas trade, imports of natural gas will also quickly increase. In addition, the supply could also be enhanced by developing coal-based synthesis gas on an appropriate scale.

For example, CAE (2011) determined that the annual natural gas production in China should amount to 300 billion m^3 by 2030 and continue to increase up to 2050 (Table 6.4). The annual natural gas import is expected to reach 150 billion m^3 in 2035 and 250 billion m^3 in 2050 (Table 6.5). In addition, the production of coal-based synthesis gas has been rapidly expanding recently, and it is expected to amount to tens of billion cubic meters per year in the future.

In China, the demand for natural gas is very uncertain, and the predictions offered by various institutes differ markedly. On the whole, however, the supply side is dominant in the supply-and-demand relationship for natural gas. The reason is that natural gas is a clean, convenient, low-carbon fuel, and so the demand for it is very large in many sectors—such as the chemical industry, industrial fuel, urban gas, and power generation—and the supply will be lower than the demand in the long term. From the perspective of the long-term national energy strategy, the end use of natural gas will be supported. Even if the annual natural gas supply were to reach 500 billion m^3 by 2030 and 550 billion m^3 by 2050, it would account only for 13–15 % of the primary energy in China (CAE 2011), which is much less than the global average of 24 %.

It is clear therefore that natural gas will be strongly supply oriented, that is, gas consumption will be restricted by the supply. Then, the important issue becomes how to distribute the limited natural gas resource or which sectors should be preferentially selected in terms of their demand. This issue is relevant to overall energy strategy and the market competitiveness of natural gas compared with other energy resources.

Using natural gas (rather than coal, coal gas, or LPG) for city gas improves the energy service quality for end users, and it also reduces air pollution; it should therefore be accorded prime importance. Moreover, both the existing city gas supply system and the economic bearing capacity in this area provide a good economic basis for natural gas application. Therefore, in the near or medium-term future, meeting the requirements for city gas should be the highest priority for natural gas utilization in China. This would be in agreement with the findings of various foreign studies. However, from a long-term view, the natural gas demand for city gas field will eventually face a saturated market, similar to the situation in developed countries (IEA 2008).

As an alternative to gasoline and diesel, natural gas offers benefits for vehicle fuel regarding energy security and reducing emissions, especially sulfur, and vehicle gas utilization should also be given preferential support. In the vicinity of gas fields, natural gas is more competitively priced than petroleum-derived fuels. Gas prices in different regions differ greatly, so this pricing advantage does not apply to all regions, and the oil price is also subject to fluctuation. With the example of CNG vehicles, their development is restricted by such infrastructural issues as the availability of gas filling stations. This problem becomes greater in big- or medium-sized cities with their shortage of land. However, with long-term effective policy support, particularly with regard to prices and infrastructure support, vehicle natural gas could be widely developed.

In the area of power generation, constructing natural gas power plants of appropriate scale would be of great benefit to the stability and flexibility of power grid operations, especially with large-scale renewable power generation. In cities, the development of natural gas-based projects for the coproduction of power and heat is also good for the environment. In the case of LNG, constructing natural gas power plants can guarantee its early place in the market. Currently, natural gas power generation is rather more expensive than coal power generation. However, this situation will change with improvements in the economy, stricter environmental standards, and lowered costs of combined cycle power plants. On the whole, therefore, the prospects are good for the rapid increase in natural gas power generation in China. Worldwide, power generation has been the largest sector of natural gas utilization, and it will be the main impetus for the long-term increase in natural gas demand (IEA 2008).

In the chemical industry, natural gas is a high-quality feedstock and important hydrogen resource. Using natural gas instead of oil is good for energy security, but it is disadvantaged by the lower costs of coal-based chemicals. With regard to industrial fuels, switching oil and coal usage to natural gas would be beneficial for energy security and environment protection, and it should therefore be supported. However, on the whole, in these two fields the gas price bearing power is weaker, and there is no great necessity to optimize the power source structure through natural gas power generation. Compared with urban gas and power generation gas, the consumption in these two fields has been increasing slowly; the rise in the use of industrial gas has been slower than that of industrial fuel gas. This trend may continue in the future.

CAE (2011) predicted that annual natural gas consumption in China in 2020, 2030, and 2050 would be 280, 450, and 500–550 billion m^3, respectively, corresponding to the gas supply. The forecast for China's natural gas consumption structure in 2030 shows that city gas (including vehicle gas), power generation, industrial fuel, and chemical use would account for 41, 22, 22, and 15 %, respectively, and that this structure would continue until 2050. However, no forecast was made for vehicle gas alone.

6.2.5.2 Potential for Natural Gas as Automotive Energy Source up to 2050

The future development of vehicle gas largely depends on energy prices and policies in addition to some uncertain factors. In this section, several possible scenarios are present with the purpose of obtaining an approximate estimate. The three scenarios are as follows:

1. Low scenario. According to gas consumption in the transport sector, the vehicle gas share of total gas consumption was 4 % in 2008. This scenario assumes that this share will remain at 4 % until 2050.
2. Intermediate scenario. In line with current policy, vehicle gas and city gas are supported as priority areas. This scenario assumes that the increase in vehicle gas consumption rises in step with that of city gas, that is, the scale of vehicle gas consumption can be calculated by predicting that for city gas.
3. High scenario. Currently, the vehicle gas proportion of city gas consumption is 15 %. Assuming that policy support for this is increased and developing conditions are favorable, this figure will amount to 20 % in 2020, 25 % in 2030, and continue at this level until 2050.

Table 6.6 Prediction of annual natural gas consumption and city gas proportion in China until 2050

Years	2005	2010	2020	2030	2040	2050
Annual natural gas consumption (billion m^3)	46.76	110[a]	280	450	487.5	525
City gas proportion	25.5 %	33 %[b]	38 %	40 %	40 %	40 %

[a]*Source*: National Energy Administration (2011)
[b]The number is estimated because of a lack of data

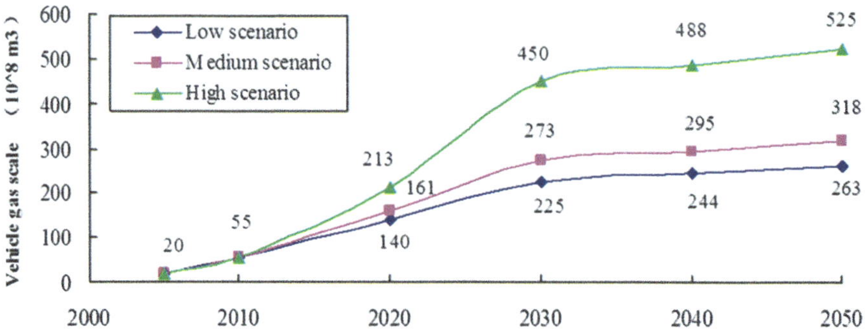

Fig. 6.5 Scenario analysis results for vehicle gas consumption in China

In this estimate, total gas consumption is based on CAE (2011), and the city gas (excluding vehicle gas) proportion is as listed in Table 6.6. The results with the three scenarios appear in Fig. 6.5. In these scenarios, the annual vehicle gas supply will be about 22.5–45 billion m^3 in 2030 and 26.3–52.5 billion m^3 in 2050.

6.3 Development of CNG Vehicles

6.3.1 Technical Background

Vehicle CNG is stored in high-pressure tanks, where the pressure is around 20 MPa. CNG vehicle engines have become refined after several generations of the following technological advances. The fuel supply system has developed from a mechanical mixer to an electronically controlled injection system. The electronic fuel injection system has changed from a single-point and open-loop control to a multipoint and closed-loop control system. Injection has developed from premixed injection outside the cylinder to combined direct injection within the cylinder. And the fuel has changed from a double to a single fuel (Li 2008). Vehicle CNG demands extra filling stations, and it is possible to build integrated oil-gas stations. CNG filling stations require purification devices, gas-storage devices, air compressors, and gas feeders (Zhou and Li 2009).

6.3.2 *Development in China and Overseas (Sidebar 6.7)*

Sidebar 6.7: Development of CNG vehicles in the United States and Italy

Currently, 40 American states have issued supporting policies for clean energy cars and filling stations. In total, there are 1,600 filling stations (mostly on the West Coast) and 147,000 natural gas vehicles in the United States. They are primarily used as large public transport vehicles, such as urban buses, airport commuter buses, and tour buses as well as heavy trucks.

By the end of 2004, Italy had 400,000 natural gas vehicles, which accounted for 1 % of all domestic cars; in some regions, the share was as high as 10 %. Annual sales of vehicle CNG are about four billion m^3. To promote the use of natural gas, the Italian government does not impose tariffs on imported gas, and value-added tax is also less than 4 %. Thus, the market price of CNG is only about 25 % that of gasoline and 50 % that of diesel. In addition, the government provides 25 % subsidies for investments in CNG filling station infrastructure. These moves have been very conducive to the promotion of CNG vehicles.

In the early twentieth century, Italy pioneered the development of CNG vehicles, and CNG vehicles gradually become the mainstream natural gas vehicles. The oil crises of the 1980s greatly boosted the development of natural gas vehicles around the world. The literature indicates that the total number of natural gas vehicles around the world has surpassed ten million. In 2009, the top six countries in terms of number of CNG vehicles were, in order, Pakistan, Argentina, Brazil, Iran, India, and Italy; China ranked seventh.[2] In recent years, the Asia-Pacific region has also experienced very rapid growth in the number of CNG vehicles, having achieved an average annual growth of about 50 % from 2000 to 2007 (Zhuang 2009; Li 2008).

As early as 1999, China initiated the Cleaner Vehicles for Cleaner Air Program, and to promote the use of natural gas vehicles, demonstrations with such vehicles were held in Beijing, Shanghai, Chongqing, Sichuan, and other 12 cities and regions. In 2005, the number of such demonstration cities and regions was expanded to 19, and the main target vehicles were buses and taxis. Through this program, the production of CNG vehicles has begun, filling stations have been built, and the technology is mature. In terms of emissions, current CNG vehicles meet Euro III standards, and China is now developing European IV standard CNG vehicles. At present, the major constraints are fuel supply, high prices, and lack of filling stations.

By the end of 2007, over 80 cities in China were promoting natural gas vehicles, including 16 key cities and regions, such as Sichuan, Chongqing, Urumqi, Xi'an,

---

[2]http://www.iangv.org/tools-resources/statistics.html

and Lanzhou. By the end of 2007, the CNG vehicle population had reached 265,000—an increase of 80,000 over 2006; 555 filling stations had been built—an increase of 75 over 2006. Proximity to natural gas sources and low prices (only 40–50 % of the oil price) are the main propelling forces for the development of CNG vehicles in those cities. According to statistics from the International Association for Natural Gas Vehicles, the CNG vehicle population in China was 4.5 million in 2009 and the number of filling stations amounted to 1,350, with annual increase rates of 12.5 and 35 %, respectively (Sidebar 6.8).

Sidebar 6.8: Development of CNG Vehicles in Shanghai and Chongqing

CNG vehicles in Shanghai are still at a stage of limited application; they are mainly used in urban public transport, but the number is small. The population of natural gas buses in Shanghai is about 281, and they account for only 1.6 % of the total number of buses; their annual consumption of natural gas is ten million m^3, which accounts for 1.5 % of the total energy consumption of buses in that city. The city has only four filling stations, so the current chances for sudden expansion are small.

The main development constraints for CNG vehicles in Shanghai are as follows. (1) CNG vehicles in Shanghai were purchased too early: all parts for these CNG vehicles were imported, and the total cost is greater than that of comparable diesel vehicles by about RMB 14 million. In addition, maintenance costs are higher by about 26 %, and it is difficult to carry out maintenance on these vehicles. (2) There is a shortage of natural gas owing to the priority given to residential, commercial, and industrial users. (3) After 2007, the price of natural gas rose from RMB 2.15 m^3 to RMB 3.58/m^3, and CNG vehicle fuel costs increased significantly. These increases amounted to almost RMB 2,000 a month on average more than diesel vehicles; thus, the price advantage compared with diesel fuel no longer exists. (4) The city's four filling stations are inconveniently located far from downtown areas, and the small business volume of these filling stations leads to long-term losses. Thus, CNG vehicles and filling stations are trapped in a vicious cycle. (5) There is also inadequate policy support: there is only a one-time RMB eight million subsidy per CNG vehicle given to bus companies on purchase (Zhuang 2009).

CNG vehicles have shown remarkable development in Chongqing for almost 12 years. Currently over 90 % of the city's taxis and buses run on natural gas energy in addition to a large number of private cars and motorcycles. The natural gas vehicle population in Chongqing totaled about 40,000; among those vehicles, the CNG consumption by buses and taxis accounted for about 70 million m^3. Annual vehicle consumption of CNG in Chongqing is 340 million m^3.

(continued)

(continued)

The main reasons for the rapid development of CNG vehicles in Chongqing are as follows. (1) There exists localized technology and strong support by local technical services. Chongqing has established Asia's largest natural gas vehicle industrial base and the domestic gas vehicle industry's only national engineering research center (National Engineering Research Center of Gas Vehicle). Chongqing has also set up the most complete supply chain in the natural gas vehicle industry. The local gas market share of cylinders, compressors, CNG buses, gas devices, and automotive products is over 30 %. (2) Chongqing's province of Sichuan is one of China's major natural gas-producing regions. The plentiful, cheap natural gas supply greatly reduces the operating costs of buses and taxis. (3) With 61 filling stations in operation in Chongqing, gas filling is very convenient. (4) There is strong policy support. The Natural Gas Cars Group Office, which was established by the Chongqing Municipal Science and Technology Commission and is under the supervision of 16 municipal departments, developed a series of industrial standards, preferential policies, and research and development policies to support the development of CNG vehicles.

Source: China Science and Technology Information, 2008, 1

6.3.3 Technological Performance

Compared with gasoline and diesel from oil refining, the main advantages of CNG are that it has low pollutant emissions and lower carbon emissions and is more economical. From rough estimates, CNG vehicles can theoretically reduce carbon monoxide emissions by 90 %, hydrocarbons emissions by 72 %, and nitrogen oxide emissions by 39 % compared with gasoline vehicles. At present, China's domestically produced CNG vehicles already meet Euro III emission standards; Euro IV standard cars are in research and development and will be launched on the market in the near future.

The main drawback with CNG is the low storage rate owing to high-pressure storage and transportation methods: the weight of the compressed natural gas is only one-tenth to one-fifth that of the storage tank itself. This leads to a short battery life and also produces filling and traffic-safety hazards. The energy density per unit volume of CNG is only 23–28 % that of gasoline, which limits the driving range to just about 60 % that of gasoline cars (Chen et al. 2009). Therefore, the application of these vehicles faces great limitations, and they are currently only used for short-distance taxis and buses.

6.3.4 Support

Years of research and development have provided good technical support conditions for the development of CNG vehicles. The equipment in CNG filling stations is basically produced domestically, and most stations feature such equipment. Dual-fuel gasoline–CNG vehicles and diesel–CNG vehicles have been produced in addition to CNG single-fuel vehicles in China. In 2007, China produced 57,000 natural gas vehicles (including chassis), with 58 domestic makers of natural gas vehicles and 18 makers of natural gas engines (Li 2008).

The Chinese government has provided strong policy support for the development of natural gas vehicles. In August 2007, the NDRC promulgated the Natural Gas Utilization Policy, which designated natural gas vehicles as a priority area in natural gas utilization. In addition, the National Clean Vehicle Action lists CNG as the preferred alternative fuel for vehicles.

The main problem for CNG vehicles at present is industrial policy: there are too many investors and operators in the CNG market, but there is lack of coordination with downstream consumption and energy supply. This has resulted in increasingly fierce competition in the downstream market. With the gradual lifting of the preferential policy for CNG filling stations in some cities, these stations are facing growing difficulties in land acquisition and construction (Zhou and Li 2009). In addition, the price volatility of natural gas is an important concern.

The availability of natural gas resources is also a major constraint in the development of CNG vehicles, but this problem will ease to a certain extent with the rapid development of China's natural gas supply. However, the gradually rising natural gas prices pose a great challenge to this fuel's current economic advantage. The large-scale development of CNG vehicles requires the establishment of a mother filling station for CNG supply in each city. At present, in the development of the gas-storage and transportation system, there is a lack of overall coordination among the CNG mother filling stations, distribution, and CNG filling stations.

To sum up, market conditions, centralized CNG supply and pricing mechanisms, and filling station availability are the major policy factors that will affect the future development of CNG vehicles (Sidebar 6.9).

Sidebar 6.9: CNG Vehicles in a Small City in Sichuan

It was known that there were two CNG filling stations in this city. One investigated station was built in 2000 with a service life of 20–30 years; the equipment (gas-storage capacity, 2,000 m^3) was produced by local enterprises, and the initial investment was about RMB 12 million. Business was poor in the early years because drivers were reluctant to spend about RMB 3,000 to retrofit their cars. Later, the gas filling station decided to pay

(continued)

(continued)
the retrofitting costs for taxi drivers and then to recoup part of those costs each time they refueled their vehicles. Thereafter, business began to improve.

In May 2010, the wholesale and retail prices of natural gas at this station were RMB 1.8/m^3 (RMB 1.3/m^3 for city gas) and RMB 2.7/m^3 (20 MPa), respectively, and about 13,000 m^3 was sold each day. Recently, most taxis and buses and some light trucks in this city have been retrofitted to use CNG, but among the total vehicle population, this is still very small. One taxi driver said that the fuel cost of a CNG vehicle was about RMB 20–30/day, which was much lower than that of gasoline car (RMB 50–60/day), and that the saving in fuel expenses was very effective. However, retrofitting the car engine influences the dynamic performance, and placing the large-volume gas tank in the rear affects the drivability. In addition, according to staff at the filling station, the CNG filling industry is a high-risk business and enterprises have to pay greater attention to security issues.

For this particular enterprise, the original aim in constructing the CNG filling station was to be in keeping with national policies, receive local government support, and participate in what it saw as a potentially big market. Although the local government helped it publicize the venture, the promised preferential CNG vehicle-maintenance cost policy had not been implemented at the time of investigation.

Local natural gas extraction plays an important role in developing CNG vehicles in this city. A Sinopec-owned gas field is located about 100 km from the city: it generates 260,000 m^3 a day, of which 240,000 m^3 is supplied to that city. The selling price of gas at the well is just about RMB 1/m^3, and the gas pipeline is owned by CNPC. The city has never been affected by a shortage in CNG supply.

Almost all the counties in Sichuan Province have CNG filling stations; almost all taxis in the province have been retrofitted to burn CNG, as have many buses and a few private vehicles. It is basically possible to drive a CNG vehicle anywhere in Sichuan. Generally speaking, the development of CNG vehicles in Sichuan has depended on spontaneous market behavior, although national policies have also played a part.

Source: The source is the interviews with the enterprise operating one CNG station in this town in May 2010.

6.3.5 Policy Suggestions

The advantages of CNG vehicles are mature technology and strong policy support. They should therefore be the main element in the future development of natural gas vehicles. However, owing to range and safety limits, the only likely application of CNG vehicles is in urban public transport. With the rapid development of city gas, the planning and construction of CNG mother stations and CNG filling stations should be coordinated. Further research and development are needed to improve safety, efficiency, and environmental performance.

6.4 Development of LNG Vehicles

6.4.1 Technical Background

LNG vehicles were initiated along with the development of the global trade in LNG. Unlike CNG, which needs to be stored at high pressure, LNG can be stored and transported at atmospheric pressure at −162 °C. LNG vaporizes with the higher temperature inside a vehicle. This allowed specialized LNG vehicles to be developed.

The volume of liquid LNG is only one-third that of CNG, which greatly increases the energy density of this form of natural gas. LNG density is about 60 % that of gasoline (Chen et al. 2009). Therefore, compared with CNG vehicles, LNG vehicles can significantly reduce the size and weight of the vehicle fuel system and the number of fillings and improve the range of the vehicle (Li 2008). LNG vehicles are thus especially suitable for fixed-line, long-distance, heavy vehicles.

The development of LNG vehicles also demands the installation of small natural gas liquefaction equipment at LNG filling stations and CNG filling stations. A specialized LNG tanker fleet is required for LNG storage and transportation.

6.4.2 Development in China and Overseas

With the growth in the global LNG trade, LNG vehicles have been developed in recent years. For example, in Canada and the United States, LNG vehicles using high-pressure direct injection technology have been put into commercial operation; they are equipped with two 680-L LNG storage tanks, the total load of the vehicle is about 45 tons, and continuous traveling distance is 800 km (Li 2008).

LNG cars have been introduced in China, and LNG demonstration and application programs for city bus systems have been conducted in Beijing, Urumqi, Guiyang, and Changsha. Since 2003, China has also embarked on research and development of LNG vehicles and accessories. With the construction of coastal LNG receiving terminals, the rapid development of LNG cars may be anticipated (Li 2008; Zhou and Li 2009).

6.4.3 Technological Performance

LNG vehicles have a greater range and lower transport costs than CNG vehicles. However, there are high costs involved with the purchase of LNG vehicles as well as the maintenance and renovation of the storage and transportation system. In addition, LNG vehicles may cause pollution, and there are safety issues if LNG is leaked into the open air. However, LNG vehicles have an unparalleled advantage in cold-energy recycling (such as refrigerator trucks, air-conditioned cars). LNG vehicles are still at an early stage of development, and their overall performance remains to be verified.

6.4.4 Support

The technological difficulties for LNG vehicles are mainly natural gas liquefaction, gasification, storage, and distribution. The United States and Canada are leaders in LNG liquefaction equipment, vehicle fuel tanks, and engine development and production. China's domestic technology in LNG cars and filling stations cars is still at an early stage. However, China has initiated the development of LNG trucks, and some cities have also launched demonstration programs for LNG buses. In addition, China's LNG storage tank technology is becoming increasingly mature.

As natural gas vehicles, LNG vehicles enjoy the same policy support as CNG vehicles, but LNG vehicles do not as yet constitute a complete industry. Market and technological development are the main factors that will affect the growth of LNG vehicles. With rising LNG imports in coastal regions and increased LNG transportation in inland regions, the development of LNG vehicles in these regions will be promoted by the available fuel supply.

6.4.5 Policy Suggestions

LNG vehicles have greater range than CNG vehicles, but the technical and policy conditions for LNG vehicles are not yet mature. It is to be expected that LNG vehicles will be promoted for fixed-line heavy vehicles (e.g., trucks, intercity buses) in coastal regions and inland regions close to natural gas fields. Technological development and policy research for LNG vehicles, filling facilities, and safety should be enhanced. New technology and policies need to be implemented in demonstrating and promoting these vehicles in appropriate regions.

6.5 Development of GTL

6.5.1 Technical Background

Natural gas GTL can be produced in two ways—direct and indirect synthesis. The latter process is more technically mature, and it consists of three parts—gas production, Fischer–Tropsch synthesis, and product refinement. GTL fuel (with a composition close to diesel) is very high in quality and very low in sulfur content. In addition, GTL fuel does not entail any changes to conventional vehicles.

6.5.2 Development in China and Abroad

Since oil supply security issues have become increasingly prominent, GTL has attracted the attention of natural gas producer countries and countries that are low in oil resources. To date, over 10 GTL facilities are in existence around the world, three of which have been put into operation (in Qatar, Malaysia, and South Africa). The largest facility is in Mossel Bay, South Africa; it was built by Mossgas in 1999 using technology supplied by Sasol Company. This facility uses natural gas from shallow gas production platforms as its raw material, and it has a capacity of 130 million tons of refined products per year (Qian 2002; Lin et al. 2008). According to IEA projections, the natural gas consumption from GTL will rise from five billion m^3 in 2006 to 50 billion m^3 by 2030, though this will be mainly accounted for by natural gas-exporting countries (IEA 2008).

Owing to the small natural gas production volume, the imbalance in supply and demand, and lack of technical breakthroughs, the development of GTL is not yet at a mature stage in China. However, indirect coal liquefaction has undergone substantial development, and its technical fundamentals are similar to those of GTL, as noted in the chapters on coal-based liquid fuels.

6.5.3 Technological Performance

Investments in GTL projects are great, but increases in the price of oil have increasingly underlined the economic advantages of this technology. In terms of vehicle usage, GTL can help improve fuel economy and reduce emissions. However, owing to energy consumption and emission in the conversion process, the full life cycle of energy consumption and emission is considerably higher than with oil from petroleum refining (Zhang et al. 2009). Therefore, GTL's sole advantage lies in terms of oil security. However, GTL itself faces supply security issues over the long term. However, through polygeneration with other refined products and chemical products, the economic merits of GTL can be greatly enhanced.

6.5.4 Support

The production of synthetic gas and refining of products are technologically mature. The problem lies in the Fischer–Tropsch synthesis. Although GTL has not been demonstrated or commercialized in China, small-scale commercial demonstration projects using indirect coal liquefaction have been set up, and this process also uses the Fischer–Tropsch synthesis. Indirect coal liquefaction of natural gas development has provided strong support for the development of GTL technology.

With China's current policy on utilizing natural gas, GTL projects are limited. However, with the increasing prominence of natural gas and oil security issues, the policy situation is expected to improve.

6.5.5 Policy Suggestions

The advantage with GTL is that it does not require any changes to vehicles or the infrastructure; its products are also of high quality. However, GTL consumes more energy and emits more CO_2 on a full life cycle basis. Hence, GTL can serve only as an alternative technical route in the future development of vehicle natural gas. With the maturation of indirect coal liquefaction, GTL should be commercially demonstrated so as to advance the technology. Its commercialization may appear later than 2030, by which time the supply constraints of coal may have accelerated the development of GTL.

6.6 Conclusions and Suggestions

6.6.1 Natural Gas Supply Will Experience Rapid, Sustained Growth

In China, the reserves and production of natural gas have grown swiftly. By 2009, the annual gas supply had already reached 90 billion m^3, almost four times the 2000 level (24 billion m^3). Considering production increases in conventional gas, the exploitation of unconventional gas reserves, the import of natural and synthetic natural gas, the annual supply capacity is expected to amount to 220, 450, and 525 billion m^3 in 2020, 2030, and 2050, respectively. Therefore, the resource constraint issue for CNG development will be reversed.

6.6.2 CNG Vehicles Will Generally Undergo Rapid Development

With the ongoing development of natural gas in China, the gas supply and pipeline coverage rate will increase rapidly. This will bring new opportunities for CNG/LNG vehicles, reduce pollution emissions, and improve fuel economy to a certain extent. The overall gas supply will continue to fall short of demand with the fast growth in gas demand in other areas; thus, the development of gas-fueled vehicles will be influenced by fuel prices and relevant policies. Depending on different policy and market conditions, the total amount of CNG consumed by vehicle is expected to amount to about 20–50 billion m^3 between 2030 and 2050, thereby making a greater contribution as an alternative to oil.

6.6.3 Pricing Mechanisms and Infrastructure for Vehicle Natural Gas

Practice has shown that resource availability, appropriate pricing of vehicle CNG, and the construction of gas filling stations are the key factors that will influence the future development of CNG vehicles. With great improvements in the resource supply, the latter two factors have become the major block to the development of CNG vehicles. Rationally determining the price ratio between vehicle CNG and vehicle oil and advancing the planning and construction of CNG mother stations and CNG filling stations in appropriate regions will guarantee the sound development of CNG vehicles. The development of LNG vehicles faces much the same problems. In addition, the localization of LNG filling stations, technological development, and policy conditions are areas that need to be addressed for the development of these vehicles.

6.6.4 CNG Public Transport and Demonstrating LNG Heavy Vehicles

Considering the restrictions on traveling range and security, the emphasis with CNG vehicle development should be placed on public transport systems, although the conditions for development in various other areas are fairly mature. The range of LNG vehicles is greater, but the technology is not yet fully mature. It follows that LNG vehicle technology should be demonstrated with heavy vehicles (such as heavy trucks and intercity buses) in appropriate regions before being promoted in other areas at a later time. Since GTL is a strategic reserve technology for oil substitution, research should be conducted and small-scaled demonstrations initiated in the near future.

Glossary

CCS CO_2 capture and sequestration
CNG Compressed natural gas
GTL Gas to liquids
LNG Liquefied natural gas
LPG Liquefied petroleum gas
SNG Synthetic natural gas

References

BP (2010) Statistical review of world energy. BP, London

CAE-Chinese Academy of Engineering (2011) Research on the mid- and long-term (2030/2050) energy development strategy of China: volume of electricity, oil and gas, nuclear energy, and environment. Science Press, Beijing (in Chinese)

Chandel M, Williams E (2009) Synthetic natural gas (SNG): technology, environmental implications and economics. Climate Change Policy Partnership, Duke University, Durham

Chen J, Li C, Chen X (2009) Review on oil substitution. China Petrochemical Press, Beijing

China Science and Technology Information (2008). Technology News: Chongqing's Natural Gas Vehicles Achieved Remarkable Results

CNPC News Center (2011). Central Asia-China Gas Pipeline acts as an important strategic channel (http://news.cnpc.com.cn/system/2011/04/27/001331940.shtml)

Doug Huxley (2006) Dakota Gasification Company CO2 sequestration verification project—a case study of greenhouse gas reduction verification and marketing. Fuel Process Technol 87: 179–183

Gray D, Challman D, Geertsema A, Drake D, Andrews R (2007) Technologies for producing transportation fuels, chemicals, synthetic natural gas and electricity from the gasification of Kentucky coal. University of Kentucky, Kentucky

Hua B (2007) Collected papers on LNG utilization. Petroleum Industry Press, Beijing

IEA (2008) World energy outlook 2008. IEA, Paris

IEA (2010) World energy outlook 2010. IEA, Paris

Li G (2008) Natural-gas vehicles in China: developments and trends. Int Petroleum Econ 7:69–74

Lin H, Han S, Wu X, Gao F (2008) Developmental prospect of GTL in China. Shanghai Chem 33(7):16–19

Lu J (2009) Current situation and proposals for the development of natural gas industry in China. Nat Gas Ind 29(1):8–12

National Energy Administration (2011) Energy and economy outlook 2010 and prospect of 2011. http://www.sdpc.gov.cn/jjxsfx/t20110128_393341.htm. Accessed 22 Jan 2013

NBS – National Bureau of Statistics of China, and NEA – National Energy Administration (2011). Chinese Energy Statistical Yearbook 2010. China Statistical Press, Beijing

NDRC (2007) Notice of the publication of the natural gas utilization policy by NDRC. http://www.ndrc.gov.cn/zcfb/zcfbtz/2007tongzhi/t20070904_157244.htm. Accessed 22 Jan 2013

NETL (2007) Industrial size gasification for syngas, substitute natural gas and power production. Report No. DOE/NETL-401/040607

Qian B (2002) Gas-to-liquid (GTL) technology: one of hot points of 21st century in the oil and natural gas industry. Natural Gas and Oil 20(1):16–19

Wang Q (2010) Reference of research on financial and economic policy of energy sustainable development: 2010 energy data. The Energy Foundation, Beijing

Xinhua Net (2009a). China Finds Onshore Natural Gas Hydrate with Prospective Reserves of over 35 Billion Tons Oil Equivalent (http://news.xinhuanet.com/newscenter/2009-09/25/content_12111977.htm)
Xinhua Net (2009b). China Officially Initiates Exploitation of Shale Gas as a New Form of Energy (http://news.xinhuanet.com/politics/2009-10/05/content_12183830.htm)
Yin J, Huang H (2009) Discussion on natural gas safety in China from natural gas dispute between Russia and Ukraine. Natural Resource Economics of China 5:33–35
Zhang K, Wang H, Li X et al (2009) Life cycle analysis of GTL fuel for city bus. Automot Eng 31(1):69–73
Zhou S, Li G (2009) Reflections on PetroChina CNG sector. Petroleum Planning & Engineering 1:20–22
Zhuang S (2009) Developing CNG in Shanghai. Energy Technol 30(1):27–29

Chapter 7
Coal-Derived Liquid Fuels

Liu Pei, Ma Linwei, Liu Guangjian, Pan Lingyin, and Li Zheng

Abstract In this chapter, after a brief introduction, we first suggest the potential of resource supplies for coal-derived fuels based on a review of coal production, utilization, and supply constraints in China. The technological development of coal-derived fuels is the main focus of this chapter. We first present the current status of the development of four technology pathways—direct and indirect coal liquefaction, coal-derived methanol, and coal-derived dimethyl ether—and compare the technical performance of their conversion processes based on published data. Then, we conduct a well-to-tank analysis of their technical performance considering three routes in the supply chain design, and we discuss the influence of carbon tax. A six-dimension method is also applied to compare the advantages and disadvantages of the four technology pathways. Finally, by identifying the barriers and future potential of these technology pathways, we propose policy suggestions for their future development.

Keywords Coal • Coal-derived liquid fuel • Coal liquefaction • Methanol • Dimethyl ether

7.1 Introduction

China's energy supply has been largely characterized by the dominance of coal based on domestic production, and this also represented the country's main energy-development strategy. In recent years, however, China's dependency on oil imports has rapidly grown, and the international price of oil has remained high. Thus, it has become increasingly necessary for China to develop its coal industry to produce

Liu Pei • Ma Linwei (✉) • Liu Guangjian • Pan Lingyin • Li Zheng
State Key Laboratory of Power Systems, Department of Thermal Engineering, Tsinghua-BP Clean Energy Center, Tsinghua University, Beijing 100084, China
e-mail: malinwei@tsinghua.edu.cn

CAERC, Tsinghua University, *Sustainable Automotive Energy System in China*,
DOI 10.1007/978-3-642-36847-9_7, © Springer-Verlag Berlin Heidelberg 2013

alternatives to oil-based gasoline and diesel, using instead coal-derived liquid fuels, for which there are relatively rich and inexpensive resources. In addition, coal-derived liquid fuels have gained attention as an important technology toward helping to assure national energy security. At the same time, though, the development of coal-derived liquid fuels is highly controversial because of the low energy efficiency, high CO_2 emissions, and high water consumption; this is therefore a problematic area for the national energy strategy. Before coal-derived liquid fuels are fully developed, the whole of society needs to arrive at a coherent understanding of their future and the various issues involved.

This chapter assesses the technological development of coal-derived liquid fuels from a variety of aspects. Through data collection and analysis, we attempt to offer a comprehensive assessment and formulate preliminary policy suggestions.

Regarding chapter content, Sect. 7.2 summarizes the present situation of coal use and development in China; it considers the availability of coal resources and analyzes potential issues regarding the resource supply of coal-derived liquid fuels. Section 7.3 is a comprehensive evaluation of various coal-derived liquid fuels. The current state of coal-derived liquid fuel technologies is presented, including those of direct coal-to-liquid, indirect coal-to-liquid, coal-derived methanol, and coal-derived dimethyl ether (DME). The technical, economic, and environmental performances of these processes are compared. In addition, a well-to-tank (WTT) analysis and comparison of energy consumption and emissions is made; the effect of a carbon tax on coal-derived liquid fuels is examined. Further, a preliminary analysis is made on the major advantages and disadvantages of several coal-derived liquid fuels by means of a six-dimension analysis method. Section 7.4 presents a development policy analysis of coal-derived liquid fuels in China. The current situation, present obstacles, and future potential of coal-derived liquid fuels are evaluated, and development policy suggestions are provided. Section 7.5 offers a preliminary conclusion and recommendations on research analysis and policy implications.

7.2 Potential of Resource Supply for Coal-Derived Liquid Fuels

Despite its rich coal resources, the sustainable coal supply in China is severely constrained by mining security, water resources, ecosystem preservation, and transportation capacity. Under these restrictions, the coal for liquid fuel production is affected by the demand for coal by other sectors and industries. This chapter therefore describes the present situation regarding coal resources, production, transportation, importation, and use to summarize coal development in China. Subsequently, a preliminary analysis on potential coal resources for liquid fuel production will be made.

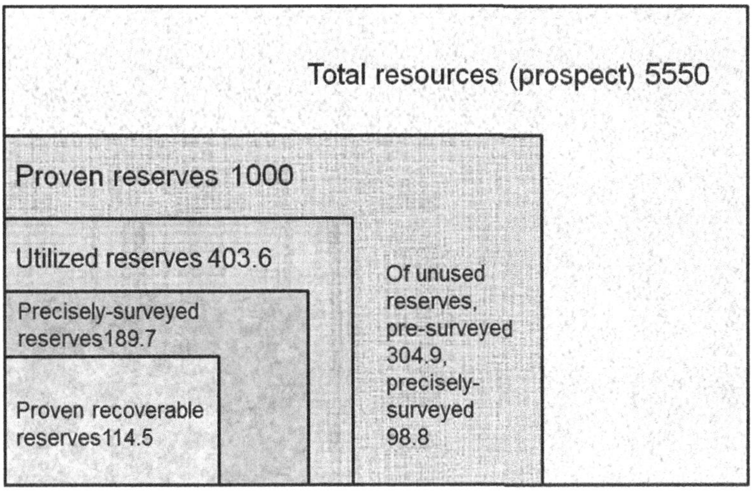

Fig. 7.1 Coal resources and coal reserves in China (2010)

7.2.1 Coal Production and Use

7.2.1.1 Coal Resources

Compared with the country's oil and natural gas resources, China's coal resources are relatively rich. According to data from BP (2010), China's proven recoverable coal resources accounted for 13.9 % of the world total in 2009, whereas its oil and natural gas resources accounted for only 1.3 and 1.1 %, respectively.

In general terms, China has great potential for increasing its coal resources (Fig. 7.1). According to the coal resource assessment in 1999 (Mao 1999), China's total coal resources (prospect resources) amount to 5,550 billion tons, and the country ranks first in the world for this commodity. The total proven reserves are about 1,000 billion tons (Pu 2010), of which the utilized reserves are 403.6 billion tons, precisely surveyed utilized reserves are 189.7 billion tons, and proven recoverable reserves are 114.5 billion tons. Of the unused reserves, the estimated reserves are 304.9 billion tons and precisely surveyed reserves are 98.8 billion tons (The Chinese Academy of Engineering 2011a, b; BP 2010). With improvements in technology and further development of coal exploration and mining, China's proven recoverable coal reserves are certain to continue increasing.

China is rich in coal resources, but they are unevenly distributed, being found more in the north and west and less in the south and east (Fig. 7.2):

1. In terms of the north–south distribution, coal resources are mainly located to the north of the Kunlun Mountains–Qinling–Dabie Mountains area: the total amount of coal resources in the provinces north of this line accounts for 93.1 % of the national coal resources, and the proven reserves account for over 90 % of the national reserves.

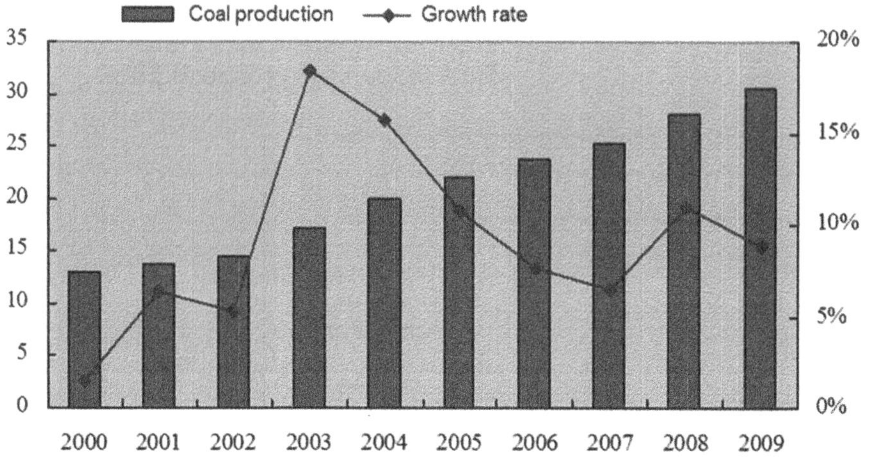

Fig. 7.2 Coal production in China (2000–2009, unit: 10^4 tons) (*Statistics source*: BP 2010)

2. In terms of the east–west distribution, coal resources are mainly located to the west of Daxinganling–Taihang Mountains–Xuefeng Mountains: in the 11 provinces west of this line, including Inner Mongolia Autonomous Region, Shanxi, Sichuan, and Guizhou provinces, the total amount of the coal resources accounts for 91.8 % of the national coal resources, and the proven reserves account for 89 % of the national reserves (Sidebar 7.1).

Sidebar 7.1: Definitions of Coal Resources and Coal Reserves

1. Prospective reserves: according to geological forecasts, coal resources above a vertical depth of 2,000 m are defined as the total coal resources.
2. Recoverable reserves: accumulated proven coal resources, including both utilized and unused reserves. Unproven resources are defined as forecast resources.
3. Utilized reserves: utilized resources for production mines and those under construction can be divided into completely surveyed and incompletely surveyed resources. The completely surveyed reserves can form the basis for mine design and construction.
4. Unused resources: among recoverable reserves, these are reserves that have yet to be utilized. They can be divided as follows: completely surveyed, incompletely surveyed, standard-level surveyed, and unsurveyed reserves.
5. Proven recoverable reserves: among utilized completely surveyed reserves, after deducting the recovered resources, the remaining recoverable resources are defined as proven recoverable reserves.

In terms of coal quality, China possesses most varieties of coal, ranging from lignite to anthracite of different coalification levels, but their amounts and distribution are uneven. Because of the high ash and sulfur content, China has limited coking coal resources, and the supply of high-quality coking coal is even more restricted. The main disadvantage with highly metamorphosed coal (lean coal, anthracite) is the high percentage of sulfur. The average calorific value of Chinese coal is below that of internationally traded coal, but it is equal to that of the coal in many countries. The average calorific value of China's coal is 5,400 kcal/kg, while that of coal from the United States is 5,600 kcal/kg (IEA 2007; Mao 1999).

7.2.1.2 Coal Production

The amount of coal production in China is huge, and the country ranks first in the world, accounting for 45.6 % of global output in 2009. In the last 30 years, coal's share among China's energy production has remained around 70 %. With the rapidly increased energy demand in recent years, there has been a swift increase in China's annual coal production—from 1.3 billion tons in 2000 to 3.05 billion tons in 2009; the average annual growth rate was 9.9 % (Fig. 7.2).

Coal production in China is mainly concentrated in Shanxi, Inner Mongolia, Shaanxi, Henan, Shandong, Guizhou, and Heilongjiang provinces. Newly increased production in recent years has mainly occurred in Shanxi, Shaanxi, Inner Mongolia, and Ningxia. According to the regional raw coal production statistics presented in the China Energy Statistics Yearbook 2010 (NSB and NEB 2011), the total annual coal production in those four provinces accounted for 34.6 % of national production in 2000; it increased to 52.4 % in 2009 and contributed to 60 % of the coal production increase during that period.

Although China's coal output has increased rapidly, it is now being produced beyond sustainable capability, and this has given rise to a number of problems (CAE 2011a; Pu 2010):

1. Low resource recovery rate: only 49 % of coal production (from state-owned key coal mines) can achieve recovery rates of over 50 %; 16 % of coal production (from state-owned local coal mines) achieves recovery rates of 30 %; the remaining 35 % of coal production (privately owned local coal mines) attains recovery rates of only 10 %.
2. Mining security problems: though in recent years, the death rate for every million tons of coal mined in China has been decreasing, the annual number of deaths in this industry is still the highest in the world. In 2008, the death rate for every million tons of coal mined was 1.184, which is 30 times higher than the number in the United States. There were 3,215 deaths in coal mining accidents in China in 2008.

3. Land subsidence problems: 95 % of coal production in China takes the form of underground mining, and mining subsidence is the main cause of land destruction by the coal industry. According to statistics, in 2005, land subsidence caused by coal mining affected more than 700,000 hm^2, and the land collapse was over 800,000 hm^2, resulting in economic losses of over RMB 50 billion. Currently, land subsidence caused by coal mining is increasing at a rate of 30,000–50,000 hm^2 per year. The problem is more severe in eastern areas of China, where land resources are relatively scarce.
4. Water resources and ecological environment problems: coal mining causes severe water pollution and destruction of water resources. According to estimates, 2.2 billion m^3 of groundwater resources are destroyed each year by coal mining in China. Coal resources and water resources are inversely distributed in China, and coal mining is constrained by scarce water resources. For example, in Shanxi, Shaanxi, Inner Mongolia, and Ningxia provinces, proven coal resources account for 64.4 % of the national resources, whereas water resources amount to only 2.6 %. At 13 large-scale mining sites in these and nearby areas, the daily water demand is about 900,000 m^3, but the local water supply can meet only half of that. At the same time, the areas of China in the north of Qinling–Dabie Mountains suffer from droughts and lack of rainfall; vegetation cover rate is low, and the ecology is fragile. Almost 90 % of China's coal resources are located in these areas, and the local ecology significantly restricts the development of coal.

7.2.1.3 Coal Transport

As noted above, China's coal resources are mainly distributed in western and northern areas of the country; however, coal demand is mostly in the economically developed coastal areas in the east and south. Long-distance, large-scale transport of coal is thus inevitable, and the transport pattern is one of western coal being transported east, and northern coal being transported south.

In 2010, the total amount of transported coal was two billion tons (NEA 2011). China's coal transport mainly relies on railway with assistance from road transport, sometimes combined with shipping. For example, coal is initially transported by rail to coastal or riverside ports and then shipped to markets in southern areas. Rail transport handles almost half of the total coal production (Yue 2008). Further, coal is the chief commodity transported by rail, accounting for over 40 % of the rail transport load. In recent years, transport capacity has become one of the major constraints in China's coal supply. Long-distance, large-scale transport and transit inevitably result in high coal prices in southeastern markets.

7.2.1.4 Coal Imports

In China's southeastern coastal areas, coal imports have been increasing in recent years as a result of scarce coal resources and limits in the coal transport capacity.

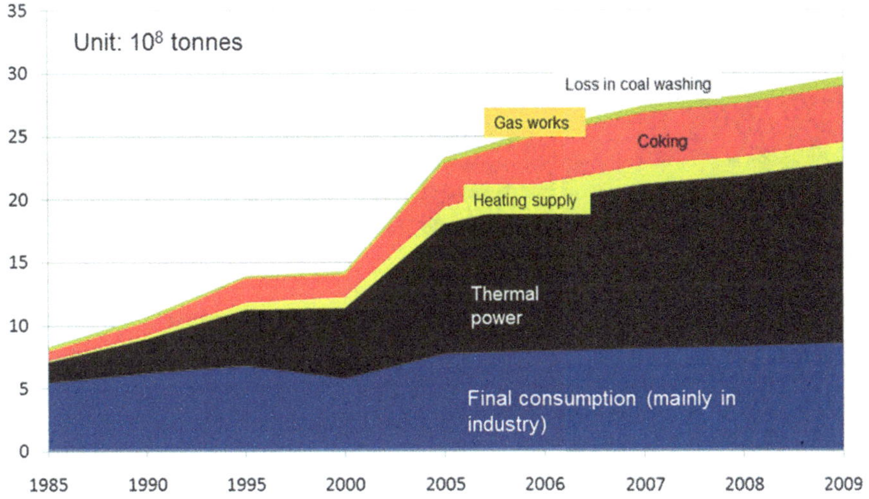

Fig. 7.3 China's coal consumption (*Statistics source*: *China Energy Statistics Yearbook 2010*)

Thus, there has been a rapid rise in national coal imports. According to the Coal Balance Sheet in the China Statistics Yearbook 2010, China's annual coal imports grew from 2.18 million tons in 2000 to 126 million tons in 2009 (there was an increase of 40 million tons in 2008). In 2010, when the coal demand in international markets was relatively weak and prices were low, power plants in southeastern coastal areas purchased large amounts of coal on international markets; national coal imports thus dramatically increased, and the accumulated net annual coal import that year was 146 million tons (NEA 2011).

7.2.1.5 Coal Utilization

In recent years, coal has maintained a level of 70 % of the national primary energy consumption. In 2009, China's annual coal consumption was 2,950 million tons—an increase of 210 million tons over the previous year. China's coal use includes the following (Figs. 7.3 and 7.4):

1. Concentrated power and heat generation: coal consumption by this sector has grown rapidly in recent years, and it currently accounts for over half of total consumption, with power generation being the major area. In 2009, 49 % of coal was combusted for power generation, and 5 % was for heat generation (heating supply).
2. Coking and gas works: this sector accounted for 15 % of total coal consumption in 2009. Until 2009, this sector was rapidly increasing, but since then it has been more stable. Coking is the dominant usage.

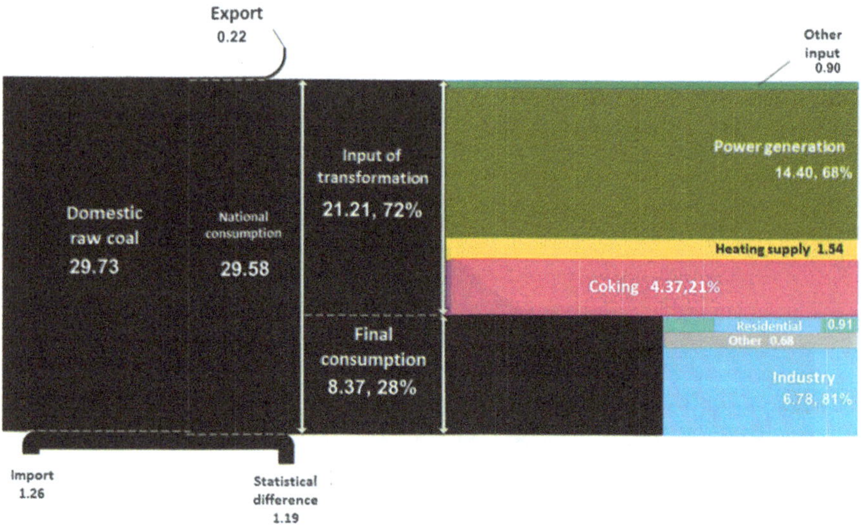

Fig. 7.4 Coal flow in China, 2009 (unit, 10^8 tons) (*Statistics source: China Energy Statistics Yearbook 2010*)

3. Final consumption: this sector includes fuel and raw materials for the chemical industry. In 2009, this sector accounted for 28 % of total coal consumption, mainly through combustion in industrial boilers or furnaces. In addition, 150 million tons of coal was consumed by the chemical industry (raw materials and production) in 2009, which compares with the 2000 figure of 88 million tons. The chemical products derived from coal include synthetic ammonia, methanol, dimethyl ether, and calcium carbide in addition to coal-derived liquid fuels.
4. Loss in coal washing: this amounted to 3 % of total coal consumption in 2009.

The large amounts of direct coal combustion result in severe environmental pollution in China. In 2007, China's SO_2 emissions reached 24.68 million tons, and NO_x emissions were in excess of 20 million tons; coal combustion produced 80 and 70 %, respectively, of these emissions.

Coal combustion produces about 80 % of China's CO_2 emissions. With stricter environmental protection policies and the development of clean coal technology, conventional coal pollution may be reduced in future, but CO_2 emissions remain a problem.

The only way to achieve large-scale reductions in CO_2 emissions produced by coal is to implement carbon capture and storage (CCS) technologies. These capture the CO_2 emissions produced by coal and transport and store them in designated sequestration sites such that the CO_2 undergoes long-term isolation from atmospheric systems. However, owing to the considerable technical difficulties and the effect on energy efficiency and costs, there has been little chance thus far for this technology to be implemented on a large scale. However, China urgently needs to promote a complete CCS deployment project to examine its feasibility. Compared with the post-capture of low-concentrated CO_2 in the flue gas of coal-fired power plants, the pre-capture of high-concentrated CO_2 in the coal chemical industry offers advantages in terms of economic performance; it will thus be the preferred approach toward complete CCS deployment in China.

7.2.2 *Constraints and Potential for Coal-Derived Vehicle Fuels*

Despite rich coal resources, China's coal mining industry faces problems with respect to mining security, destruction of water resources and the environment, and transport capacity. For example, the Chinese Academy of Engineering (CAE 2011a) comprehensively analyzed China's resource volume, mechanical mining, mining security, water resources, ecological preservation, and transport capacity, and it concluded that China's sustainable coal production cannot exceed 3.8 billion tons. If the above constraints are factored in, the actual level of sustainable coal production would be smaller (CAE 2011b): if only mining security and mechanical mining are considered as constraints, sustainable coal production would be 3.5 billion tons; if mining security, water resources, and ecological preservation are considered, sustainable coal production would be 2.9 billion tons; if transport constraints in Shanxi, Shaanxi, Inner Mongolia, and Xinjiang are also considered, sustainable coal production would be less than 2.5 billion tons.

In a situation of limited sustainable coal production, the rational allocation of coal resources remains a problem. At present, power generation, the steel industry, construction materials, and the chemical industry account for over 85 % of national coal consumption. According to forecasts by these major coal-consuming sectors, future coal demand will be 3.7–4.0 billion tons in 2030 and 3.5–3.8 billion tons in 2050. The chemical industry will require 0.35–0.4 billion tons of coal in 2030 and 0.4–0.45 billion tons in 2050 (CAE 2011a). Except for coal used to produce synthetic ammonia, calcium carbide, coal-to-olefin, and residential liquid fuels (e.g., dimethyl ether), coal used to produce vehicle liquid fuel can amount to no more than approximately 0.1–0.2 billion tons before 2050. If stricter sustainable coal production capacity constraints and CO_2 emission limits are considered, the actual coal resources for liquid fuel production would be less.

Local production of liquid fuels in coal-rich areas can help solve the coal transport capacity problem. Meanwhile, since coal-derived liquid fuel production consumes large volumes of water, depletion of local water resources will be a significant limiting factor. Referring to the interview with industrial experts, the average water consumed in producing 1 ton of coal-derived liquid fuel is about 10 tons.

Coal imports can partly solve the problems of domestic coal production capacity, but total coal consumption has to take CO_2 emission limits into consideration. Scenario analysis (CAE 2011b) has showed that to maintain CO_2 emissions below nine billion tons, China's coal consumption would have to be 3.2–3.4 billion tons by 2020 and thereafter decrease. It should be possible for China's increased energy demand to be met by nuclear power, natural gas, and renewable energy sources (CAE 2011b). CCS technology is not considered in the above scenario since its development is at an early stage, and it will probably not be commercially applicable in the near- to mid-term.

7.3 Technological Development of Coal-Derived Liquid Fuels

Coal-derived liquid fuel conversion technologies are advanced, clean technologies that convert solid coal into vehicle liquid fuels. Through various technical routes, coal can be converted into vehicle fuels, such as gasoline, diesel oil, methanol, and dimethyl ether. Fuels derived in this way can partly replace conventional petroleum-based gasoline and diesel that are currently in wide use.

This section will briefly introduce the development status of the above technologies. It will focus on an analysis of energy consumption, environmental emissions, and economic performances. It will also consider the influence of a carbon tax in direct coal liquefaction, indirect coal liquefaction, coal-derived methanol, coal-derived dimethyl ether and present a WTT analysis. In addition, a six-dimension comprehensive evaluation method is developed to evaluate the advantages and disadvantages of coal-derived liquid fuel development.

7.3.1 Current Status of Technology

7.3.1.1 Direct Coal Liquefaction

China started studying direct coal liquefied technology in the 1970s. Bituminous coal and lignite are appropriate types for direct liquefaction, and the products are mainly gasoline and diesel. Though technologies developed over the last 30 years are different in many ways from older technologies, the basic principles and industrial methods are the same or similar. Following research and development efforts in recent years and the advent of new scientific approaches, technical development is now much more feasible (Du 2006).

From 1997 to 2000, the China Coal Research Institute cooperated with government departments and companies in Germany, Japan, and the United States toward the completion of large-scale pilot plants for Shenhua Coal, Yunnan Xianfeng Coal, and Heilongjiang Yilan Coal. The institute also conducted a pre-feasibility study of a direct liquefaction demonstration plant using three different types of coal. In August 2001, China's first coal liquefaction demonstration project—the Shenhua direct coal liquefaction project—was finally licensed with a total designed production capacity of 500 million tons of oil. In 2009, the project achieved production of one million tons/year. By the first quarter of 2011, the plant was operating at higher than 80 % of the designed capacity, generating over 200,000 tons of fuel products.

7.3.1.2 Indirect Coal Liquefaction

Bituminous coal, lignite, and anthracite are appropriate types for indirect liquefaction, with the products being mainly fuels such as gasoline, diesel, naphtha, and jet fuel. In 1937, Japan introduced the Fischer–Tropsch (F-T) synthesis technology from Germany, whose core technology is a cobalt catalyst, to the sixth factory of China Jinzhou Oil. In 1943, an indirect liquefaction plant with a production capacity of about one million tons of crude oil per year was built by China Jinzhou Oil. After the founding of the New China, that plant made handsome profits. Following the discovery of the Daqing Oilfield, the Jinzhou synthetic oil plant ceased operations in 1967.

In the early 1980s, in the aftermath of the world oil crises and also taking the country's rich coal resources into account, China restarted research and development efforts into indirect coal liquefaction technology. The Institute of Coal Chemistry in the Chinese Academy of Sciences made some achievements in improving technology and catalyst development, and it conducted a number of industrial tests. In recent years, other domestic research institutions have carried out considerable research work in F-T synthesis. The Dalian Institute of Chemical Physics has examined F-T synthesis using a supported iron catalyst. Nanjing University and Nanjing Institute of Chemical Industry have developed two processes for the conversion of synthetic gas to gasoline using oxygenated hydrocarbons. Tsinghua University, the former Beijing Institute of Chemical Technology (currently named as Beijing University of Chemical Technology), and other institutions conducted a considerable amount of research into F-T synthesis on a small laboratory scale.

Many large enterprises and research institutes jointly funded a new company named Synfuels China, which was established in 2006 and aiming to commercialize the technology of indirect coal liquefaction. Shanxi Luan Mining Group built an indirect coal liquefaction polygeneration demonstration plant in 2008, whose annual production capacity is 160,000 tons of liquid fuels, in Changzhi, Shanxi, using Synfuels China's technology. Synfuels China has also cooperated with the Mongolia Yitai Group and Shenhua Group in the construction of two indirect coal liquefaction polygeneration demonstration plants whose annual production capacity is 160,000 and 180,000 tons in Erdos, Inner Mongolia. By 2009, the Luan Changzhi and Yitai Erdos projects had carried out successful test runs and were producing qualified oil.

7.3.1.3 Coal-Derived Methanol

China's methanol industry has developed very rapidly in recent years: the total annual production capacity of methanol increased from under 1 million tons in

1990 to 16 million tons in 2007. According to the interview with industrial experts in 2010, the annual production capacity of methanol (using coal as the main raw material) in 2009 amounted to 25–26 million tons.

However, compared with advanced production equipment employed around the world, the unit capacity of methanol production plants in China is generally small, the technology is backward, and the energy consumption and production costs are high. Currently, China has about 200 methanol production plants, and most of them produce under 0.05 million tons; foreign methanol production plants are mainly large scale and use gas as the main raw material; 80 % of them produce 0.3 million tons per year. In addition, the cost of foreign methanol production plants is lower than in domestic plants.

In the area of automotive energy, China has conducted methanol car demonstrations in several provinces and cities. Among these, Shanxi has carried out methanol-gasoline closed pilot schemes all over the province; the proportion of gasoline in the mix was tested at 15 % (M15 gasoline) to 85 % (M85) and then at 100 % (M100). The heat equivalent of the different mixtures of methanol-gasoline was lower than with gasoline. On July 2, 2009, the Standardization Administration of China announced that the methanol-gasoline standard for vehicle use (M85) (GB/T 23799-2009) had been officially approved and that it would be implemented on December 1, 2009. However by April 2011, the standard for M15 was still under discussion.

7.3.1.4 Coal-Derived Dimethyl Ether

In 2008, China's dimethyl ether (DME) production capacity had reached about 4.1 million tons/year, but the yield was only 2.05 million tons. This rate is quite low, though this production project is still in the initial phase. During the period of the Eleventh Five-Year Plan (2006–2010), proposed and under-construction DME projects had a total estimated production of about 5–8 million tons. Referring to the interview with industrial experts in 2010, it was estimated that China's DME productivity amounted to over ten million tons in 2009.

Currently, Shanghai Jiao Tong University, Xi'an Jiaotong University, and Jilin University are investigating DME vehicle fuels. The tests of DME vehicle development have been carried out in Shanghai, Shanxi, and Xi'an. From March 2006 to July 2007, the Shanghai DME car project had completed DME fuel car production with test runs using ten cars.

7.3.2 Comparison of Technical Performance

This section will compare the technical performance of different coal-derived fuels from three aspects: technological indicators in the production of coal liquefaction (e.g., resource consumption, energy efficiency), economic indicators (e.g., investment, product costs), and direct environmental emissions (mainly CO_2, SO_2, and NO_x).

Table 7.1 Technological, economic, and environmental indicators for various production technologies for coal-derived fuels

		Direct coal liquefaction	Indirect coal liquefaction	Coal-derived methanol	Coal-derived dimethyl ether
Technological indicators	Annual production (10,000 tons/year)	250	250	60	20
	Production technology	HTI	High-temperature /low-temperature synthesis	–	–
	Coal consumption (tce/GJ)	0.065	0.076	0.064	0.067
	Electricity consumption (kWh/GJ)	12.80	18.53	19.2	24.11
	Water consumption (tons/GJ)	0.145	0.255	0.31	0.37
	Energy efficiency (%)	50.31	41.41	50.22	47.455
Economic indicators	Investment (RMB/GJ)	169.76	199.74	144.12	196.79
	Product cost (RMB/GJ)	41.41	53.45	46.54	62.23
	IRR (%)	10.16	8.615	14.91	10.63
Direct environmental emissions	CO_2 (kg-C/GJ)	20.09	35.17	26.55	29.815
	SO_2 (kg/GJ)	0.01	0.004	0.003	0.004
	NO_x (kg/GJ)	0.09	0.160	0.17	0.17

Sources: Liu et al. (2009), Yu et al. (2006), Li (2003), Zhang et al. (2006), and Kuang (2009)
Notes:
1. Values in the table are average results derived from a comprehensive comparison of data in literature
2. IRR, internal rate of return
3. The heat value of direct and indirect coal-to-liquid fuels is 41.868 GJ/tons, methanol is 22.7 GJ/tons, and DME is 29.7 GJ/tons

Many related studies have been conducted in this field; data from the literature and average data chosen as performance parameters for coal-derived fuel production processes are presented in Table 7.1. A comparison of the resource consumption, energy efficiency, investment, production costs, and environmental emissions in direct coal liquefaction, indirect coal liquefaction, coal-derived methanol, and coal-derived DME production process are shown in Figs. 7.5, 7.6 and 7.7. In those figures, the technological, economic, and environmental indicators for direct coal liquefaction are used as a comparative basis (shown as 1); the indicators for other

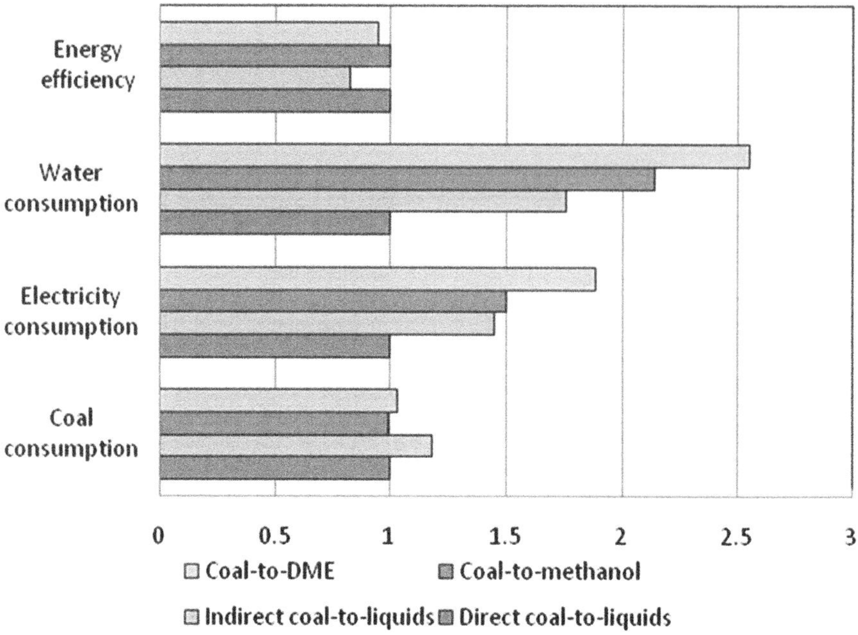

Fig. 7.5 Comparison of technological indicators in the production of coal-derived fuels

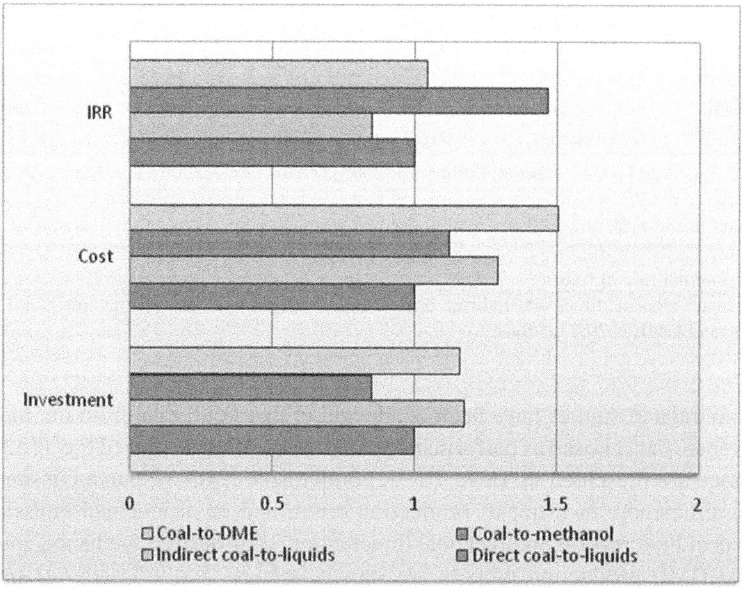

Fig. 7.6 Comparison of economic indicators of coal-derived liquid fuel production

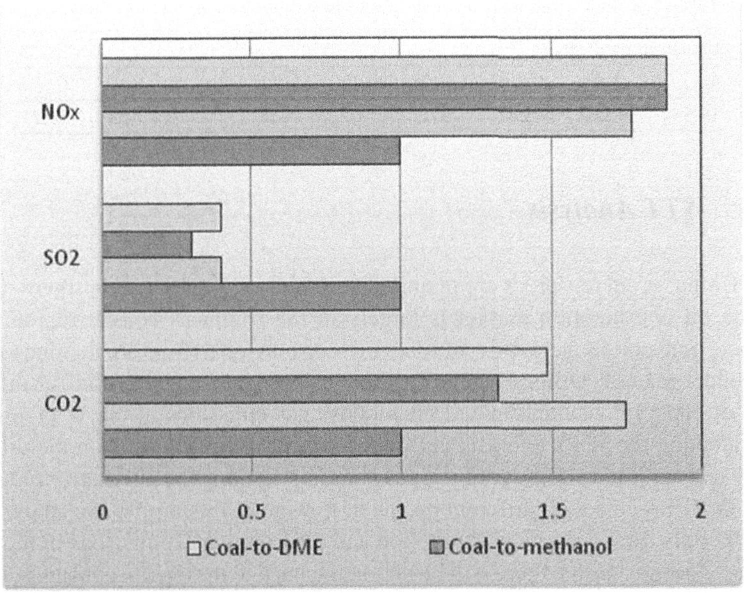

Fig. 7.7 Comparison of environmental indicators in coal-derived liquid fuel production

production technologies are expressed as relative values. For example, in Fig. 7.5, indirect coal liquefaction of coal is 1.18-fold that of the direct coal liquefaction.

It is evident in Fig. 7.5 that coal consumption in indirect coal liquefaction is higher and leads to low efficiency in the corresponding production process. Overall, energy efficiency in the production process of coal-derived fuels is mainly in the range of 40–60 %. Electricity and water consumption are closely related to such factors as the specific process, type of equipment, and scale of production; however, it can be broadly stated from an examination of Fig. 7.5 that power and water consumption are higher with coal-derived DME. In the production of coal-derived fuels, the consumption of large quantities of water is a significant problem. Water consumption for coal-derived fuel is about 0.145–0.37 tons/GJ, which amounts to about 6.07–15.49 tons to produce 1 ton of oil.

Figure 7.6 shows the investment, product cost, and internal rate of return (IRR) of coal-derived fuel production. The investment and cost with indirect coal liquefaction and coal-derived DME are high, but the IRR is low; the investment and cost of coal-derived methanol are low, but the IRR is high. Therefore, in terms of economic performance, coal-derived methanol is superior to other coal-derived fuels.

Figure 7.7 shows the direct emissions of CO_2 and the air pollutants SO_2 and NO_x in the process of coal-derived fuel production. CO_2 and NO_x emissions in direct coal liquefaction are lower than with indirect coal liquefaction, coal-derived methanol, and coal-derived DME when producing 1 GJ oil; however, the SO_2 emissions are more than double those with the other three routes.

Table 7.2 Industrial analysis of lignite from Xilinhot Shengli Coal Mine

Industrial analysis					
M_{daf}/%	A_d/%	V_{daf}/%	Calorific value (MJ/kg) ($Q_{gr,daf}$)	S_d/%	C_{ar}/%
19.56	17.03	43.47	26.3	1.52	39.71

7.3.3 WTT Analysis

Since China's coal resources are mainly distributed in western and northern regions and the oil consumption market is largely in the southeast coastal region, long-distance transport is generally needed to transport coal from coal mines to the oil product market. Different combinations of transport modes result in different levels of energy consumption and greenhouse gas emissions in the WTT process. In conducting the WTT analysis in this section, the main focus is on the different energy consumption and greenhouse gas emissions among the alternative fuels used in the WTT process with different means of transport. To simplify the analysis, we consider only direct energy consumption and environmental emissions in the WTT process. This means that there will be no tracing back to the energy consumption and carbon emissions in the infrastructure, production, transport equipment, and other materials.

7.3.3.1 Basic Parameters and Routes Setting

Bituminous coal and lignite are both appropriate for the production of coal-derived fuel. Among China's coal resources, lignite accounts for about 70 %, and it is mainly distributed in Inner Mongolia, Xinjiang, and Yunnan. In the analysis presented in this chapter, lignite from the Shengli Coal Bed in Xilinhot, Inner Mongolia, is used as the source of coal. Industrial analysis of this type of coal is presented in Table 7.2.

The coal energy consumption factor in the mining process is about 0.016 tce/tons[1] and the emission factor is 131.9 g/kWh (SBQTC 2007; Ma 2002).

In this analysis, the oil consumption is centered in Shanghai. For the various coal-derived fuels, three production-transportation pathways are designed, as indicated in Fig. 7.8.

With current technological levels, it is difficult to transport methanol by pipeline, so only Pathway 2 and Pathway 3 are investigated with respect to methanol. The unit transport energy consumption and CO_2 emissions appear in Table 7.3.

[1]The energy consumption limit of raw coal production in Shandong Province in 2008 was 5.2–15.58 kg tce/tons; the upper limit is used and rounded off for the purpose of this analysis as 16 kg tce/tons.

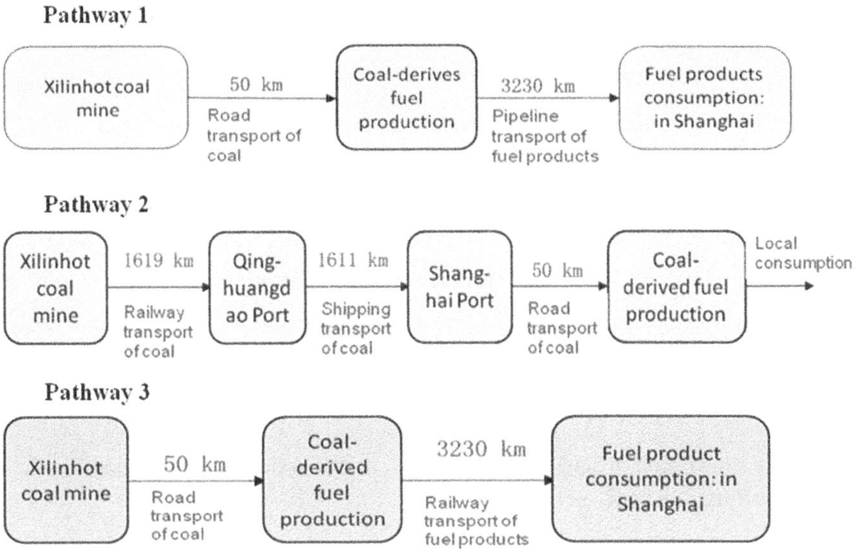

Fig. 7.8 Production-transport pathway design for WTT analysis of coal-derived liquid fuels

Table 7.3 Unit energy consumption and CO_2 emission by transport in China

Transportation mode	Rail transport		Road transport	Water transport	Pipeline transport
Energy consumption (GJ/tons · km)	0.000260		0.002147	0.000246	0.000405
CO_2 emission (kg/tons · km)	Electric locomotives 0.0695	Diesel locomotives 0.0192	0.159	0.0183	0.108

Sources: China Economic Herald (2008) and Wu et al. (2008)

Notes:

1. In this table, diesel locomotives in Rail transport, Road transport, and Water transport are all powered by diesel; electric locomotives in Rail transport and Pipeline transport are all powered by electricity
2. The data in the table were converted by the author from data in the literature. Among them, the CO_2 emission factor for diesel combustion in the internal combustion engine is 74.1 g/MJ and the CO_2 emission factor for thermal power is 267.753 g/MJ. CO_2 emissions in the transport process = energy consumption × CO_2 emission factor (kg/tons · km)

7.3.3.2 Calculating Method

Assuming a loss of 0.6 % during transport, the energy consumption and CO_2 emissions of various coal-derived fuels can be calculated for coal mining, transportation, and oil production. The calculating method is as follows.

1. Amount of raw coal required to obtain 1 GJ liquid fuel:

$$M_c = 1/(1 - \text{losses during transport}) \times \text{coal consumption in the production of 1 GJ liquid fuel}/\text{conversion factor in standard coal and raw coal} \tag{7.1}$$

2. Energy consumption and CO_2 emissions in the coal mining process:

$$E_1 = M_c \times \text{power factor of coal mining} \tag{7.2}$$

$$P_1 = M_c \times \text{calorific value of raw coal} \times \text{equivalent } CO_2 \text{ emission factor of coal mining} \tag{7.3}$$

3. Energy consumption and CO_2 emissions in the transportation process: when transporting raw coal,

$$E_{21} = M_c \times \text{transport energy consumption factor} \tag{7.4}$$

$$P_{21} = M_c \times \text{transport equivalent } CO_2 \text{ emission factor} \tag{7.5}$$

when transporting liquid fuels,

$$E_{22} = 1/(1 - \text{losses during transport})/\text{calorific value of liquid fuel}^2 \times \text{transport energy consumption factor} \tag{7.6}$$

$$P_{22} = 1/(1 - \text{losses during transport})/\text{calorific value of liquid fuel} \times \text{transport equivalent } CO_2 \text{ emission factor} \tag{7.7}$$

4. Energy consumption and CO_2 emissions in the production process:

$$E_3 = 1/(1 - \text{losses during transport})/\text{energy efficiency in the production process} \tag{7.8}$$

$$P_3 = 1/(1 - \text{losses during transport}) \times \text{equivalent } CO_2 \text{ emission factor in the production process} \tag{7.9}$$

Therefore, to obtain 1 GJ liquid fuel in the process from coal mining, transportation, and oil production to market energy consumption:

$$E = E_1 + E_2 + E_3 \tag{7.10}$$

$$CO_2 \text{ emissions } P = P_1 + P_2 + P_3 \tag{7.11}$$

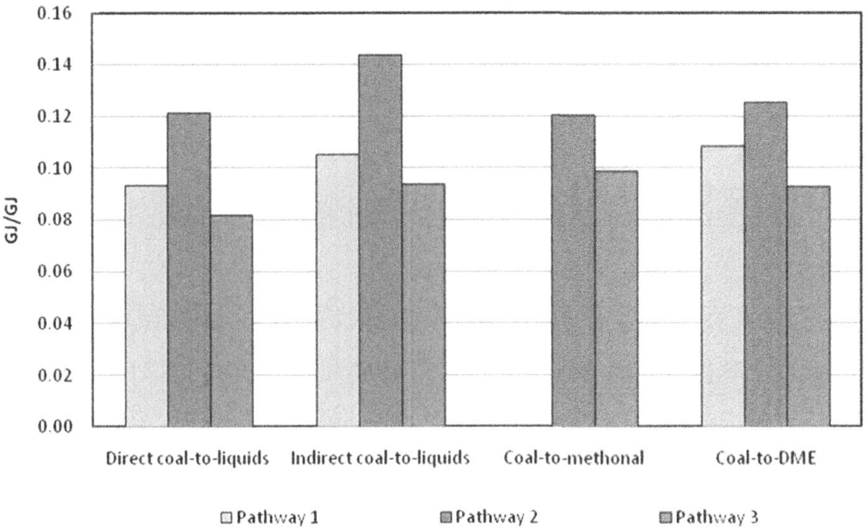

Fig. 7.9 Energy consumption for coal-derived fuels WTT with the set pathways

7.3.3.3 Results and Analysis

From the results of the WTT analysis, in addition to the conversion process, coal mining and transport consume remarkable amounts of energy, whereas the CO_2 emissions in transportation account for only about 1–7 % of total emissions. The WTT energy consumption and CO_2 emissions with coal-derived liquid fuels appear in Figs. 7.9 and 7.10.

WTT energy consumption is very different with the same kind of coal-derived fuel production technology (Fig. 7.9): it is evident that energy consumption is relatively low in Pathway 3, though it is relatively high in Pathway 2. It is clear that building a coal-derived fuel factory near a coal mining area and undertaking long-distance transport of refined oil is an energy-saving option; by contrast, long-distance coal transport and building a coal-derived fuel factory near the refined oil market increases fuel consumption. In this analysis, energy consumption by long-distance pipeline transportation of refined oil is greater than that of rail transport—at least in terms of China's current level of pipeline development. With the development of pipeline transportation technology, long-distance pipeline transport has the potential for producing further reductions in power consumption.

For the same kind of coal-derived fuel production technology using a different mining–transport–production pathway, there is no significant difference in the equivalent CO_2 emissions (Fig. 7.10). This is because CO_2 emissions are mainly concentrated in coal mining and liquid fuel production links.

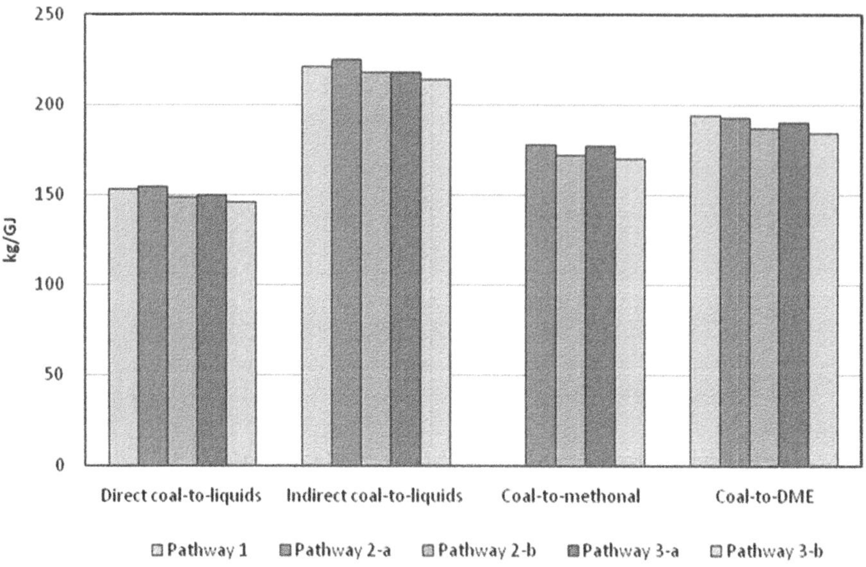

Fig. 7.10 Equivalent CO_2 emissions of coal-derived fuels WTT with the set pathways. *Note*: A represents rail transport by electric train; b represents rail transport by diesel train

7.3.4 Influence of Carbon Tax on Cost of Coal-Derived Fuel Products

Throughout the entire WTT process of coal-derived alternative fuel conversion, considerable amounts of CO_2 are emitted to the environment—largely through emissions in the production links. Therefore, there is great potential for reducing CO_2 emissions with coal-derived fuel production links. Reducing CO_2 emissions not only depends on promoting particular technologies (such as CCS) but also demands certain policy support and guidance. Here, introducing a carbon tax is one possible future policy option for reducing greenhouse gas (GHG) emissions.

A carbon tax (GHG emission price) is a kind of pollution tax. It is levied according to the production, distribution, or use of fossil fuels based on the amount of carbon emissions after burning fossil fuels. With such a tax, government departments first determine a price for carbon emissions per ton and then calculate the taxes for electricity, natural gas, or oil according to that price. The tax increases the cost of polluting fuels, and it leads to reduced fuel consumption and energy-efficient improvements among public utilities, business organizations, and private individuals. In addition, a carbon tax can improve the cost competitiveness of alternative-energy sources such that they are able to compete with low-price polluting fuels.

The costs of coal-derived liquid fuels under different levels of carbon tax appear in Fig. 7.11. The cost of fuel increases with increasing carbon tax. The more CO_2

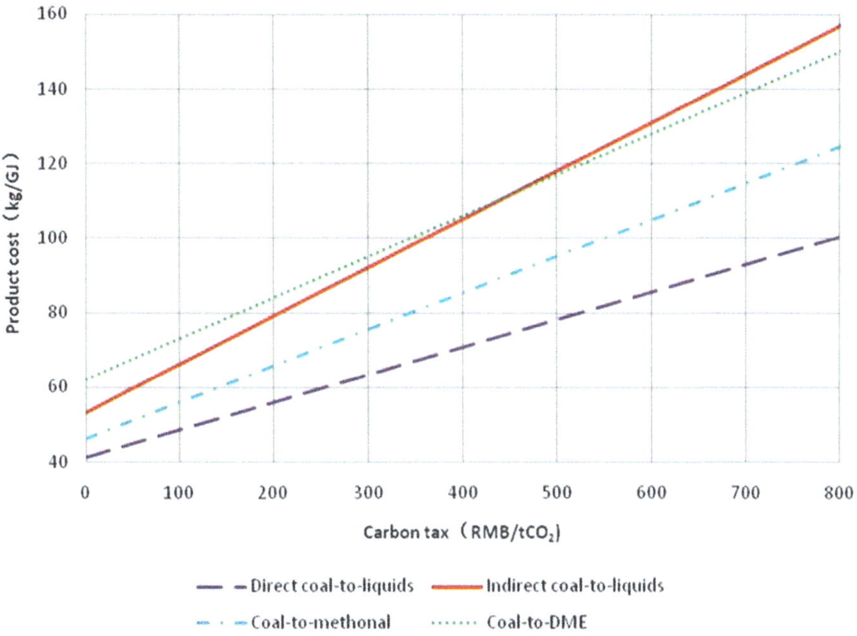

Fig. 7.11 Cost of coal-derived fuels under a carbon tax

is emitted in the production of 1 GJ, the greater the curve's gradient in the figure; thus, there is a greater increase in the price for a particular increase in the carbon tax. If the carbon tax is lower than RMB 450/tCO_2, the production cost of indirect coal liquefaction is less than that of coal-derived DME; however, if the carbon tax is higher than RMB 450/tCO_2, the production cost of indirect coal liquefaction products is greater than that of coal-derived DME. At this point, coal DME acquires a certain competitive advantage since its CO_2 emissions are lower than with indirect coal liquefaction technologies.

7.3.5 Six-Dimension Comprehensive Evaluation

The above analysis is a preliminary one that compares energy consumption, emissions, economy, and other technical factors for China's coal-derived liquid fuel. However, the development of coal-derived liquid fuels is a complex system, being influenced by energy, economic, environmental, social, and many other factors. Therefore, the following analysis represents an expansion of the original 3E (energy–economy–environment) framework by taking into account a comprehensive consideration of six dimensions—resource, environmental, economic, social, technical, and management factors; it makes a comprehensive evaluation of the benefits, costs, potential, and obstacles in the various types of coal-derived liquid fuel technologies (Ma et al. 2009).

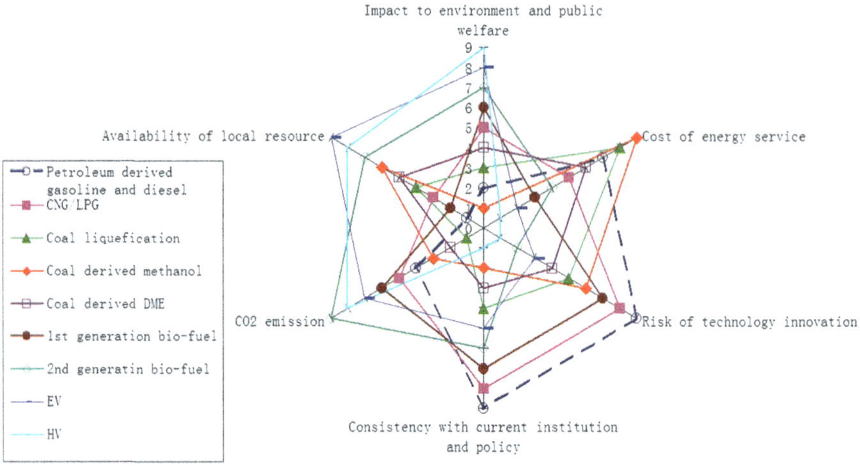

Fig. 7.12 Results of six-dimension evaluation of coal-derived fuel compared with other forms of automotive energy

7.3.5.1 Six-Dimension Evaluation Method

In making a unified evaluation and comparison of the six dimensions reflecting characteristic differences in various coal-derived liquid fuels, one core index was selected as the evaluation criterion for each dimension; these are presented in Fig. 7.12. In this analysis, other vehicle alternative fuels are also considered as references in the evaluation.

1. *Energy Security Dimension—Availability of Domestic Resources*

   The ability of alternative fuels to improve energy security is mainly related to the availability of domestic resources, such as resource reserves and energy supply capacity. In this regard, renewable energy is, generally speaking, superior to fossil energy, and coal is better than oil and gas. Biomass is considered to be inferior to wind and solar power because biomass energy consumes land and water, both of which are scarce in China, and the prospect for energy crops is uncertain.
2. *CO_2 Emissions*

   These mainly relate to the international analysis results of CO_2 emissions in the WTW life cycle of different kinds of alternative vehicle fuels (Pehnt 2006; WBCSD 2007).
3. *Economic Dimension—Energy Service Costs*

   Energy service costs mainly depend on energy production costs, such as power costs and liquid fuel production costs. These costs are also related to the compatibility of the existing infrastructure, such as with regard to power grids, gas distribution systems, and automotive diesel system. In terms of the economic efficiency dimension, the existing infrastructure needs to be fully utilized because the cost of building new infrastructure is high and perhaps cannot be supported.

4. *Social Dimension: Effects on the Environment and Public Welfare*
   This dimension is affected by many factors, and it can also be understood as public acceptance. Emissions from conventional sources are not the sole important factor here; for example, methanol toxicity, the impact of wind farms on the local ecology, and the safety of nuclear power need to be considered (technically, however, these may not present substantial problems).
5. *Technical Dimension: Technological Innovation Risks*
   Technological development generally proceeds through four stages—research and development, demonstration, promotion, and commercialization. Different obstacles occur at these different stages, such as technical barriers, engineering barriers, and cost barriers, and these produce some risk. In general, the more advanced the technology appears to be, the greater is the potential risk.
6. *Political Dimension: Accordance with Existing General Principles*
   The existing general principles can be understood as the current basic state policy, such as with respect to energy conservation and development of renewable energy. Coal-derived fuels often lack the necessary policy and regulatory support. If inconsistent with existing general principles, commercial application of alternative fuels will usually take longer and may even be postponed indefinitely.

7.3.5.2 Evaluation Results and Analysis

To better display the characteristics of the six dimensions with respect to various forms of alternative energy, this analysis adopts a uniform method of combining scores with radar diagrams in presenting the evaluation results. The rating criteria are divided into 10 levels (0–9 points), with a higher point representing better performance of that dimension; a larger coverage area of the curves represents better combined performance of the six dimensions.

The various technical routes, their corresponding dimensions, and their relative merits are presented in Fig. 7.13 (coal liquefaction in the figure includes direct and indirect coal liquefaction). It is evident that petroleum-based liquid fuel is the best in terms of technical and economic feasibility. Alternative technologies, such as electric cars, second-generation biofuels, and hydrogen/fuel cells, are the most favorable regarding long-term energy security and climate change, and they are also in accordance with political requirements and the public need. However, in economy, they are far behind more expedient alternative technical routes, such as coal-derived fuels, first-generation biofuels, and natural gas vehicles.

It is evident from this analysis that the advantages of general coal-derived liquid fuels are in their benefits through energy security and economy. Their common disadvantages are high CO_2 emissions, divergence from existing political objectives, and poor social acceptance. Achieving large-scale development of coal-derived liquid fuels in the future demands considerable efforts to be made in CCS, energy saving, publicity, and education.

It is also possible to make observations about the individual merits of several coal-derived liquid fuels (coal liquefaction, coal-derived methanol, coal-derived

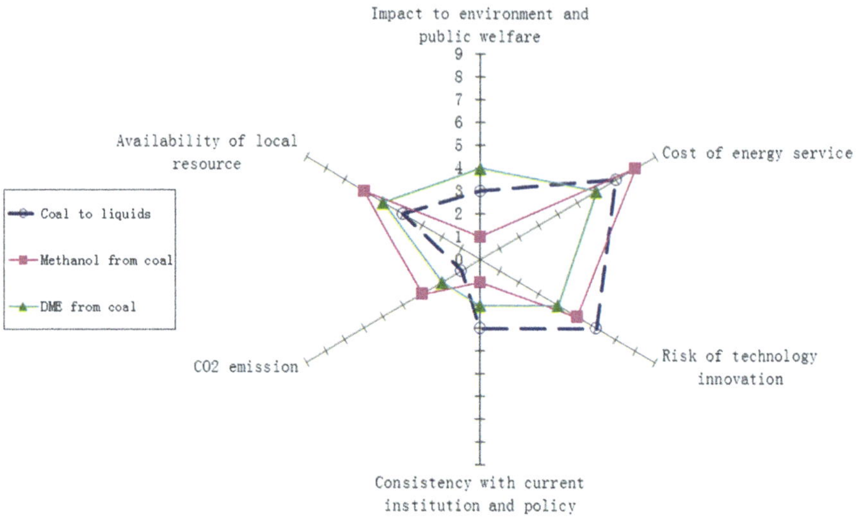

Fig. 7.13 Results of six-dimension evaluation of coal-derived fuel

DME) from the results in Fig. 7.13. It is evident there that the main problems relating to coal-derived methanol are in the political and social dimensions; this fuel performs well in the other dimensions. Apart from its high carbon emissions, the main problem with coal liquefaction is that coal-derived DME has no obvious advantages except for better social acceptance. Hence, we can conclude that vehicles powered by coal-derived methanol should be promoted by focusing on their political and social acceptance, and coal liquefaction needs to effectively solve the problem of CO_2 emissions. There are no advantages in promoting coal-derived DME vehicles except for their social acceptability.

7.4 Development Strategy for Coal-Derived Liquid Fuel

This section aims to briefly review the development status of various coal-derived liquid fuels, focus on the obstacles and potential benefits facing their future development, and provide appropriate policy recommendations (Ma et al. 2008).

7.4.1 Analysis of Different Technology Pathways

7.4.1.1 Coal-Derived Methanol

Current situation: undergoing rapid expansion, surplus production capacity; local demonstration and promotion projects.

Major obstacles: mainly industrial policy issues. There is a lack of clear guidance policies and industry standards (M15 standard) at the national level, and there are some public acceptance issues. Some minor technical issues regarding engine corrosion, exhaust gas treatment, and other aspects remain to be solved.

Future potential: recent applications are mainly seen in gasoline blending and the development of the M85/M100 motorcade. Long-term potential development lies in methanol flexible-fuel automobiles and using the large amount of methanol storage to accommodate short-term disruptions in the oil supply or rise in oil prices.

7.4.1.2 Coal-Derived DME

Current situation: undergoing rapid expansion, surplus production capacity; Shanghai is promoting demonstration of DME in public transport.

Major obstacles: mainly engineering problems. Research and development of vehicle technology are required in addition to construction of the supporting infrastructure.

Future potential: has the possibility of being an alternative fuel to diesel in the medium and long term, depending on the development and promotion of vehicle technology.

7.4.1.3 Direct Coal Liquefaction

Current situation: in the commercial demonstration stage; the economy of such vehicles still needs to be proved in industrial practice.

Major obstacles: mainly engineering problems—how to achieve large-scale, stable, continuous industrial production; quality problems with synthetic oil. High energy and water consumption and large emissions of CO_2 in the production process will affect the long-term promotion of coal liquefaction.

Future potential: an alternative to diesel fuel; also, there is a promising market for liquefied oil in the short to medium term as an alternative to oil in non-transport use (fuel oil and petrochemical raw materials).

7.4.1.4 Indirect Coal Liquefaction

Current situation: in the demonstration stage; independent indirect coal liquefaction technology has been developed. The economic efficiency still needs to be evaluated in large-scale industrial demonstrations.

Major obstacles: mainly engineering problems and the need to accumulate experience in industrial production and operation. It faces similar problems to direct coal liquefaction regarding resources and environmental aspects.

Future potential: as transport oil, it is compatible with existing vehicles and the existing infrastructure; there is also the possibility of cogenerating some chemical products.

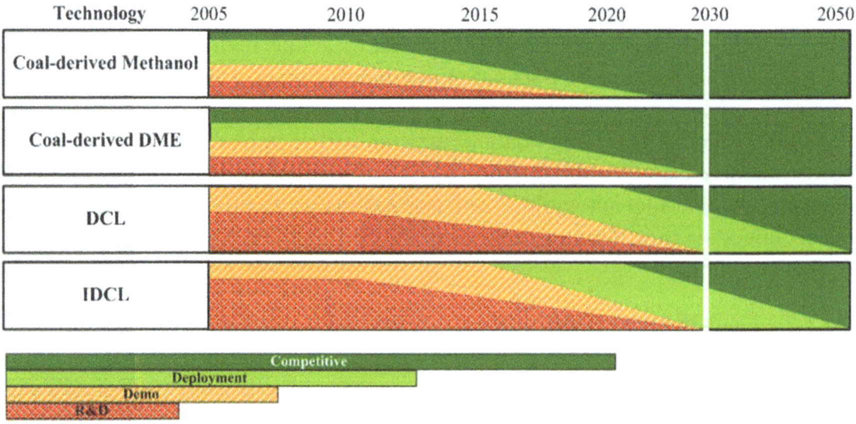

Fig. 7.14 Policy recommendations for the development of coal-derived liquid fuels (policies to be adopted at different times)

7.4.2 Policy Recommendations for Comprehensive Development

Development policies for different kinds of coal-derived alternative fuels for 2050 are made using the above analysis. As shown in Fig. 7.14, a recent key point has been the promotion of coal-derived methanol and ongoing commercial demonstrations of coal-oil fuels. After 2020, the development of coal-derived alternative fuels will depend on the energy security situation and overall national considerations regarding climate change. However, in light of the constraints facing sustainable coal production capacity and the current state of development of CCS, it is expected that the total size of the coal-derived liquid fuel market will not be very large.

Based on the above analysis, we make the following main policy recommendations for the development of coal-derived liquid fuel.

1. Coal-derived methanol: provide clearer industry guidance policies and industrial standards; expand the range of the project demonstrations; focus on the development of a full-fuel closed-running fleet and flexible-fuel vehicles. Industrialization should basically be achieved by 2015.
2. Coal-derived DME: focus on promotion by means of DME bus demonstrations; the industrialization of DME as a replacement for diesel should basically be achieved by 2020. As a substitute for diesel, coal-derived DME is an important way to deal with the possible problem of a future shortage of diesel.
3. Direct/indirect coal liquefaction: continue to promote business demonstrations and take further steps in improving product quality and reducing production costs, accumulate operating experience regarding industrialization, and make this a strategic technology reserve for energy security. After 2015, a modest promotion of this technology can be made depending on energy security needs.

7.5 Conclusions and Suggestions

7.5.1 Conclusions

7.5.1.1 Limited Potential Resource Supply of Coal-Derived Liquid Fuels

Although China has abundant coal resources, coal mining faces many problems, such as safety in production, limited water resources, and environmental protection. Current production capacity has already reached the sustainable level, which is only three billion tons per year. This means that increased coal production will become more and more difficult. It is possible to ease domestic coal production capacity constraints by importing coal and generating electricity in coastal regions. However, the total consumption of coal will still be restricted by overall CO_2 emission control, and it will be limited to about three billion tons a year owing to the unclear future regarding CCS technology. In addition, given that various other sectors and industries also have a huge demand for coal, it is estimated that coal resources that can be used to produce coal-derived liquid fuel will be 200 million tons or less per annum before 2050.

7.5.1.2 Improving Technical Performance of Coal-Derived Liquid Fuels

Coal-derived liquid fuels have recently undergone rapid technological development in China. With some technologies, China is at the forefront of world development efforts. However, in general, the technical performance of coal-derived liquid fuels needs to undergo further improvement. An analysis of the literature shows that the energy efficiency of conversion of coal-derived liquid fuel is only 40–60 %, and 6.1–15.5 tons of water are consumed in the production of 1 tons of oil products. Except for coal-derived methanol, direct and indirect coal liquefaction projects and coal-derived DME projects have an IRR of under 12 %; this means that they have poor ability to resist such fluctuations as increases in the price of coal and falling oil prices.

The result of WTT analysis indicates that the production of liquid fuel near coal mines results in lower energy consumption and lower emissions. The development of coal-derived liquid fuels also faces the problem of scarce water resources. In addition, a carbon tax will probably lead to higher costs of coal-derived liquid fuel production—especially for indirect coal liquefaction with higher CO_2 emissions. Overall, the technical performance of coal-derived liquid fuels need to be further improved.

7.5.1.3 Controlling the Total Production Capacity of Coal-Derived Liquid Fuels

The evaluation results of the integrated six-dimension analysis (resource, environmental, economic, technological, political, and social factors) indicate that coal-derived liquid fuel is stable in terms of technological and economic performance and is conducive to energy security. However, the development of coal-derived liquid fuels should not proceed too quickly owing to their poor public acceptance and high GHG emissions, which would make them a political target. To promote the large-scale development of coal-derived liquid fuels, more attention should be paid to CCS, energy saving, and appropriate consumer education.

A policy analysis of the development of coal-derived liquid fuels up to 2050 shows that it is necessary to promote coal-derived methanol and continue the construction of coal liquefaction commercial demonstration projects. The development of coal-derived liquid fuels after 2020 depends on China's status in terms of energy security and considerations regarding climate change. Overall, an excessively large capacity of coal-derived liquid fuels should be avoided because of coal production limits and lack of clarity regarding the future of CCS technology.

7.5.2 Suggestions

7.5.2.1 Steady Development of Coal-Derived Liquid Fuels

At a certain scale, the development of coal-derived fuels can contribute to national energy security. However, considering their negative effects and poor public acceptance, it would be inappropriate to develop coal-derived liquid fuels in a rapid fashion and on too great a scale. When choosing local production projects, coal resources, water resources, and environmental capacity need to be carefully considered. At the same time, more attention should be paid to education and publicity to achieve a better public understanding of coal-derived liquid fuels.

7.5.2.2 Promote Utilization of Methanol as Vehicle Fuel, Continue Commercial Demonstrations of Coal Liquefaction

Under the premise of normal, safe, long-term operations, coal-derived methanol for vehicle use offers advantages in terms of assured production capacity and lack of risk. Thus, coal-derived methanol can be considered an important development direction for coal-derived liquid fuel technology. Regarding public acceptance, technological improvements, and infrastructure construction, the focus in the development of coal-derived liquid fuels should be on their promotion in certain regions of the country rather than nationwide.

On the basis of existing technical and commercial demonstrations of coal liquefaction, demonstration projects should be further established; technical, economic, and environmental performance need to be further improved so that coal liquefaction can develop as a national energy security reserve. More research projects on vehicle technology and lower fuel costs should be the directions of development for coal-derived DME. As a substitute for diesel, coal-derived DME is an important way of dealing with the possible problem of future diesel shortages.

7.5.2.3 Improve Research and Development of Energy-Saving, Emission-Reduction Technologies of Coal-Derived Liquid Fuels

More efforts are needed in the research and development of energy and water conservation as well as emission-reduction technologies for the production of coal-derived fuels. Under the constraints of national GHG emissions, CCS would appear to be a key technical solution for the large energy consumption and CO_2 emissions of coal-derived liquid fuel production. Meanwhile, CO_2 capture in coal-derived liquid fuel production has more technological and economic advantages than CO_2 capture in a coal-fired power plant. Therefore, application of coal-derived liquid fuel CCS technology should be promoted.

References

BP (2010) BP statistical review of world energy. BP, London

CAE-Chinese Academy of Engineering (2011a) Research on the mid- and long-term (2030/2050) energy development strategy of China: volume of energy-saving and coal. Science Press, Beijing (in Chinese)

CAE-Chinese Academy of Engineering (2011b) Research on the mid- and long-term (2030/2050) energy development strategy of China: volume of integration. Science Press, Beijing (in Chinese)

CAH-China Economic Herald (2008) The coal industry risk analysis report 2008. China Economic Herald, Beijing (in Chinese)

Du M (2006) Special forum on coal's liquefaction. China Coal 32(2):10–12 (in Chinese)

IEA (2007) World energy outlook 2007. IEA, Paris

Kuang S (2009) High oil price brings opportunities for development of coal-derived energy resource technology in China (part 2). Mod Chem Ind 9(29):1–9

Li D (2003) Analysis and evaluation on technology of oil making from coal. Coal Chem Ind 2(1):17–23 (in Chinese)

Liu F, Hu M, An Y et al (2009) Coal liquefaction technology progress and discussion. Chem Eng Equip 11:106–110 (in Chinese)

Ma, Z (2002) Comparative evaluation and research for the emission coefficients of several main energy greenhouse gases in China. China Institute of Atomic Energy, Ph.D. thesis 6, p 42

Ma L, Fu F, Li Z, Zhang X, Ni W (2008) Analysis on the development of advanced coal-derived energy conversion technology in China. Coal Convers 31(1):82–88

Ma L, Li Z, Fu F, Zhang X, Ni W (2009) Alternative energy strategies for China towards 2030. Front Energy Power Eng China 3(1):2–10

Mao J (1999) China's coal resources predicting and evaluating. Science Press, Beijing (in Chinese)

NEA-National Energy Administration of China (2011) 2010 energy economic situation and outlook for 2011. http://www.sdpc.gov.cn/jjxsfx/t20110128_393341.htm. Accessed 22 Jan 2013 (in Chinese)

NSB-Energy Statistics Office of National Statistics Bureau, NEB-Comprehensive Office of National Energy Bureau (2011) China energy statistics 2010. China Statistical Press, Beijing

Pehnt M (2006) Dynamic life cycle assessment (LCA) of renewable energy technologies. Renew Energy 31(1):55–71

Pu H (2010) Sustainable development and utilization of coal and environment policy in China. China University of Mining and Technology Press, Xuzhou (in Chinese)

SBQTC-Shandong Bureau of Quality and Technical Supervision (2007) Comprehensive energy consumption limit of tons of raw coal production. Local Standards in Shandong Province DB37/832

WBCSD-World Business Council for Sustainable Development (2007) Mobility 2030: meeting the challenges to sustainable mobility. http://www.wbcsd.org/web/publications/mobility/mobility-full.pdf. Accessed 22 Jan 2013

Wu W, Fan Y, Li L et al (2008) Energy efficiency, energy saving potential and countermeasures in transport sector. Macroeconomics 6:28–33 (in Chinese)

Yu Z, Chen G, Yang L (2006) Assessment of coal-derived liquid fuel production technologies. China Energy 2(28):14–18 (in Chinese)

Yue F (2008) Annual report on coal industry in China (2006–2010). Social Sciences Academic Press, Beijing (in Chinese)

Zhang L, Huang Z (2006) Analysis of life cycle energy consumption and greenhouse gases emission of coal-derived vehicle fuels. J China Coal Soc 10(5):662–665 (in Chinese)

Chapter 8
Liquid Biofuels

Chang Shiyan, Zhao Lili, Zhang Ting, and Zhang Xiliang

Abstract Over the last 10 years, the development of biofuels in China has undergone three distinct stages. By the end of 2010, the utilization of fuel ethanol reached 1.86 million tonnes in China and that of biodiesel was about two million tonnes. This chapter analyzes the biomass resource potential in China and reviews conversion technologies and policies. A scenario-based analysis on projecting the use of biofuels by 2050 is carried out. The major conclusions include the following: (1) Biofuel production will continue to grow in China until 2050, and the actual supply capacity will be about 32.4–79.7 million tonnes of oil equivalent (mtoe) in 2050; (2) biodiesel will continue to rise, accounting for over 50 % of total biofuel production after 2030; and (3) second-generation biofuels will serve as important alternatives in the long term. Several suggestions regarding biofuels are also proposed.

Keywords Biofuel • China • Policy

Throughout the world, biofuels developed rapidly during the period of 2000–2010, with the yield increasing 6.25-fold from 16 to 100 billion L (IEA 2011). According to projections in the *Biofuels Technology Roadmap* published by the International Energy Agency (IEA) in 2011, 27 % of global transportation fuels by 2050 will derive from biofuels, resulting in a CO_2 reduction of 2.1 billion tonnes. It is predicted that biofuels will play a crucial role in the global transportation fuel mix and that they will contribute greatly to reductions in carbon emissions.

Bioenergy accounts for a large share in the primary energy mix in China, thanks to the country's rich biomass resources. Over the past 10 years, China's liquid biofuels industry has gone from nonexistent to then undergoing rapid growth;

Chang Shiyan (✉) • Zhao Lili • Zhang Ting • Zhang Xiliang
China Automotive Energy Research Center, Tsinghua University, Beijing 100084, P.R. China
e-mail: changshiyan@mail.tsinghua.edu.cn

DOI 10.1007/978-3-642-36847-9_8,

however, that was followed by a period of stagnation. Biofuels have been included among national development strategies for science and technology and for industrial optimization.

8.1 Current Status

8.1.1 Fuel Ethanol Grows Slowly: 2010 Utilization Volume Far Less Than Planned

Over the last 10 years, the development of biofuels in China has undergone three distinct stages. At the start year of 2001, the Chinese government approved the construction of four fuel ethanol plants, with an initial capacity of 1.02 million tonnes. In 2002, five cities (Zhengzhou, Luoyang, and Nanyang in Henan Province and Harbin and Zhaodong in Heilongjiang Province) were chosen to launch the Vehicle-Use Ethanol Gasoline Pilot Testing Program. In 2004, the National Development and Reform Commission (NDRC) and seven other departments expanded the test areas for fuel ethanol to cover the whole areas of five provinces (Heilongjiang, Jilin, Liaoning, Henan, and Anhui) and partial areas of four provinces (Hubei, Shandong, Hebei, and Jiangsu). From 2004 to 2006, the use of biofuels grew rapidly. In December 2006, to promote the entry regulations for biofuels, the NDRC and Ministry of Finance (MOF) issued the "Circular on Strengthening Construction Management of Fuel Ethanol and Promoting Healthy Development of Industries." After the regulations were enforced, the diffusion of biofuels slowed significantly and the increase in production came mainly from expanding the capacity with existing projects. By the end of 2010, the utilization of fuel ethanol amounted to 1.86 million tonnes in China (Zhao et al. 2011) (Fig. 8.1); that was only 18.5 % of the goal targeted in the Medium- and Long-Term Development Plan for Renewable Energy in China (2007).

8.1.2 Biodiesel Utilization Exceeded the Goal: Volume Far Behind Accumulated Capacity

In China, biodiesel is produced mainly from waste oil. The annual production capacity for biodiesel is estimated to be two million tonnes (China Renewable Energy Society 2011). Plants using waste oil as feedstock face many difficulties in terms of feedstock collection and marketing, and so the actual utilization of biodiesel is only 400,000 tonnes (Zhao et al. 2011) (Fig. 8.1). Following

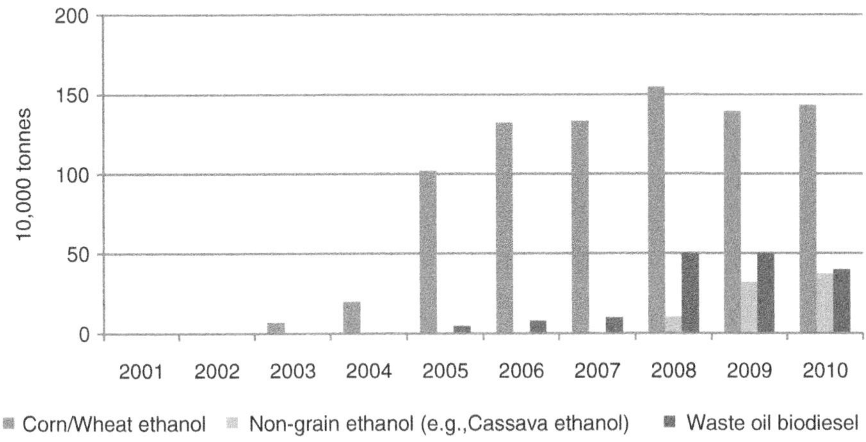

Fig. 8.1 Status of biofuel development in China

research and demonstration efforts, some projects have undergone expansion using oil-bearing energy crops, such as *Jatropha*, as feedstock. In 2007, three *Jatropha* biodiesel projects were approved by NDRC, including those of Petrochina Nanchong Petrochemical Co., Ltd. (60,000 tonnes/annum), SINOPEC Guizhou Oil Products Company (50,000 tonnes/annum), and CNOOC Hainan Biodiesel Project (60,000 tonnes/annum). The Hainan Biodiesel Project has been put into operation.

8.2 Biomass Resource Potential

Biomass resource is the basic input for liquid biofuel conversion. Although biomass is renewable, there is uncertainty about the volume of the resource available for large-scale, cost-effective use within a certain time frame. It is very important to carry out an assessment of biomass resource. In this context, biomass resource can be categorized as either a plantation resource or a non-plantation[1] resource, according to the input factors of production (especially land). Plantation resources consist of plants used for energy purposes, whereas non-plantation resources consist of all kinds of residues and wastes during planting and processing, mainly cellulosic biomass.

[1]The concept of the non-plantation resource comes from Bhattacharya et al. (2005) and Li et al. (2005). The scope of non-plantation resources is very large, and this chapter does not examine such areas as animal waste and industrial organic waste since the focus here is on liquid biofuels.

Table 8.1 Biomass resource inventory

Resource	Feedstock	Fuel products
Plantation resources	Non-grain agricultural energy crops	Fuel ethanol
	Cassava, sweet sorghum, sweet potato, etc.	
	Oil-bearing trees	Biodiesel
	Jatropha, Chinese Pistache, etc.	
	Lignocellulosic energy crops	Cellulosic ethanol, biodiesel or bio-oil
	Fast-growing trees and grasses	
Non-plantation resources	Agricultural residues	Cellulosic ethanol, biodiesel, or bio-oil
	Primary residues	
	Secondary residues from agricultural processing	
	Forestry residues	Cellulosic ethanol, biodiesel, or bio-oil
	Primary residues: logging and forestation residues, logging and slash of firewood forest, stump refreshing and rejuvenation residuals of shrub forest, fostering and intermediate cutting residuals, updating and trimming residuals of municipal greening	
	Secondary residues: residues from forestry processing	
	Tertiary residues	
	Residues from municipal solid waste	
	Others	Biodiesel
	Used waste oil	

8.2.1 Biomass Resource Inventories

The General Office of the State Council and NDRC issued a succession of notices in 2007 about the production of oilseed crops and healthy processing of maize. These notices emphasized stricter regulations being imposed on the conversion of rapeseed for biodiesel and projects for processing maize into fuel ethanol would no longer be established. It was stated in the Medium- and Long-Term Development Plan for Renewable Energy (2007) that no new fuel ethanol projects using edible feedstock would be part of near-term planning efforts. Instead, the rational use of non-grain raw materials for producing fuel ethanol was encouraged. In the near term, the priorities for producing fuel ethanol were placed on non-grain feedstock, such as cassava, sweet potato, and sweet sorghum; such oil-bearing plants as *Jatropha* and Chinese Pistache were to be used for producing biodiesel. It was stated that the used-oil-recovery system in the catering industry should be gradually established. For the long term, cellulosic biomass-derived biofuels would be actively promoted. In accordance with the above policies, the plantation resources discussed in this chapter do not include such crops as corn and wheat nor do they include oilseed crops, such as rapeseed (Table 8.1).

Table 8.2 Collectable volume of crop straw (2010)

Agricultural product		Production/ 10,000 tonnes[a]	Coefficient/ tonne/ tonne[b]	Generation of straw/ 10,000 tonnes	Collectable coefficient/ tonne/ tonne[c]	Collectable volume/ 10,000 tonnes
Grain	Rice	19,576	1.00	19,576.00	0.75	14,682.00
	Wheat	11,518	1.17	13,476.06	0.74	9,972.28
	Corn	17,725	1.04	18,434.00	0.95	17,512.30
	Bean	1,897	1.50	2,845.50	0.88	2,504.04
	Tuber	3,114	0.50	1,557	0.8	1,245.6
Cotton		596.1	2.91	1,734.65	0.9	1,561.19
Oil-bearing crop	Peanut	1,564.4	1.14	1,783.42	0.88	1,569.41
	Rapeseed	1,308.2	2.87	3,754.53	0.88	3,303.99
	Sesame	58.7	2.01	117.99	0.88	103.83
Fiber crop		31.7	1.73	54.84	0.87	47.71
Sugar crop	Sugarcane	11,078.9	0.1[d]	1,107.89	0.88	974.94
	Beetroot	929.6	0.43	399.73	0.88	351.76
Tobacco		300.4	0.71	213.28	0.8[e]	170.63
Total				65,054.89		53,999.68

Notes:
[a]Data are from the National Bureau of Statistics of China (2011)
[b]The definition of coefficients comes from Xie et al. (2011a, b). National average values for the straw coefficients are applied in this context
[c]The collectable coefficients are quoted from Cai et al. (2011)
[d]Data are from the Research Group of China's Renewable Energy Development Strategy (2008)
[e]Estimated value

8.2.2 *Non-plantation Resources Have Large Potential with Various Competitive Uses*

8.2.2.1 Non-plantation Resource Potential

Agricultural Residues

The rough volume of agricultural residues in China may be estimated based on the output of the main crops, straw coefficients, and collection coefficients. The equation is as follows:

$$\mathrm{CA}_j = \sum_{i=1}^{n} P_{ij} \times \mathrm{PRR}_{ij} \times \alpha_{ij} \tag{8.1}$$

where i is the crop type, which equals 1, 2, 3, …, 13 in the context; j signifies years; CA_j is the collectable volume of crop straw; P_{ij} is the crop output; PRR_{ij} is the straw coefficient; $P_{ij} \times \mathrm{PRR}_{ij}$ is straw generation; and α_{ij} is the straw collection coefficient. The overall crop straw volume of China in 2010 was about 651 million tonnes, and the collectable volume amounted to 540 million tonnes (Table 8.2).

Forestry Residues

Forestry residues come from multiple resources, including logging and forestation residues, residues from the forestry processing industry, fostering and intermediate cutting residuals, residues from economic forest cultivation and cutting, and logging and processing residues of bamboo forests. The volume of forestry residues of China is roughly 855 million tonnes at present, among which 461 tonnes are collectable (Table 8.3).

Other Non-plantation Residues

In addition to agroforestry residues, waste oil is another important feedstock used to produce liquid biofuels in China (Table 8.4).

8.2.2.2 Competitive Uses of Non-plantation Resources

Crop residues are important organic and energy resources. In the agricultural production cycle, they play a crucial role in maintaining soil fertility, avoiding soil erosion and supporting continuous crop production (Xie et al. 2011a, b). In addition to being returned to the field, crop straw can be put to many uses. Crop residues have been long used as primary forage for raising stock and as fuel for heating and cooking in rural China. Two projects encouraged the utilization of crop residues as part of China's industrial adjustment policy were Returning Crop Straw to Fields and Comprehensive Use of Crop Straw (including silage, ammoniated straw for raising cattle, edible fungi breeding, man-made board from crop straw, lignocellulosic fuel ethanol from crop straw, non-grain exploitation and development of forage, straw biogas, straw pyrolysis, gasification, and pelleting) and Processing and Product Development of Low-Quality Wood Fuel and Its Residues. In utilizing forestry residues, the stumping residues of shrubs can be used in such areas as livestock feeding, weaving, and boarding production. The residues produced from logging in mills (including laths, sawdust, wood shavings, blocks, and fragments) can be used in man-made board production, animal feeding, and papermaking. Figure 8.2 presents an analysis of competitive uses of agroforestry residues. Because of competitive uses, the volume of crop straw for energy use[2] is only 81 million tonnes, and that for forestry residues is 124.41 million tonnes. If other uses of biomass (direct combustion, gasification and power generation of agroforestry residues, centralized gasification of straws, and straw briquettes[3]) are considered, the residue resources for liquid biofuels production amount to only 185 million tonnes.

[2]Following the definition of Cai et al. (2011), the traditional use of residues as burning fuels is not included.

[3]In 2010, the volume of agroforestry residues for modern bioenergy use was about 21 million tonnes.

Table 8.3 Forestry residue potential

Type		Production	Unit	Coefficient[a]	Unit	Generation of residues	Collectable coefficient[a]	Collectable volume/10,000 tonnes
Logging residues	Commercial timber	1,5769.7[b]	10,000 m^3	0.47[b]	tonne/m^3	7,380.22	0.7[c]	5,166.15
	Noncommercial timber	9,045.8[b]	10,000 m^3	0.59[b]	tonne/m^3	5,291.79	0.7[c]	3,704.26
Forest tending and thinning residues		19,545.22[d]	10,000 hm^2	2.2[e]	tonne/ha	43,021.81	0.25[c]	10,755.45
Firewood-harvesting material		174.73[d]	10,000 hm^2	16.00[c]	tonne/ha	2,795.68	0.80[c]	2,236.54
Shrub stubble and rejuvenation residues		5,365.34[d]	10,000 hm^2	3.33[f]	tonne/ha	17,866.58	1[c]	17,866.58
Urban greening update and pruning residues		–	–	–		5,500[c]	0.5[c]	2,750
Processing residues		–	–	–	–	3,600[c]	1[c]	3,600
Total		–	–	–	–	85,456.08	–	46,078.99

Notes:

[a]The definitions of generation volume and collectable volume of forestry residues are the same as with agricultural residues. The coefficients of generation and collection are estimated empirically

[b]According to the forest logging limits set by the State Council, the quota for forest logging in China is 248 million m^3 annually, including 158 million m^3 of commercial timber and 90 million m^3 of noncommercial timber. Based on the data of various forest zones, the volume of logging and processing residues may be calculated by the biomass ratio of forest woods, with 40 % being commercial timber and 50 % noncommercial timber. The biomass is defined as 1.17 tonnes/m^3

[c]Data come from Zhang and Lv (2008)

[d]Data come from the Seventh National Forest Resource Inventory

[e]The coefficient is estimated based on the residue volume of 340–403 million tonnes concluded by Wang et al. (2006)

[f]Most shrub forests are rejuvenated by stumping. The biomass obtained from shrub stumping amounts to 8–12 tonnes/hm^2 (10 tonnes/hm^2 is taken in the context), with stumping taking place every third year

Table 8.4 Collectable waste oil

Type	Feedstock	Consumption	Coefficient	Generation of residues/ 10,000 tonnes	Collectable coefficient	Collectable volume/ 10,000 tonnes
Waste oil	Food oil	2,680[a]	0.2–0.3	670[b]	0.5[c]	335

Notes:
[a]The population in 2010 was 1.34 billion (National Bureau of Statistics of China 2011), and the assumed per capita consumption of food oil was 20 kg/annum
[b]The value of 0.25 is taken as the residue coefficient
[c]Estimated value

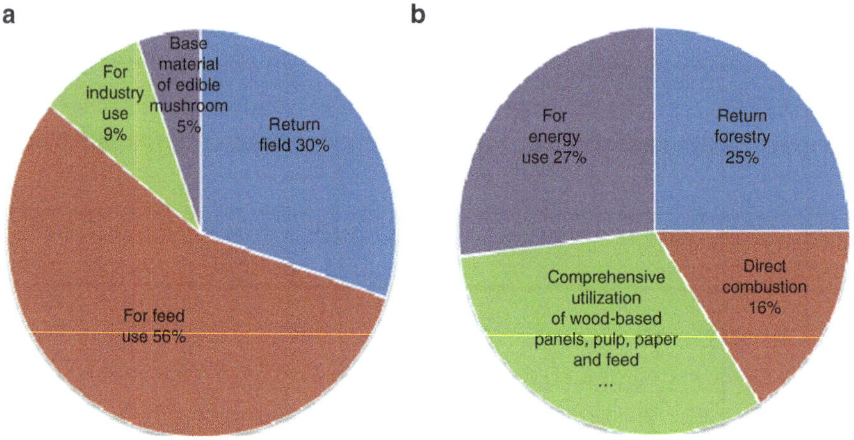

Fig. 8.2 Competitive uses of (**a**) agricultural residues and (**b**) forestry residues. *Notes*: The data for agricultural residues are derived from Cai et al. (2011), including the ratio of residues returned to the field, the volume of residues used as forage, industrial feedstock, and base material for edible mushroom growing and also the volume for rural energy use (direct combustion of residues); the ratios of forestry residues come from Chang et al. (2009)

8.2.2.3 Non-plantation Resources Potential and Increase

The increased use of non-plantation resources is uncertain. More and more residues will be used for energy conversion in China through the industrialization of China's bioenergy development, technological advances in converting agroforestry residues, and reduction in the cost of residue collection. Through related policies and planning, research results indicate that a number of positive factors will facilitate the long-term increase in non-plantation resources (Table 8.5).

Agricultural residues consist primarily of grain crop residues. Taking 2010 as an example, the grain yield accounted for 77 % of several main agricultural products. It is estimated that the collectable straw from grain crops amounted to 85 % of the total collectable straw. Therefore, the long-term volume of collectable crop straw may be estimated based on long-term grain production:

$$CA_{kj} = P_{kj} \times \beta_{kj} \tag{8.2}$$

Table 8.5 Policies and research results on potential growth of non-plantation resources

Type	Index	Reference	Content
Crop straw	Food yield	State Council (2008)	The cultivated area is greater than or equal to 1.8 billion mu (*120 million ha*), which is the binding target by 2020. Grain yields will increase from 316.2 kg/mu in 2007 to 350 kg/mu in 2020. And the overall grain production capacity should be greater than 540 million tonnes in 2020
		Institute of Agricultural Economics and Development, Chinese Academy of Agricultural Sciences (2007)	Projections indicate that China's food production will continue to grow and will reach 597 million tonnes in 2050 (Medium Scenario)
	Straw volume	Ministry of Agriculture (2007)	It is projected that China's main crop straw production will reach 900 million tonnes in 2015, of which about half can be used as the feedstock of bioenergy
		Shi (2011)	It is projected that crop straw production in 2030 will increase by 64 % compared with 2007, i.e., 137 million tonnes
Forestry residues	Forest coverage	State Forestry Bureau (2009)	China's forest industry sets up three targets in addressing climate change as follows: by 2010, the national forest coverage rate amounts to 20 %; by 2020, the national forest coverage increases to 23 %; by 2050, the forest coverage rate will amount to and stabilize at more than 26 %
	Logging residues	State Council (2011)	Forest cutting quota (excluding bamboo cutting quota) during the period of the Twelfth Five-Year Plan is 271.054 million m^3 per year, of which 218.359 million m^3 is for commercial timber harvesting and 52.695 million m^3 for public welfare forest harvesting (rearing, rejuvenation, and others)

where $k = 1, 2$ (agricultural residues, forestry residues); j signifies years; CA_{1j} is the collectable volume of crop straw; P_{1j} is the production of crops; and β_{1j} is the coefficient of comprehensive collectable crop straw. The collectable crop straw volume will be roughly 599 million tonnes by 2050 if "1" is applied, which is the

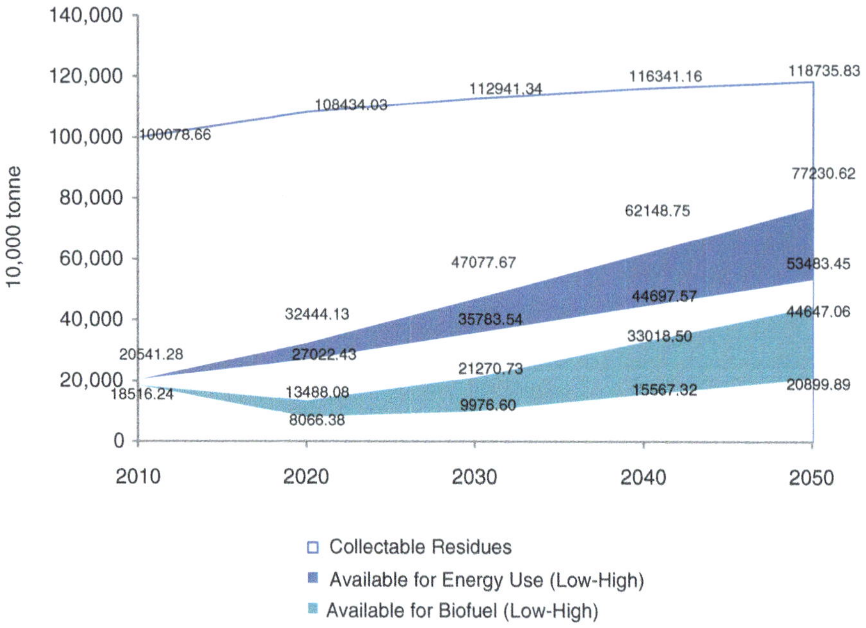

Fig. 8.3 Agroforestry resource potential in China

coefficient of comprehensive collectable crop straw for 2010. Similarly, the volume of forestry residues may be calculated based on the expected forest area. P_{2j} is the area of forest, and β_{2j} is the coefficient of comprehensive collectable forestry residues.

It is estimated that the potential of China's agroforestry residues will be roughly 1.187 billion tonnes by 2050. However, only 535–772 million tonnes could be put into energy use. The amount available for liquid biofuels will be about 209–446 million tonnes, taking into account the competitive uses of residues (Fig. 8.3, Appendix 8.1).

8.2.3 *Plantation Resource Potential Uncertain Owing to Multiple Constraints*

The plantation resources for liquid biofuels consist of non-grain sugars, starch plants, oil-bearing plants, lignin cellulosic plants, and oil-bearing microalgae (Xie 2011). Non-grain sugars and starch plants include sweet sorghum, cassava, and sweet potatoes. Oil-bearing plants include *Jatropha* and Chinese Pistache and so on. Lignin cellulosic plants consist of short-rotation woody crops, such as poplars, and fast-growing herbaceous plants, such as switchgrass.

8.2.3.1 Potential of Plantation Resource

Method

The main difference between energy crops and non-plantation resources is the requirement for land and possible resulting land-use changes. The potential for energy crops can be determined by the prospective area for crop plantation and the average crop production per unit land (Japan Institute of Energy 2007; Haberl et al. 2010). The direct equivalent method (Sidebar 8.1) is used in this study to calculate the energy crops potential in China. For those resources whose heat values cannot be estimated directly, including non-grain sugar materials, starch plants, and oil-bearing plants, the potential is estimated according to the heat value of their fuel products. The potential of lignocellulosic plants is assessed based on their own heat values. The estimation is made based on the following equation:

$$P_{lj} = \mathrm{Pro}_{lj} \times \mathrm{ML}_{lj} \times \mathrm{LHV}_{lj} \tag{8.3}$$

where l is the type of energy crop; j signifies years; P_{lj} is the potential of energy crops; Pro_{lj} is the average production per unit of marginal land; ML_{lj} is the area of marginal land available for energy crop plantation; LHV_{lj} is the low heat value of possible fuel products if l is sugar, starch plants, and oil-bearing plants; and LHV_{lj} is their own low heat value when l is lignocellulosic plants.

Sidebar 8.1: Three Methods for Calculating Biomass Potential

The resource potential of energy crops is calculated based on bioenergy potential. The following methods are usually applied:

1. Primary energy method
   The resource potential is estimated based on the heat value of biomass itself, ignoring the energy loss in biomass conversion. The method is applied generally in estimating the global biomass resource potential. Haberl et al. (2010) assumed that all biomass resources are calculated on their dry mass and that the carbon shares in different resources are the same, i.e., 0.5 tonnes per tonne of biomass, equivalent to 18.5 MJ/kg.
2. Method of theoretical conversion ratio
   For sugar and oil-bearing plants, de Wit and Faaij (2010) proposed estimating the potential according to the heat value of sugar and oil obtained by pressing the feedstock, i.e., the volume of biomass resource is multiplied by the compression ratio and by the heat value of sugar or oil. This method is more accurate than the primary energy method.

(continued)

(continued)

3. Direct equivalent method (hybrid method)
   In estimating the renewable energy resource potential, IPCC (2011) utilized the primary energy method in the field of bioenergy. However, in the fields of wind, solar, and nuclear energy, IPCC calculated the potential based on the heat values of the products—an approach called the direct equivalent method. This method is also applied in regional analysis in the case of feedstock that can be subdivided into various types and where the manner of production with each type is different from the others. For example, in estimating China's biomass resource potential, the Research Group of China's Renewable Energy Development Strategy (2008) and Shi (2011) applied the primary energy method to calculate the potential of agroforestry residues and fast-growing energy plants; however, they employed the method of theoretical conversion ratio to calculate the potential of sugar, starch, and oil-bearing plants, and that corresponded to the results with the direct equivalent method. For the purpose of comparison with other studies, the present chapter uses the same methods as those originally adopted.

Main Energy Crops

The energy crops grown in China vary in terms of plantation conditions, marketing, and production per unit land area, as shown in Table 8.6.

Resource Potential of Energy Crops

The potential of China's energy crop resources depends on the area of available marginal land and average production per unit land area. According to the present techno-economic trend, energy crop resources will increase in the future, though with uncertainty over the volume of increasing resources. In terms of driving factors, the main factors for supply increase are the expansion of available marginal land and the increase of output per unit land. Energy crops may be improved by such methods as crossbreeding, physically induced mutation breeding, chemically induced breeding, cell engineering, and genetic engineering, which result in increased output per unit land. In the long term, the area of available marginal land for energy crop planting is the hard constraint.

There is considerable uncertainty regarding the area of marginal land in China for energy crop planting. Figures in the literature range from 83 to 203 million ha (Appendix 8.2). This chapter adopts 113 million ha as a reasonable value. Among unused land, the marginal land available for agriculture is 7.02 million hm^2, marginal land available for forestry is 37.36 million hm^2, and arable land available

Table 8.6 Main characteristics of energy crops grown in China

Energy crops		Current plantation	Current market	Yield	
				tonnes/ha	tonnes/ha
Sugar and starch crops	Cassava	Cassava was harvested over a total area of 387,400 ha in 2008, with the total output of 7.9394 million tonnes (Fang et al. 2010). Main producing areas are Guangxi, Guangdong, Hainan, Yunnan, and Fujian provinces, among which Guangxi accounts for over 60 % of the national harvest area and fresh cassava production. At present, four dominant cassava-growing zones have been primarily established: west Hainan and west Guangdong; south Guangxi, east Guangxi, and central Guangdong; west Guangxi and south Yunnan; east Guangdong and southwest Fujian (Huang et al. 2008)	The proportion of cassava used as forage (including that self-used by farmers) in 2007 was 30 %; the remaining 70 % was used for producing starch, ethanol, starch sugar, etc. (Tian et al. 2010)	20.55 (average in 2008) 45–75 (high yield)	2.94–4.11 (fuel ethanol)[a] 6.43–15 (fuel ethanol)[a]
	Sweet sorghum	Pilot plantations have been carried out in Huachuan of Heilongjiang, Urumqi of Xinjiang, Anqing of Shandong, Hohhot of Inner Mongolia, and Chaoyang of Liaoning Province, and among others. Sweet sorghum breeding and cultivation techniques have been included in national high-tech research development plans	No mature market is available. A small amount of sweet sorghums is used for making alcohol and as animal feed	60–80 (straw)	3.75–5 (fuel ethanol)[b]

(continued)

Table 8.6 (continued)

Energy crops		Current plantation	Current market	Yield tonnes/ha	Yield tonnes/ha
	Sweet potato	Sweet potato cultivation area was 4.76 million ha in 2008, with the production of 102 million tonnes (Department of Rural S&T, Ministry of Science and Technology 2009)	50 % for livestock feed, 15 % for industrial processing, 14 % as food, 6 % for breeding, and 15 % falling into decay (Jin et al. 2011)	21.47 (average value)	2.39–3.07 (fuel ethanol)[c]
Oil-bearing trees[d]	*Jatropha*	Demonstration planting bases in Panzhihua of Sichuan, Hainan, Guangxi, and other places have been established	No mature market has been established. A small portion is used for pharmaceutical purposes	1.5–3 (3–5 years after afforestation in the southwest) 6.75–9 (3–5 years after afforestation in Hainan)	0.5–1 (biodiesel)[e] 2.25–3 (biodiesel)[e]
Cellulosic plants	Switchgrass, *Miscanthus*, *Arundo donax*[f]	The United States is cultivating new varieties of switchgrass for bioenergy use through crossbreeding, molecular breeding, and other means. Research institutions in China also began the planting and cultivation of energy grass. For example, according to the Grassland and Environmental Research Development Center, Beijing Academy of Agriculture and Forestry, switchgrass is able to adapt to the barren desertification of marginal land (surface sand, coarse sand, and gravel soil) in north China, but direct seeding on marginal lands is difficult; seedling transplanting is required (Fan et al. 2010)	Can be used for biomass power generation, heating, solid fuel, and liquid fuel raw materials	3.77 (switchgrass)[g] 11.45 (*Arundo donax*)[g]	0.89 (switchgrass)[g, h] 2.7 (*Arundo donax*)[g, h] 2.67–4.5[i]

Fast-growing energy forest	The adaptability of fast-growing energy forest on natural conditions and land is lower	Can be used as feedstock for biomass power generation, briquette and liquid fuel can also be used for feed, weaving, and making plates, and other purposes	2–98.8[j]	0.47–23.4[h]

Notes:
[a]Assuming that 1 tonne of ethanol may be produced from 5–7 tonnes of feedstock
[b]Assuming that 1 tonne of ethanol may be produced from 16 tonnes of crop straw
[c]Assuming that 1 tonne of ethanol may be produced from 7 to 9 tonnes of feedstock
[d]China is rich in oil-bearing trees. According to *Approaches for Inspection and Approval of Forestry Bioenergy Feedstock Bases* (2011), issued by the State Forestry Administration, the major oil-bearing plants in China include *Jatropha*, shiny leaf yellowhorn, Chinese Pistache, *Swida wilsoniana*, *Vernicia fordii*, Chinese tallow tree, and soapnut tree. Thus far, 200,000 hm^2 of oil-bearing energy forestry base has been established
[e]Assuming that 1 tonne of biodiesel may be produced from 3 tonnes of feedstock
[f]Lignocellulosic energy grasses mainly include switchgrass, Chinese silvergrass, *Anaphalis margaritacea* var. *yedoensis*, and bamboo reed
[g]Data come from Hou et al. (2011), which were based on experiments carried out on a large-scale plantation in the land desertified by sand dredging in the Beijing suburbs
[h]Estimation is made based on the higher lignocellulosic ethanol conversion rate (300 L/tonne) (Ralph et al. 2010)
[i]Global average volume and predicted volume for 2050 of lignocellulosic ethanol produced from fast-growing energy grasses, estimated by IEA (2011)
[j]Diversified energy trees are quite different in terms of yield, though all of them are high in production, according to Zhang and Lv (2008)

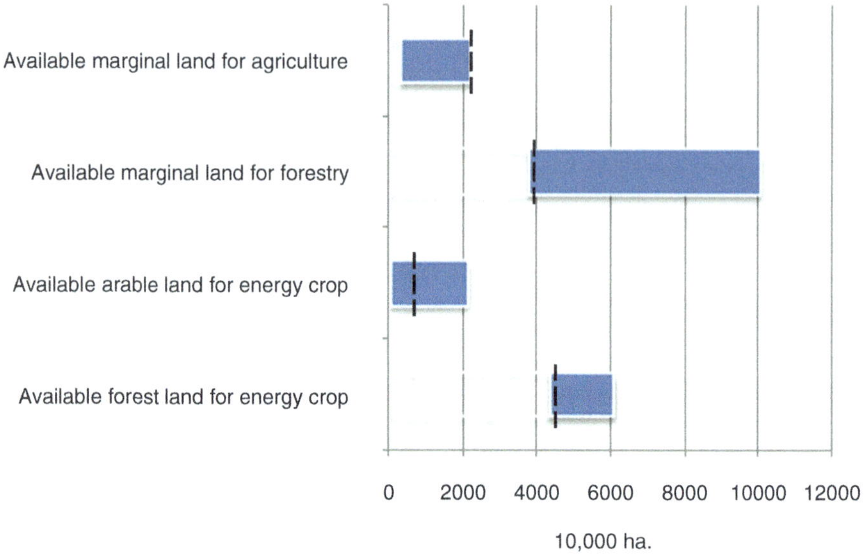

Fig. 8.4 Marginal land in China. *Notes*: the values used in this chapter are indicated by *dotted lines*

for energy crops is 20 million hm^2. In addition, forestland available for energy crop planting is 48.2414 million hm^2 (Fig. 8.4).

The potential of energy crop resources in China is roughly 17.29 EJ (Table 8.7), based on an estimation using the direct equivalent method.

8.2.3.2 Uncertainty Analysis of Plantation Resources

Estimating the biomass resources potential is complex and influenced by numerous factors. Results vary as a result of different research categories and assumptions. Taking the global biomass resources potential as an example, the estimated potential for 2050 varies between 5 and 1,272 EJ—a roughly 240-fold gap of 1,200 EJ (Haberl et al. 2010).

Land resource potential is a vague concept. It is not just a scientific notion but also a concept that involves economic and political factors. For example, an area regarded as appropriate for energy plants in terms of planting technology is different from a suitable planting area using economic criteria (e.g., opportunity cost) and different again from an appropriate area from the perspective of environmental conservation and policies. Accordingly, owing to the variety of categories, the area of available energy crop planting land potential ranges from 60 to 3,700 million hm^2—a 60-fold gap (Haberl et al. 2010) (Fig. 8.5).

Table 8.7 Resource potential of energy crops

Energy crop	Marginal land	Area (10,000 ha)	Yield (tonnes/ha)	Lower heating value (LHV)[a] (GJ/ha)	Potential (EJ)
Sugar and starch energy crops	Available marginal land for agriculture	702	3 (fuel ethanol)	80.10	0.56
	Available arable land for energy crops	2,000	3 (fuel ethanol)	80.10	1.60
Oil-bearing trees	Available unutilized land for oil-bearing trees	2,200	1 (biodiesel)	37.8	0.83
	Available forest land for oil-bearing trees	1,400	1 (biodiesel)	37.8	0.53
Lignocellulosic energy crops	Available unutilized land for lignocellulosic energy crops[b]	1,536.06	15 (feedstock)	277.50	4.26
	Available forest land for lignocellulosic energy crops[b]	3,424.14	15 (feedstock)	277.50	9.50
Total					17.29

Notes:
[a]The heat value of fuel ethanol is defined as 26.7 GJ/tonne, that of biodiesel 37.8 GJ/tonne, and that of lignocellulosic plants 18.5 GJ/tonne (Haberl et al. 2010)
[b]To avoid calculation overlaps, the area available for lignocellulosic plants equals the area of unused land available for forestry plus the area of forestry land available for energy plants minus the proportion for oil-bearing energy plants

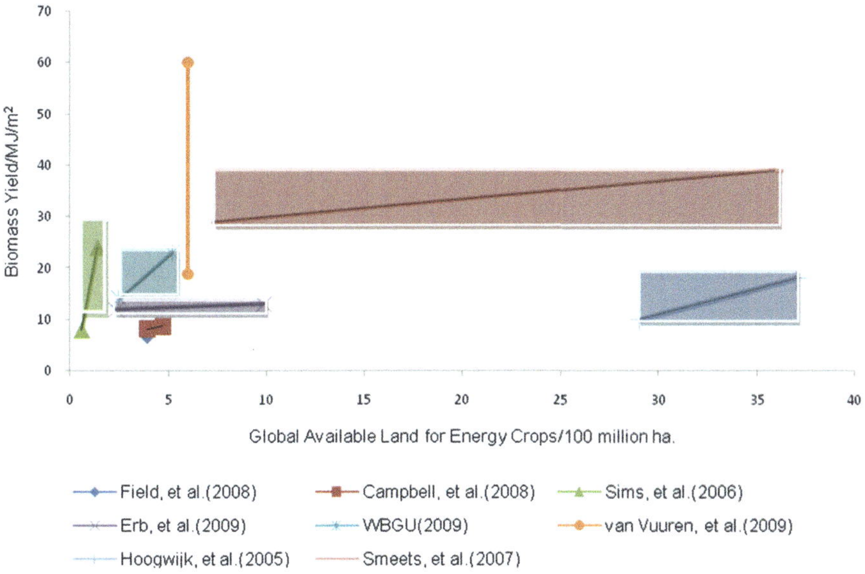

Fig. 8.5 Global marginal land and yield (*Data source*: Haberl et al. 2010)

The development of liquid biofuels depends on the volume of available marginal land. Owing to the negative impact on food security of biofuel production from grains, the principle of "neither competing for food with people, nor competing for arable land with crops" was formulated in the *Development Planning of Agricultural Biomass Energy Industry* by the Ministry of Agriculture (Sidebar 8.2). All studies on marginal land in China have taken social and environmental factors into consideration. The Research Group of China Renewable Energy Strategy (2008) defines marginal land as high-quality land in the category of unused land and lower-quality land in the category of used land. This chapter follows the classification system of the Research Group but with some modifications. Marginal land in this chapter is classified into three categories: (1) unused but usable land suitable for forestation and agricultural planting, (2) existing forestland of oil-bearing trees and potential land suitable for forestry,[4] and (3) low-grade land available for energy crop planting that grows low-yield non-grain products but can be improved by planting structure adjustment. It should be noted that there is even difficulty in utilizing unused but arable land (arable land reserves) to plant energy crops. The exploitation of arable land reserves is difficult since the reserves are mainly distributed in the northeast, northwest, and Huanghuai area. The northeastern reserves are mainly wetland, and drainage systems must be set up to exploit them, but such moves are constrained by wetland conservation policies. The northwestern area suffers from drought, and exploration is therefore constrained by lack of water resources. The largest area of reserved arable land is waste pastureland, which is mainly found in mountainous and hilly areas (Zhang et al. 2004). Accordingly, accurate figures for the area of marginal land available for energy crop planting need further investigation (Appendix 8.2).

Sidebar 8.2: National Reservation Policies for Arable Land (Forestland) and Land Classification in China

The land resources in China are quite limited. In recent years, arable land resources have been decreasing yearly owing to such factors as nonagricultural construction use, ecological restoration, natural disasters, and agricultural structure adjustment, though exploiting land resources enhances the development of China's economy (Fig. 8.6). The Law of Land Administration emphasized, "It is strictly prohibited to convert land from agricultural to construction use. The total amount of land used for construction should be

(continued)

[4]The definition of "forestland" comes from the Regulations for the Implementation of Forestry Law of the People's Republic of China: it consists of arbor forest with a crown density above 0.2, and it includes bamboo forest, shrubbery, and sparse woodland as well as land suitable for forestry planned by county-level or higher governments, i.e., wild mountains and areas suitable for forestry, sand wasteland suitable for forestry, and others.

(continued)
restricted to a certain level. The priority should be put on preserving arable land." Owing to the growing population and other factors, the total demand for food will continue to rise and will amount to 572.5 billion kg by 2020; the demand and supply will be imbalanced in the long run, according to the State Council (2008). For long-term national food security, China puts further emphasis on the strategy of arable land preservation by setting the goal of 1.8 billion mu (120 million hm^2) as the critical area for arable land.

The main policies for arable land (forestland) conservation in China are indicated in Table 8.8.

The Law of Land Administration classifies land into that for farm use, construction use, and unused land. Land for farm use refers to that directly used for agricultural production, including cultivated land, wooded land, grassland, land for farmland water conservancy, and water surfaces for breeding. Land for construction use refers to that on which buildings and structures are erected, including land for urban and rural housing and public facilities, land for industrial and mining use, land for building communications and water conservancy facilities, land for tourism, and land for building military installations. Unused land refers to land other than that for agricultural and construction uses. Adopting additional information from Current Land Use Classification and National Plan for Forestland Conservation Utilization (2010–2020), the category system of China's land is shown in Fig. 8.7.

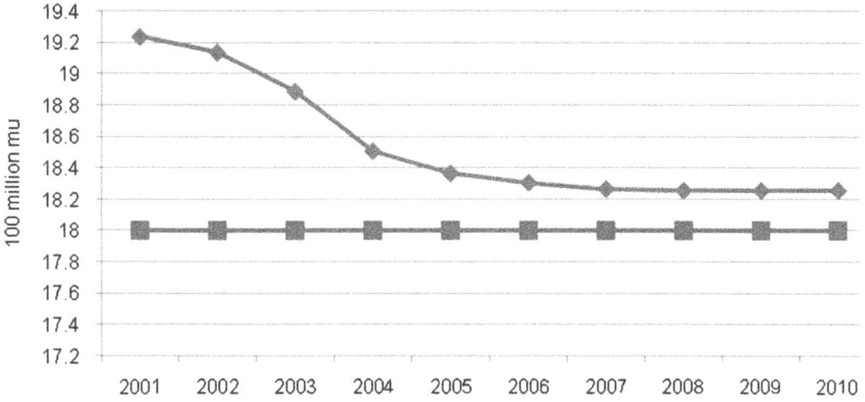

Fig. 8.6 Historical volume of China's arable land

Table 8.8 Main policies for arable land (forestland) conservation in China

Policies	Promulgated by	Date of issue	Related text
Law of Land Administration	Eleventh Session of the Standing Committee of the Tenth National People's Congress	Aug. 20, 2004	Strict control is placed on converting farmland to construction use; the total amount of land for construction use is controlled and cultivated land receives special protection
Regulations on the Protection of Basic Farmland	Decree No. 257 of the State Council	Dec. 27, 1998	The state practices a system of protecting basic farmland. Policies of overall planning, rational utilization, combination of utilization, nurturing, and strict protection shall be adhered to in protecting basic farmland
Eleventh Five-Year Guidelines for National Economy and Social Development	Fourth Session of the Tenth National People's Congress	Mar. 14, 2006	120 million hm^2 was set as a binding target for the total volume of arable land
State Council (2008)	Executive Meeting of the State Council	Jul. 2, 2008	By 2020, the area of arable land should be no less than 1.8 billion mu (12 million hm^2). The quality of basic land should be enhanced, and the area should not been reduced
National Plan for Forest land Conservation Utilization (2010–2020)	Executive Meeting of the State Council	Jun. 9, 2010	By 2020, the volume of forestland will increase to 312 million hm^2, accounting for 32.5 % of the total land area of China. By 2020, the volume of forests will reach 223 million hm^2—an increase of 40 million hm^2 compared with 2005. Forest coverage will amount to 23 % or above

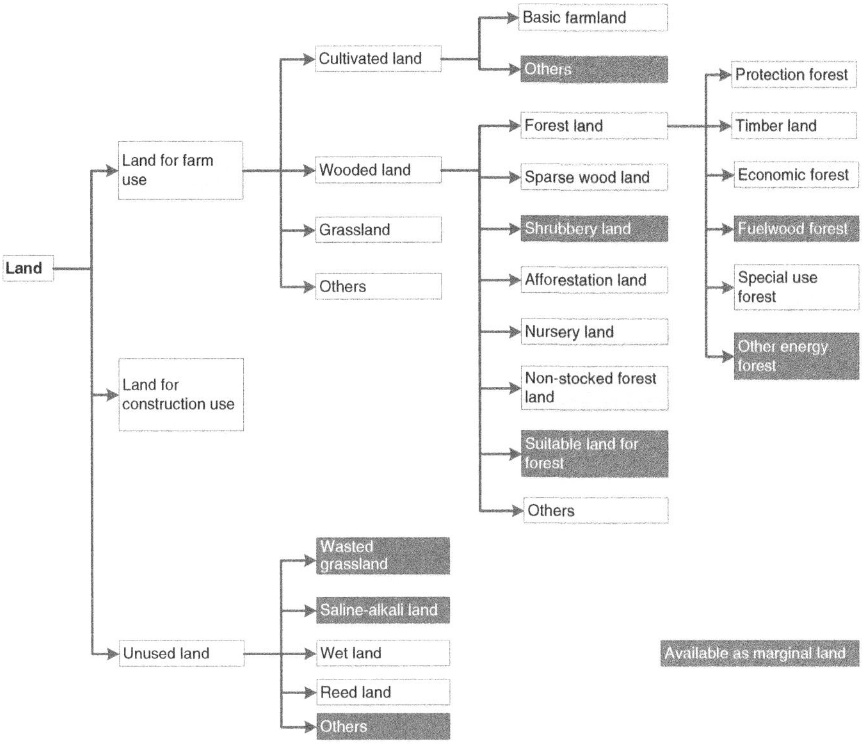

Fig. 8.7 China's land categories and potential marginal land

8.3 Conversion Technology Development and Industrial Policies for Liquid Biofuels

Biofuels may be classified in terms of feedstock, conversion processes, and fuel products (Fig. 8.8). Chinese scholars usually define the fuel ethanol produced from non-food sugar and starch energy crops and biodiesel produced from oil-bearing trees as "1.5-generation" liquid biofuels. They define cellulosic ethanol and Fischer-Tropsch (F-T) biodiesel as second-generation liquid biofuels. Biodiesels from algae are classified as third-generation liquid biofuels.

8.3.1 Non-grain 1.5-Generation Liquid Biofuels

8.3.1.1 Current Status of Non-grain 1.5-Generation Liquid Biofuels

Although the technical route is relatively mature, most raw materials for the world's fuel ethanol are derived from maize and sugarcane, which has aroused considerable

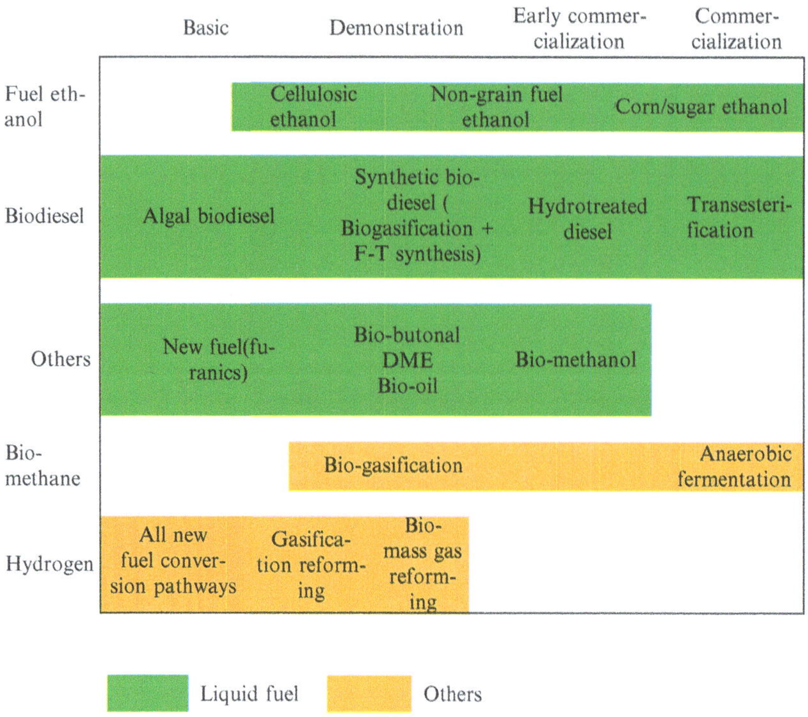

Fig. 8.8 Development stages of bioenergy technology pathways (*Source*: modified from Bauen et al. 2009)

debate in terms of food safety. To solve the problem in supplying raw materials, the global community is actively exploring high-yield non-grain crops as alternatives. Based on soil and weather conditions, China plans to grow such non-food crops as sweet sorghum and cassava on marginal land. Compared with the proven conversion technology with feedstock grain crops, such as maize and wheat, the technical chain of ethanol production using non-grain crops remains at the demonstration or application and promotion stage. Research is needed with respect not only to fuel conversion technology but also in breeding and cultivating the crops and evaluating land suitability.

A cassava fuel ethanol project with a 200,000-tonne capacity has been set up and promoted in Guangxi. In Huachuan, Heilongjiang Province, a demonstration project with a 5,000-tonne capacity for producing ethanol from sweet sorghum has been established. The project applies solid-state fermentation (SSF) technology, and it requires such simple equipment as stalk grinders, combined heating and blending machines, and beer wells. Because of the low investment and variable costs, this technology can be promoted in underdeveloped rural area (Xiao and Yang 2006). Tsinghua University developed a new SSF technology to address the difficulties in storing and transporting sweet sorghum in industrialized ethanol production. A demonstration project with a 127-m^3 capacity has been established and put into operation in Inner Mongolia.

Biodiesel is prepared mainly by transesterification, including chemical catalysis, enzyme catalysis, and supercritical methods. The chemical conversion technology has certain disadvantages, such as high investment for equipment, acidic and alkaline emissions, large-scale water scrubbing, and great energy consumption. Thus, the environmentally friendly process of biological enzyme catalysis has gained a great deal of attention. However, a technology bottleneck is hindering industrialization: the feeding methanol and byproduct (glycerine) inhibit enzyme activity and lead to fast deactivation of the enzyme. To overcome this problem, the Laboratory of Renewable Resources and Bioenergy at the Department of Chemical Engineering of Tsinghua University has developed a new technology. This has been applied by Hunan Rivers Bioengineering Co., Ltd. in establishing a set of biodiesel conversion facilities with an annual capacity of 20,000 tonnes. These were put into commission in 2006.

At present, the total production cost of biodiesel produced from energy crops, such as *Jatropha* and Chinese Pistache, is quite high owing to the expense of the feedstock (Table 8.9).

8.3.1.2 Policies for Non-grain Liquid Biofuels Industry

China has promulgated a variety of policies for the industrial development of non-grain liquid biofuels. But most of the policies related to fuel ethanol were issued before 2007. Few policies were released in the period 2007–2009. The policies for the biodiesel industry are actively implemented to a certain extent, but they have led to some undesirable effects owing to the lack of a supporting scheme for the whole industrial chain (Fig. 8.9). Details of the policies appear in Appendix 8.3.

8.3.2 Second-Generation Biofuel Technology

8.3.2.1 Current Status of Second-Generation Biofuel Technology

Second-generation biofuel technology is defined as technology that makes use of wider resources, such as lignocellulosic biomass, for liquid biofuel production. The technology consists of biochemical, thermal chemical, and hybrid processes. According to statistics of the IEA (2010), the global output of second-generation biofuels is estimated to be 680,000 tonnes, and by 2016 the production will exceed 1.6 million tonnes (Figs. 8.10 and 8.11). The United States is the world leader in second-generation biofuel development, both in terms of capacity and technological development (Fig. 8.12).

Table 8.9 Overview of non-grain 1.5-generation liquid biofuels technology and industry

Technology	Potato bioethanol	Sweet sorghum bioethanol	Waste oil biodiesel	*Jatropha* biodiesel
Development stage	Early commercialization	Demonstration and promotion	Early commercialization	Small-scale pilot projects
Current status	China currently has world's largest cassava bioethanol production facilities	Pilot project with 5,000-tonne capacity been established; NDRC approved preparation for a 100,000-tonne bioethanol project using sweet sorghum stalk as feedstock, undertaken by ZTE Energy Company in Inner Mongolia (Liu et al. 2011)	About 40 biodiesel plants use waste oil in China	3 million mu (200,000 hm^2) of energy woodland for biodiesel production planned (Wang 2011); CNOOC's project with 60,000-tonne capacity put into operation in Dec. 2009 with waste oil as main feedstock
Cost estimation (RMB/tonne)	4,887–6,802	5,444–5,968	7,000	8,000
Of which: proportion of feedstock cost (%)	55.8–68	70.5–73	79.37	75
Technological bottleneck	Stable feedstock cultivation, immature supply mode	Underdeveloped equipment and operations for continuous production	Underdeveloped collection system	Immature fostering and collection technology with underdeveloped system
Potential development in near future	Need for a sound industrial system	Need for a sound industrial system	Need for a sound industrial system	Need for better breeding and cultivating technology as well as a sound industrial system

Source: Modified by author based on Wang et al. (2010a, b)

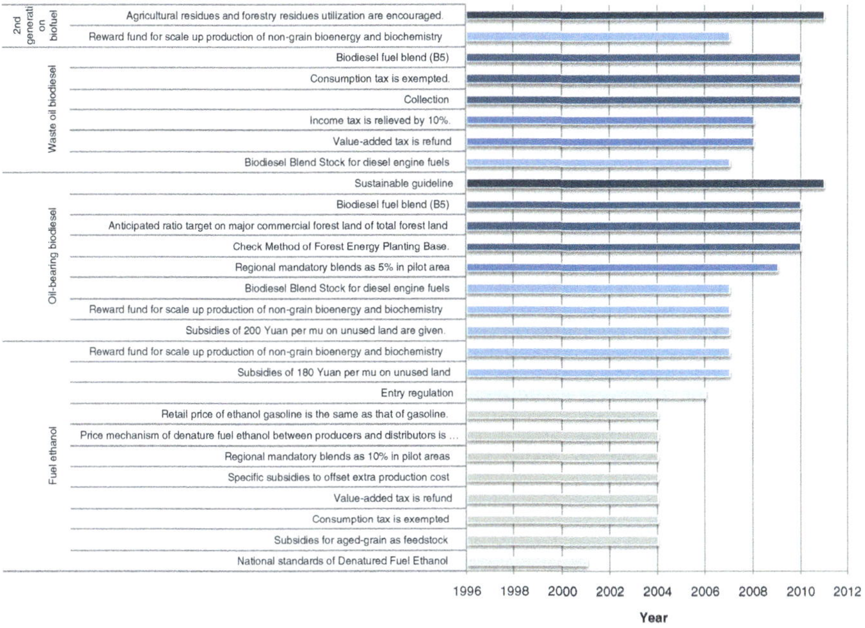

Fig. 8.9 Historical overview of biofuel policies in China

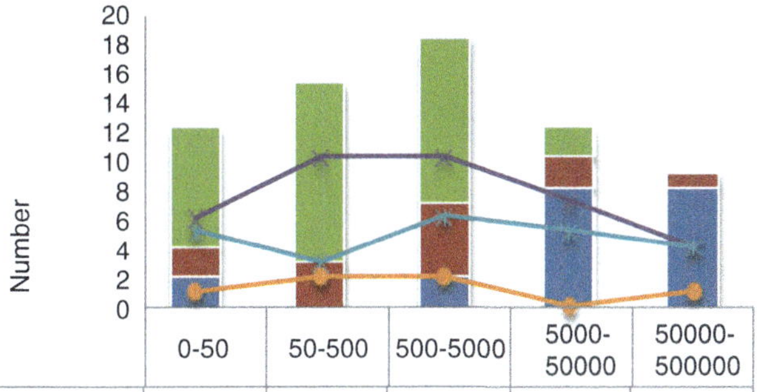

	0-50	50-500	500-5000	5000-50000	50000-500000
Operation	8	12	11	2	0
Under construction	2	3	5	2	1
Under Plan	2	0	2	8	8
Biochemical	6	10	10	7	4
Thermalchemical	5	3	6	5	4
Hybrid	1	2	2	0	1

Fig. 8.10 Global number of second-generation biofuel projects (*Data source*: IEA 2010)

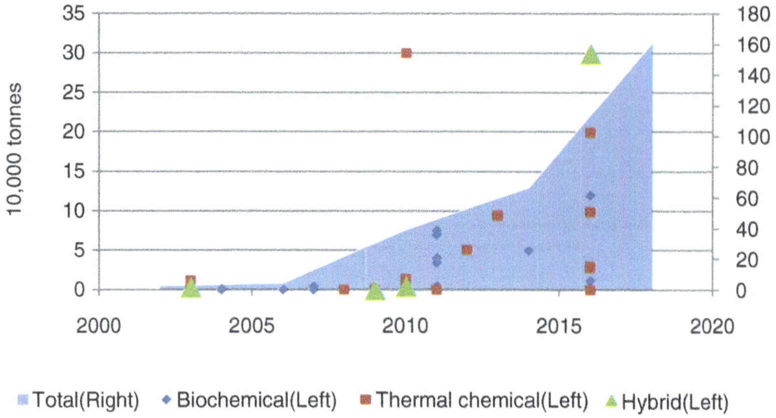

Fig. 8.11 Projection of global biofuel production based on current projects (*Data source*: IEA 2010)

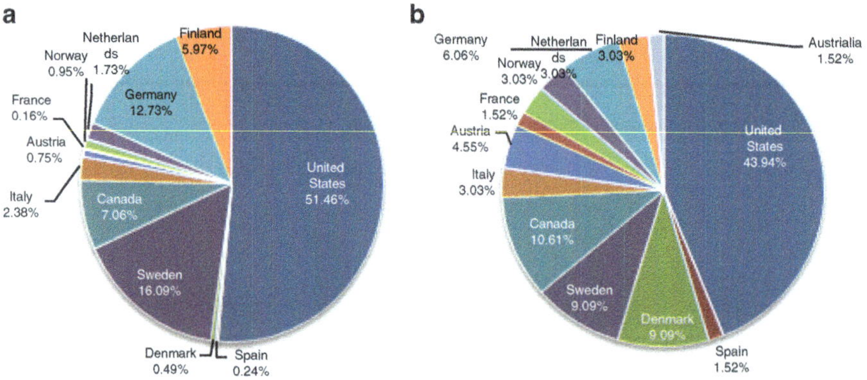

Fig. 8.12 (**a**) Projected production and (**b**) number of second-generation biofuel project by 2016. *Note*: Second-generation projects in China are excluded (*Data source*: IEA 2010)

Biochemical Process

The biochemical process is commonly used in second-generation technology for ethanol production. As seen in Fig. 8.10, 37 of 66 projects (56 %) in operation or under planning adopt the biochemical process. This process uses a hydrolysis-fermentation method to produce fuel ethanol, and it consists of acid hydrolysis and enzyme hydrolysis. The main difference between first- and second-generation biofuels lies in the front-end pretreatment and hydrolysis (Sims et al. 2010). According to the different ways of integrating the hydrolysis and fermentation, the technical pathways are classified as follows: separate (sequential) hydrolysis and fermentation, separate hydrolysis and cofermentation, simultaneous saccharification

Table 8.10 Key pilot projects for cellulosic ethanol in China

Organization	Location	Feedstock	Test capacity (tonnes)	Technology
COFCO Biochemical Co., Ltd.	Zhaodong, Heilongjiang	Corn stover	500	Enzyme hydrolysis
Henan Tianguan Enterprise Group Co.	Nanyang, Henan	Corn stover	3, 000	Acid and enzyme hydrolysis
East China University of Science and Technology	Jixian, Shanghai	Sawdust and rice straw	600	Dilute acid hydrolysis

and fermentation, simultaneous saccharification and cofermentation, and consolidated bioprocessing. In 2006, COFCO Biochemical Co., Ltd. started building a 500-tonne cellulosic ethanol demonstration plant in Zhaodong, Heilongjiang using a hybrid saccharification and fermentation process. The Institute of Nuclear and New Energy Technology of Tsinghua University and the University of Oxford have jointly developed a synchronous simultaneous multienzyme synthesis and hydrolysis and separate fermentation process (Li and Chan-Halbrendt 2009).

Despite its simple principle, the process of enzyme hydrolysis faces many technical problems in industrialization. The cost of cellulase accounts for 56–60 % of the total cost of bioethanol production. Accordingly, advances in developing cellulase for bioethanol production need to embrace two aspects—enhancing its economic performance and improving its technical performance. The Danish firm Novozymes declared they had achieved success with Cellic CTec2, which reduces the cost of cellulase to US$0.5/gal, such that biofuel producers can achieve industrialization of the process with total costs of under $2/gal (RMB 3.6/L). Based on Novozymes' cellulosic-production tests, Mogensen (2009) made predictions on cellulosic ethanol production by China for 2010–2015 based on estimates of material and energy equilibrium and equipment investment. Mogensen concluded that 74 gal of cellulosic ethanol could be produced per dry tonne of biomass at a cost of $2.59/gal in 2010; the figure for 2015 was 92 gal at a cost of $1.5/gal.

In May 2006, SINOPEC, COFCO, and Novozymes initiated a commercial fuel ethanol production project using corn stover as feedstock. The project developed a new conversion process as well as new enzyme products. In October 2006, the first batch of ethanol was produced and a threefold cost reduction over the 2007 figure was achieved. According to predictions, this decrease will continue in the future. In China, at least three pilot plants have been established for second-generation fuel ethanol production (Table 8.10).

Thermochemical Process

The commonly used thermochemical methods include biomass gasification and synthesis and direct biomass liquefaction. The principle of biomass gasification and

Table 8.11 Comparison of biochemical and thermochemical pathways

Process	Biofuel yield (L/dry tonne)	Energy content (MJ/tonne)	Energy yield (GJ/tonne)
Biochemical			
Enzymatic hydrolysis ethanol	121.3–330.8	23.3	2.5–6.9
Thermochemical			
Syngas to F-T diesel	82.7–220.5	32.9	2.9–7.6
Syngas to ethanol	132.3–176.41	23.3	2.8–3.7

Data source: IEA (2008)
Note: 1 ton = 0.907 tonne

synthesis is to produce high-quality fuel ethanol, ether, and hydrocarbons by such processes as F-T synthesis following biomass gasification. The diesel production process of gasification followed by F-T synthesis has been industrialized though mainly using fossil feedstock. The process with biomass feedstock is still under development. The German firm CHOREN utilizes such biomass as wood and crop straw to produce synthetic diesel, and it has developed a biomass gasification process called Carbon V. The company has been devoted to research and development of this process since its establishment in 1990. In 1998, a demonstration project was set up, and in 2007 a commercial plant with an annual output of 15,000 tonnes was built. At present, CHOREN is planning a plant with an annual capacity of 200,000 tonnes Biomass to Liquid (BTL). The main product, Sunfuel, has excellent performance, and it can be used in current diesel engines by being blended with conventional diesel at any ratio.

Biomass synthetic fuels have suffered from high costs in recent years. The cost of F-T diesel is 27 % higher than that of lignocellulosic bioethanol. And the future development of biomass synthetic fuels is a matter of dispute. However, some studies have shown that the process for F-T diesel will be commercialized within 10 years—before that of lignocellulosic ethanol. However, this will require breakthroughs to be achieved with key technologies for biomass pyrolysis and liquefaction (Table 8.11).

The IEA (2008) made estimates about the cost of biofuels and concluded that the production costs of both cellulosic ethanol and synthetic biodiesel would be greatly reduced after 2010 and reach a stable level by 2030 with an optimistic estimate of technological development. Adopting a pessimistic estimate, the IEA forecasts that the production cost would slowly fall to between \$0.65 and \$0.7/L of gasoline equivalent by 2050 (Table 8.12).

Hybrid Process

In addition to the biochemical process, more efficient energy-conversion technologies have been explored internationally, such as the biogasification-fermentation process. This technology represents leading-edge development in the production of

Table 8.12 Projected costs of second-generation biofuels

		Cost/RMB/tonne[a]		
Second-generation biofuel	Assumption	2010	2020	2030
Cellulosic ethanol[b] (biochemical)	Optimistic estimate	6,806	5,105	4,934
	Pessimistic estimate	6,806	5,955	5,785
BTL[c]	Optimistic estimate	10,000	6,500	6,000
	Pessimistic estimate	10,000	8,000	7,800

Source: IEA (2008)
Notes:
[a]US$1 = RMB1
[b]1 L cellulosic bioethanol equals 0.84 L oil equivalent
[c]1 L BTL equals 1 L oil equivalent

liquid biofuels. Preliminary research and evaluation results point to good economic efficiency. Coskata, an American company, has established a biogasification-fermentation process demonstration plant in Madison, Pennsylvania.

8.3.2.2 Policies for Second-Generation Biofuels

It is crucial to support the development of renewable and new-energy technologies by means of definite policies at different stages of technological development. China has been promoting first-generation biofuels since 2001 in such ways as supporting R&D, enforcing pilot operation, and providing subsidies. However, policies for second-generation biofuel production are still under discussion or initial preparation except with respect to R&D (Fig. 8.13).

8.3.2.3 Future Progress with Second-Generation Biofuels

Progress needs to be made to utilize biomass resources more extensively, promote the conversion efficiency of biomass resources, and address the conflict between demand for biomass resources and sustainable development. The Research Group of Biomass Resource Strategy of the Chinese Academy of Sciences (2009) believes that the development of second-generation biofuels should follow the trajectory indicated in Fig. 8.14.

8.3.3 Algae

Algae is a promising feedstock that features low input but high output, with production per unit area 30 times higher than with land resources. Algae do not compete for land with grains and can be grown rapidly in the natural medium of seawater. The use of algae to produce such biofuels as biodiesel and pyrolysis fuel oil by means of cell engineering and biochemical technology has become an

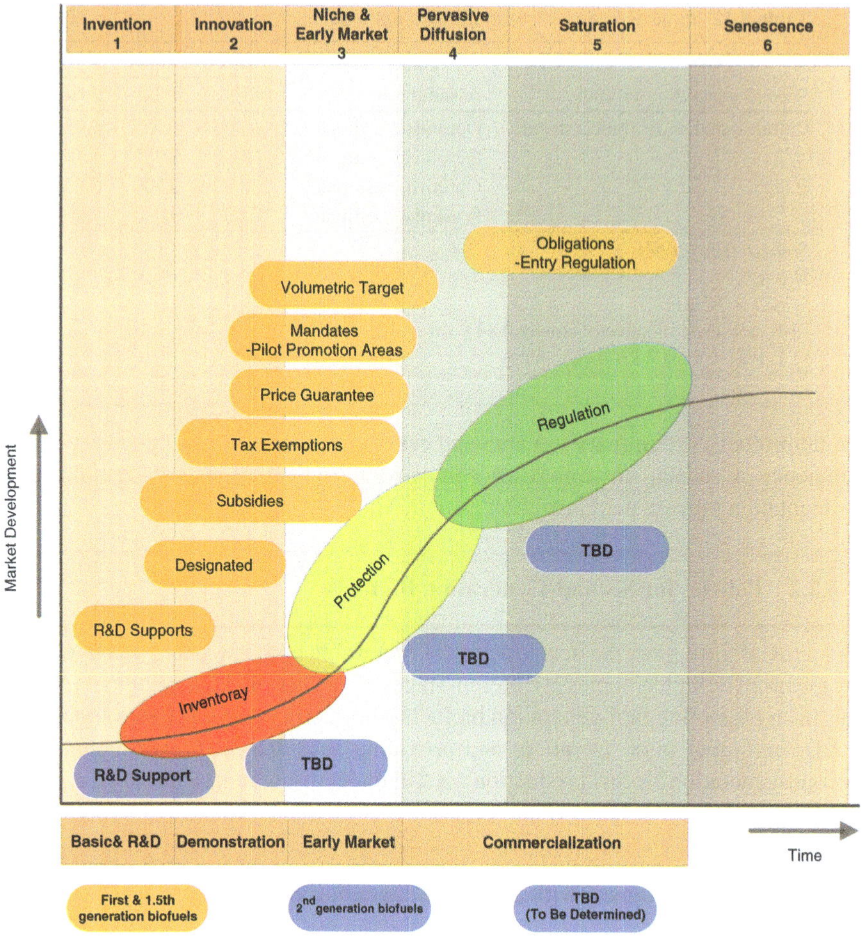

Fig. 8.13 Review of policies relating to second-generation biofuels (Adapted from Ros et al. (2006) and Bauen et al. (2009))

area of keen interest. ENN Group has started R&D in this area in China. A carbon sequestration bioenergy project has been initiated using microalgae in Daqi, Inner Mongolia.

8.4 Biofuel Development Scenario

8.4.1 Pathway Options

To investigate the specific situation in China by means of a development scenario, 18 pathways were selected by the author. The pathways included 1.5-generation

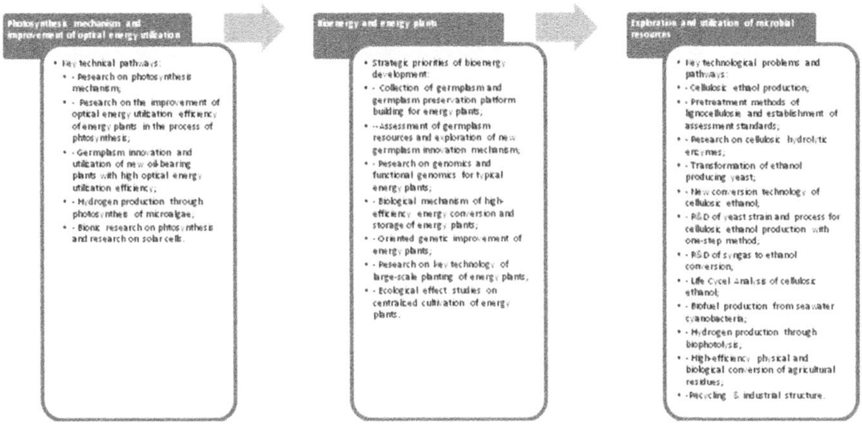

Fig. 8.14 Technical pathways for the future development of biofuels (*Source*: summarized by the author based on the Research Group of Biomass Resource Strategy of the Chinese Academy of Sciences 2009)

non-grain biofuel production from cassava, sweet sorghum, and *Jatropha curcas*, and second-generation biofuel production using broader biomass resources, e.g., agroforestry residues as feedstocks (Table 8.13).

8.4.2 Scenario Setting

8.4.2.1 Technology Development Scenario

Second-generation biofuels have been determined as one of the most promising alternative transport fuels; it is believed that they will represent an important breakthrough in addressing energy security and environmental protection while using low resources input. However, there are many uncertainties with these biofuels. Many research institutes around the world have advanced different visions regarding the success of second-generation biofuels, and the differences in the time frame for deployment vary from 10 to 20 years. As evident in Table 8.14, with the ACT and BLUE scenarios, the IEA (2008) has projected that second-generation technology will be deployed by 2012 and 2015, respectively, and commercialized by 2030 and 2035. The OPEC Fund for International Development and the International Institute for Applied System Analysis (OFID/IIASA 2009) has adopted the assumption of scenario WEO-V2/TAR-V2, whereby second-generation conversion technologies will be deployed after 2030. Therefore, two scenarios of second-generation biofuels are chosen in the present study—a slow and a fast development scenario.

Table 8.13 Biofuel pathways

No.	Resource	Feedstock	Conversion technology	Fuel product
1	Arable land	Corn/wheat	Fermentation	Fuel ethanol
2	Marginal land	Cassava	Fermentation	Fuel ethanol
3	Marginal land	Sweet sorghum	Fermentation	Fuel ethanol
4	Marginal land	Sweet potato	Fermentation	Fuel ethanol
5	Industry/service industry	Waste oil	Transesterification	Biodiesel
6	Marginal land	*Jatropha curcas*	Transesterification	Biodiesel
7	Marginal land	*Pistacia chinensis*/ *Xanthoceras sorbifolia*	Transesterification	Biodiesel
8	Agriculture	Primary agricultural residues	Hydrolysis and fermentation	Cellulosic ethanol
9	Manufacturing	Secondary agricultural residues	Hydrolysis and fermentation	Cellulosic ethanol
10	Forestry	Primary forestry residues	Hydrolysis and fermentation	Cellulosic ethanol
11	Manufacturing	Secondary forestry residues	Hydrolysis and fermentation	Cellulosic ethanol
12	Agriculture	Primary agricultural residues	Gasification and synthesis	Synthetic biodiesel
13	Manufacturing	Secondary agricultural residues	Gasification and synthesis	Synthetic biodiesel
14	Forestry	Primary agricultural residues	Gasification and synthesis	Synthetic biodiesel
15	Manufacturing	Secondary forestry residues	Gasification and synthesis	Synthetic biodiesel
16	Urban	Municipal solid waste	Gasification and synthesis	Synthetic biodiesel
17	Marginal land	Energy crops	Hydrolysis and fermentation	Cellulosic ethanol
18	Marginal land	Energy crops	Gasification and synthesis	Synthetic biodiesel

8.4.2.2 Policy Scenario

Policy support will strongly influence the development of future energy demand as China faces rigid limitations regarding land availability for the further expansion of energy crops. Two options considered here have the following characteristics: (1) The first option is to maintain the existing policy scenario and await the success of second-generation technology innovations. This is similar to the second type of biofuel policy scenario, called Moratorium, which was developed by the Food and Agriculture Organization of the United Nations (2008). That scenario advocates a 5-year moratorium for the sustainable development of biofuel technology, experience accumulation, and preventing a potentially negative impact on the environment, social community, and human rights; (2) The second option is to adopt a more active

Table 8.14 Review of the outlook for second-generation biofuels

Reference	Scenario	Description
IEA (2008)	ACT	Bring global CO_2 emissions back to current levels by 2050 (485 ppm), deployment begins by 2015, full commercialization by 2035
	BLUE	Reduce CO_2 emissions by 50 % by 2050 (450 ppm), deployment begins by 2012, full commercialization by 2030
	WEO-V1/TAR-V1	Become commercially available after 2015; deployment is gradual
OFID/IIASA (2009)	WEO-V2/TAR-V2	Owing to delayed arrival of second-generation conversion technologies, all biofuel production until 2030 is based on first-generation feedstock
	TAR-V3	Accelerated development of second-generation conversion technologies permits rapid deployment; 33 and 50 % of biofuels used in developed countries are second generation in 2020 and 2030, respectively

policy for sustainable development, increase R&D investment for second-generation technology, promote the formulation of standards for biofuel energy efficiency and greenhouse gas (GHG) emissions, and strictly supervise management of the biofuel industry. This option involves giving strong support to projects in line with the requirements of sustainable development and promoting the smooth transition from first- to second-generation technology.

In accordance with the above technology and policy information, four scenarios were designed (Table 8.15).

8.4.3 *Cost and Technology Diffusion*

8.4.3.1 Cost Calculation Method

The cost calculation method for various biofuel pathways is divided into different stages along the supply chain as follows:

$$C_{qj} = \sum_{p=1}^{n} (\text{CAPEX}_{pqj} + \text{FO}\,\&\,M_{pqj} + \text{VO}\,\&\,M_{pqj}) \tag{8.4}$$

where q is the biofuel pathway, j is the year, C_{qj} is the average cost of biofuel pathway q in the year j, p is the stage of the biofuel pathway up to n, CAPEX_{pqj} is the average unit capital cost, FO & M_{pqj} is the average unit fixed operation and maintenance (O&M) cost, and VO & M_{pqj} is the average unit variable O&M cost. The flow chart of the biofuel system is shown in Fig. 8.15.

Table 8.15 Scenario description for developing biofuels in China

	Abbr.	Scenario	Description
Slow development of second-generation biofuels	M1	Moratorium 1	Technology outlook: second-generation biofuel technology is deployed commercially by 2025. Policy option: little special R&D support for 1.5- and second-generation biofuels; awaiting the global deployment of second-generation technology; strict land-use policy with the upper limit of unused land utilization proportion as 20 %; prohibited use of newly added arable land for biofuels.
	L1	Moderate development 1	Technology outlook: second-generation biofuels technology is deployed commercially by 2025. Policy option: encourage investment in biofuel projects, especially for 1.5- and second-generation biofuels; formulation of standards for biofuel energy efficiency and GHG emissions; strict management of the industry; permitted utilization of marginal agroforestry land for biofuel production; moderate land-use policy with upper limit of unused land utilization proportion as 50 %.
Fast development of second-generation biofuels	M2	Moratorium 2	Technology outlook: second-generation biofuel technology is deployed commercially by 2015. Policy option: little special R&D support for 1.5-generation biofuels, awaiting the global deployment of second-generation technology; strict land-use policy with upper limits of unused land utilization proportion as 20 %, prohibited use of newly added arable land for biofuels.
	L2	Moderate development 2	Technology outlook: second-generation biofuel technology is deployed commercially by 2015. Policy option: encourage investment in biofuel projects, especially for 1.5- and second-generation biofuels; formulation of standards for biofuel energy efficiency and GHG emissions; strictly supervise management of the industry; permitted utilization of marginal agroforestry land for biofuel production; moderate land-use policy with upper limits of unused land utilization proportion as 50 %.

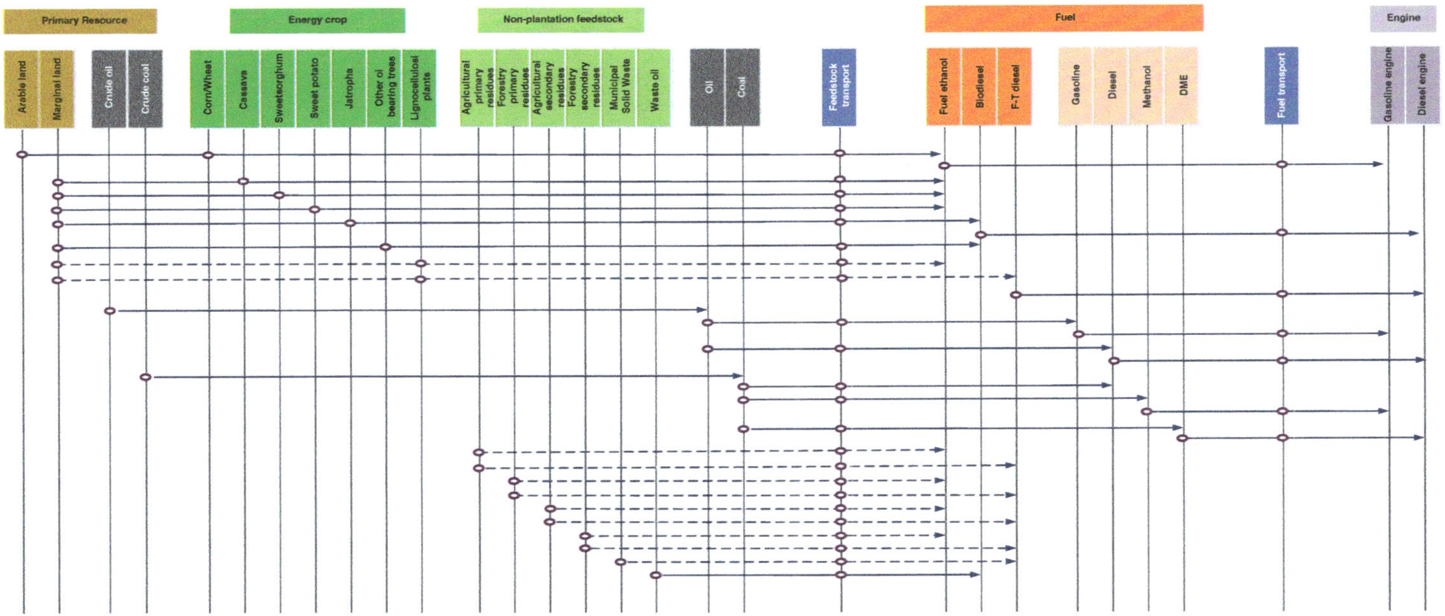

Fig. 8.15 Flow chart for biofuels (the fossil fuel pathway is included for reference). *Notes*: *Solid lines* indicate that the technology is at the stage of promotion or early commercialization; *dotted lines* denote that the technology is at the R&D or demonstration stage

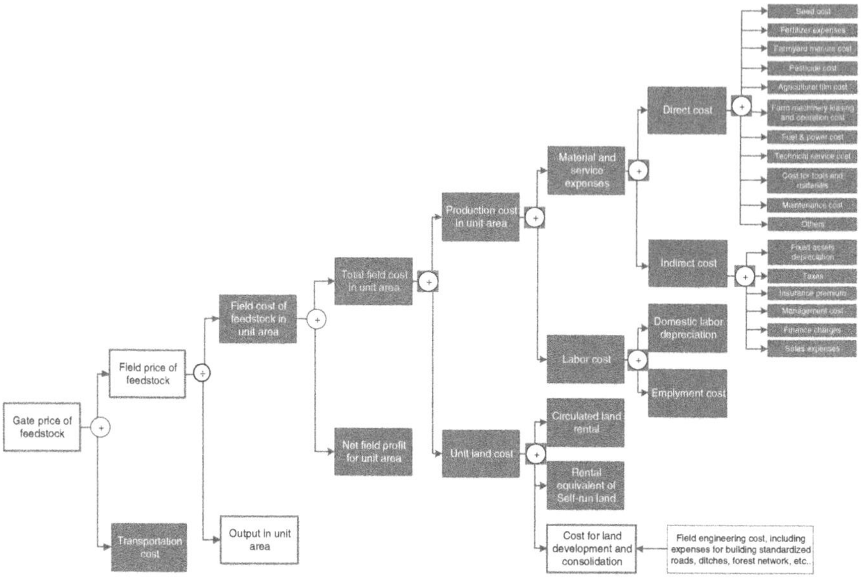

Fig. 8.16 Cost accounting system for agricultural products in China

8.4.3.2 Cost Assumption of Various Biofuel Pathways

The 1.5-generation non-grain technology has the following features: (1) Feedstock cost accounts for a large proportion of the total cost, (2) long-term cost reduction mainly relies on improving the conversion efficiency and decreasing the conversion cost, and (3) there is little potential for unit cost reduction. In terms of the cost accounting system of China's agricultural products, the major cost of energy plants derives from materials and services, labor, and land resources. In addition, for energy plants growing on marginal land, the cost of land exploitation and development also has to be considered (Fig. 8.16). With China's food crops, the main cost involves material and services, followed by labor. The land cost is relatively low. From 1978 to 2009, the average land cost underwent a 50-fold increase from RMB 2.23/μ to RMB 114.6/μ (Fig. 8.17). In the long term, there is great potential for increasing the cost of land for food crop cultivation, though currently it is relatively low. The land cost will be relatively high for biofuel development in China.

The cost reduction with second-generation biofuels depends on technological innovation in the near term and on economies of scale and technology learning in the long term, as shown in Fig. 8.18.

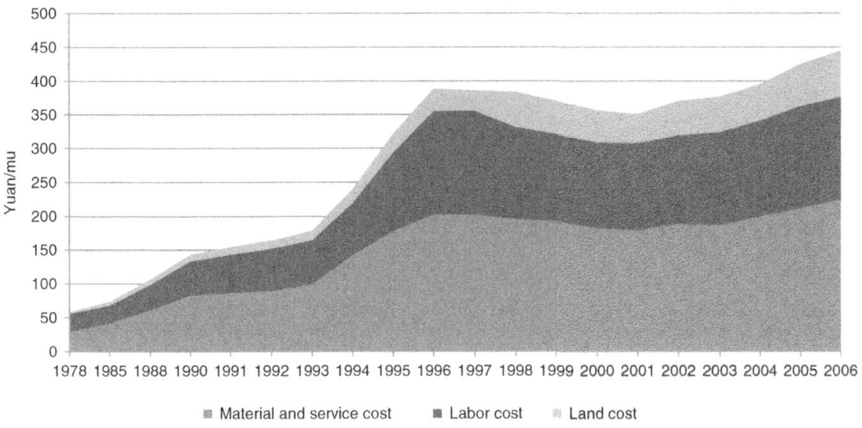

Fig. 8.17 Historical costs for China's food crops (*Source*: Department of Pricing of National Development and Reform Commission 2004, 2010)

8.4.4 Constraints

Regarding the long-term prospect for biofuels, four kinds of constraints need to be considered: those relating to balancing energy flow, capacity, dynamic technical change, and the environment (Fig. 8.19).

8.4.5 Analysis and Evaluation of Development Potential of Biofuels in China

8.4.5.1 Automotive Energy Alternative

As evident in Fig. 8.20, biofuel production will continue to grow in China up to 2050, with the actual supply capacity then being about 32.4–79.7 mtoe. The oil substitution effect is lowest in the M1 scenario; the yield differs from that in L2 by 47.31 million tonnes in 2050. Owing to strict regulation, industrial development of biofuels lacks incentives in the M1 and M2 scenarios. The yield of fuel ethanol in 2020 in M1 and M2 is only 3.82 million tonnes and 4.17 million tonnes, respectively. That cannot achieve the expected target of 10 million tonnes as determined in the Medium- and Long-Term Development Plan for Renewable Energy (2007). In all four scenarios, the yield of biodiesel exceeds three million tonnes, which

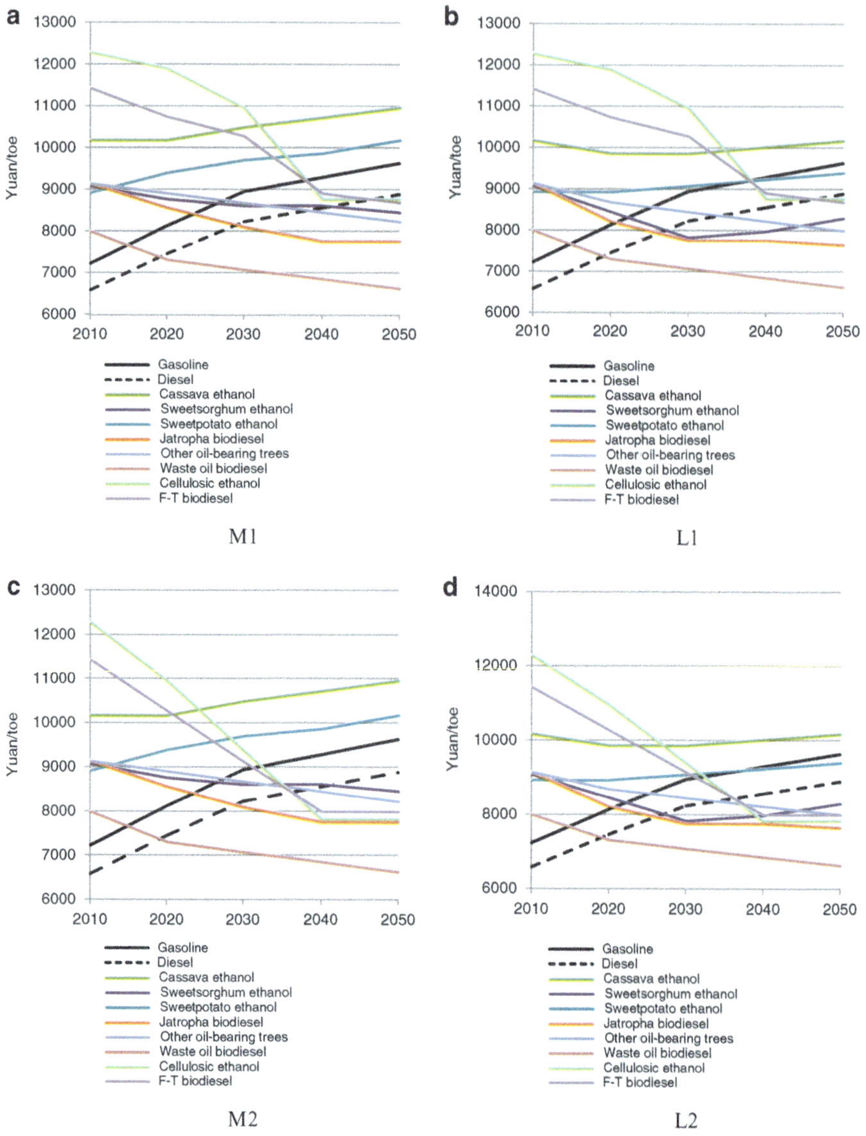

Fig. 8.18 Costs of biofuels

surpasses the identified target of two million tonnes. In the near term, the growth of biofuel production depends largely on the diffusion of cultivation technology of 1.5-generation energy crops, reduction in feedstock cost, and related industry policies. In the medium term, it depends largely on innovation breakthrough with

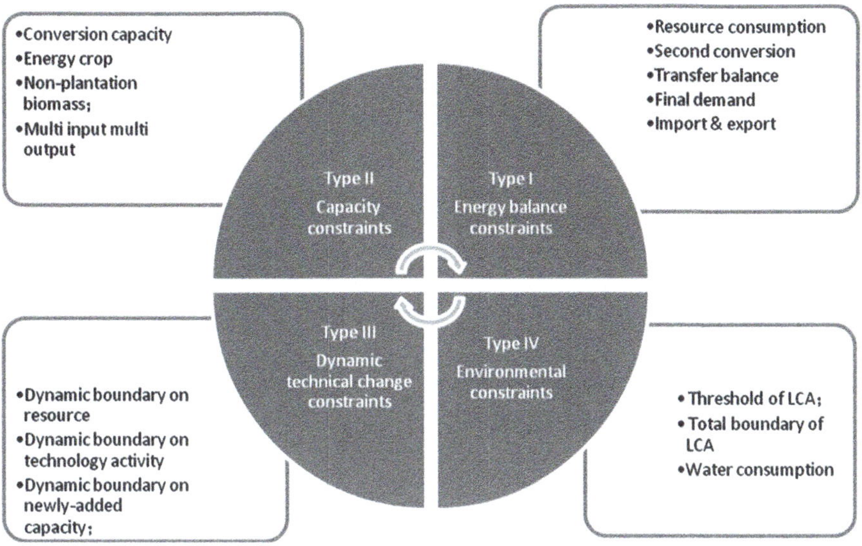

Fig. 8.19 Constraints on biofuel development

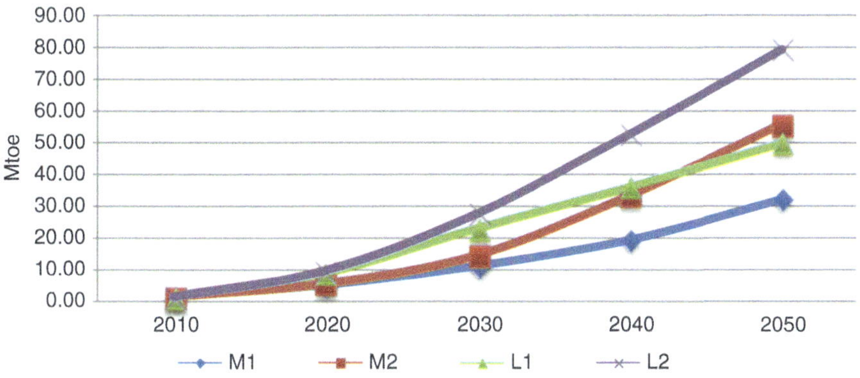

Fig. 8.20 Scenario projections for biofuels

second-generation biofuels. And in the long term, it depends on the biomass resource potential and market potential.

Owing to high compatibility with the existing transport infrastructure and high market demand, the demand and production of biodiesel will continue to rise. It will account for over 50 % after 2030 (Fig. 8.21).

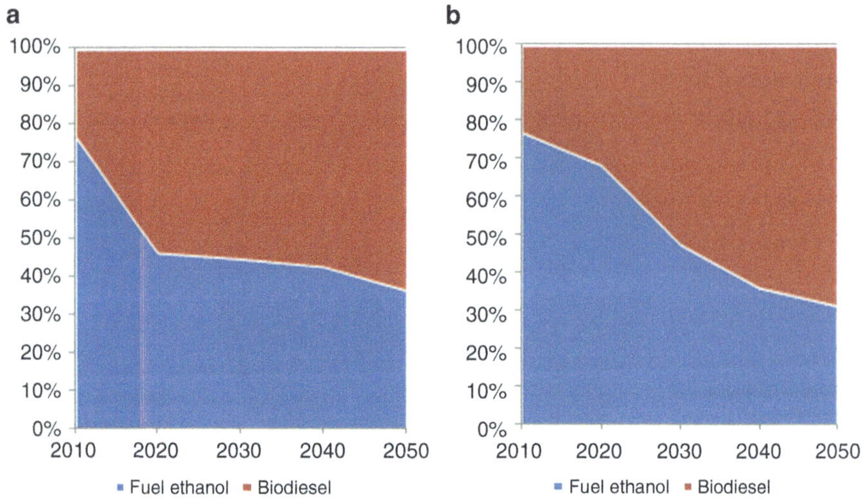

Fig. 8.21 Proportions of fuel ethanol (including cellulose ethanol) and biodiesel (including F-T synthetic diesel) in the (**a**) M1 and (**b**) L2 scenarios (based on energy value)

8.4.5.2 Second-Generation Biofuels

The difference in production under the various scenarios is great; however, the proportion of second-generation biofuels displays a significant upward trend in each scenario. By 2050, the production in three of the four scenarios is greater than 50 % (Figs. 8.22 and 8.23).

8.4.5.3 Feedstock Portfolio

The feedstock portfolio under the various scenarios is different. Biofuels derived from starch and sugar crops will decrease, those derived from oil-bearing trees will account for a large proportion in the medium term, and biofuels derived from agricultural and forestry residues will play an important role after 2030. Biofuels obtained from cellulosic energy crops will gradually increase after 2030 (Fig. 8.24).

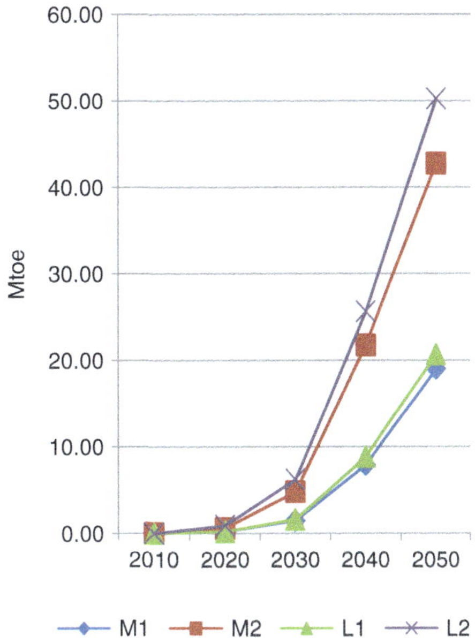

Fig. 8.22 Production of second-generation biofuels

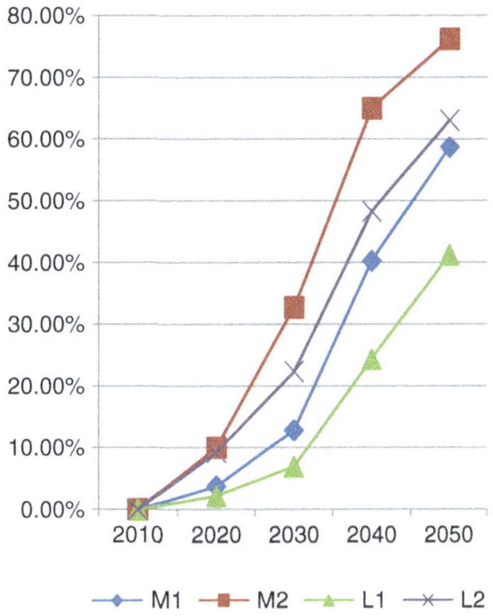

Fig. 8.23 Proportion of second-generation biofuels

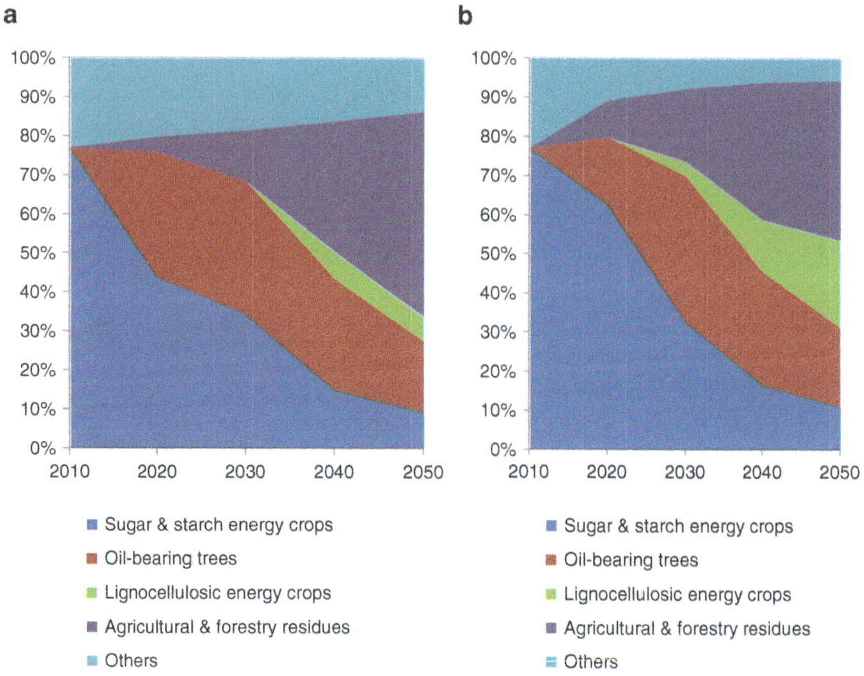

Fig. 8.24 Feedstock portfolio under scenarios (**a**) M1 and (**b**) L2

8.5 Conclusions

8.5.1 Major Conclusions

8.5.1.1 Biofuel Production Will Amount to 32.4–79.7 mtoe by 2050

The scenario analysis shows that biofuel production will continue to grow in China up to 2050, and the actual supply capacity will be about 32.4–79.7 mtoe by 2050. In the near term, the growth of biofuel production depends largely on the diffusion of cultivation technology of 1.5-generation energy crops, reduction in the feedstock cost, and related industry policies. In the medium term, it depends largely on innovation breakthroughs with second-generation biofuels. In the long term, it depends on the biomass resource potential and market potential.

8.5.1.2 Resource Potential

The collectable volume of agricultural and forestry residues will be 1.187 billion tonnes in 2050. Of that, about 535–772 million tonnes can be used for energy production, and about 210–446 million tonnes can be used for liquid biofuel production considering many other uses of biomass. The resource potential of energy crops in 2050 in China amounts to 17.29 EJ based on the direct equivalent method of calculation.

8.5.1.3 Fuel, Technology, and Feedstock Portfolios

Fuel portfolio: Owing to high compatibility with the existing transport infrastructure and high market demand, biodiesel will continue to rise. After 2030, it will consume over 50 % of total biofuel production.

Technology portfolio: The radical technological changes required for 1.5-generation non-grain ethanol and biodiesel fuel are almost impossible to achieve. However, incremental innovation will contribute to cost reduction, such as increased energy crop yield, increased conversion efficiency, and comprehensive utilization of wastewater and solids. China does not have any cost advantage with several mature-market energy crops, such as cassava; growing energy crops, such as sweet sorghum and *Jatropha curcas*, on marginal land will add to land development costs. Therefore, feedstock cost will continue to be a problem for the large-scale development of 1.5-generation biofuels in China over the next 10–20 years, during which time second-generation biofuels are expected to become a revolutionary innovation. The 1.5-generation non-grain biofuel technology will play an expedient role in the long term, and its overall trend will become stable after 2030 as its growth rate slows down. This technology will make a contribution mainly in the period from 2020 to 2040. Second-generation technology will grow during the period from 2020 to 2030, and it is expected that second-generation biofuels will serve as important long-term alternatives.

Feedstock portfolio: The feedstock portfolios are different under the various scenarios. Biofuels derived from starch and sugar crops will decrease; those derived from oil-bearing trees will assume a large proportion in the medium term and those derived from agricultural and forestry residues will play an important role after 2030. Biofuels derived from cellulosic energy crops will gradually become prominent after 2030.

8.5.1.4 Technology R&D and Industrial Policy

China's cautious biofuel policies would appear to be appropriate for the short term. However, it is expected that more supportive policies will be implemented to promote the long-term development of liquid biofuels.

8.5.2 Suggestions

1. Since they are of strategic significance, priority should be given to the future development of second-generation biofuels. It is imperative that China lends greater support to R&D of second-generation technology. Technical support should be extended to cover demonstration projects with scales of over 1,000 or even 10,000 tonnes. Demonstration and subsidy policies urgently need to be formulated and implemented in accordance with China's technology development status.
2. China has a large potential for developing biodiesel. Since biodiesel production from waste oil is quite mature, both technologically and industrially, it is critical that a well-organized recovery management system for waste oil be established. Owing to the important role it will play in the medium term and its rich feedstock resources, there has to be greater policy support for biodiesel derived from oil-bearing trees. An upward adjustment can be made to the biodiesel development target for 2020, and guidance and incentives can promote the utilization of biodiesel in automobiles.
3. Three types of energy crops—non-grain energy crops, oil-bearing trees, and lignocellulose energy crops—will become important in, respectively, the short, medium, and long term. R&D efforts into biofuel production from non-grain energy crops have been conducted over a number of years, and it is now necessary to implement policy support for demonstration and expansion projects. Breeding and cultivating technology for oil-bearing energy plants is not at an advanced state of development. Over the next 10–15 years, therefore, accumulated experience in technological and industrial development needs to lead to the establishment of a sound feedstock cultivation system. Lignocellulosic energy crops will become important in the middle and long term; accordingly, it is suggested that appropriate plans be made in advance and pilot projects promoted.
4. It is suggested that the priority be given to biofuel production using waste oil and agroforestry residues as feedstock. This is less controversial in terms of sustainability and is more compatible with existing industrial and environmental policies.
5. Greater efforts need to be made into analyzing the macro- and microlevel impact of biofuels. At the macrolevel, it is necessary to explore the economic and environmental capacity of biofuel development under various scenarios. It is essential to determine an appropriate boundary between biofuel development and land use. The interaction among biofuel development, energy substitution, and reducing GHG emissions likewise demands investigation. Coupling studies need to be conducted on the complex interaction of energy, economy, the environment, and land use. At the microlevel, it is necessary to promote life-cycle analyses of energy consumption and carbon emissions with biofuel production to facilitate the formulation of sustainable development standards for liquid biofuels.

Appendices

Appendix 8.1 Related Parameters in Estimating Non-plantation Resources

Parameter	Unit	2010	2020	2030	2040	2050
Food	10,000 tonnes	53,830	56,202[a]	58,439[a]	59,572[a]	59,703[a]
Forestry	10,000 ha	19,545.22[b]	22,080[c]	23,040[d]	24,000[d]	24,960[c]
Comprehensive coefficient of agricultural residues	tonnes/tonnes	1.04	1.04	1.04	1.04	1.04
Comprehensive coefficient of forestry residues	tonnes/ha	2.49	2.49	2.49	2.49	2.49
Biomass power capacity using agricultural and forestry residues as feedstock[e]	10 MW	232.00	1,769.00	2,193.00	2,628.00	3,083.00
Biomass briquettes	10,000 tonnes	280.00	5,000.00	8,000.00	8,000.00	8,000.00
Centralized supply of biomass gas	10^3 m^3	456,250.00	4,562,500.00	8,668,750.00	12,775,000.00	6,881,250.00

Notes:

[a]Data from Institute of Agricultural Economics and Development of CAAS (2007)

[b]Seventh Forest Resources Inventory Data.

[c]According to the Action Plan to Address Climate Change of the National Forestry Administration, three-phased target should be achieved: national forest coverage of 20 % by 2010; national forest coverage increasing to 23 % by 2020; stability of forest coverage above 26 % by 2050

[d]Linear interpolation between 2020 and 2050

[e]Including direct fired power generation, gasification, and power generation and IGCC (Integrated gasification combined cycle). Bagasse-fired power generation is not included in the power capacity for 2010

Appendix 8.2 Overview of Studies on China's Marginal Land Resources

Type of marginal land	Area	Overview of study results
Land suitable for agricultural use in the category of unused land	702[a](304.36–2,136)	Research Group of China Renewable Energy Strategy (2008) concluded that land suitable for forestation in the category of unused land was 7.344 million hm^2, based on Survey and Evaluation of Arable Land Reserve Resources (Wen and Tang 2005)
		According to Xie et al. (2007), Marginal Land I includes arable land reserve resources as well as barren hills and wasteland with a slope of less than 15°, accounting for 30 % of the total area, i.e., 21.36 million hm^2
		The Editorial Board of Sustainable Development of Energy Crops in the People's Republic of China (2009) used the data of arable land reserve resources released by the Ministry of Land and Resources as the basis for its studies. The board did not think it appropriate to reclaim sparsely wooded land and shrubbery for crop cultivation owing to the existing national policy on converting farmland to forest. Further, in terms of ecological protection, it was not desirable to reclaim marshland and reed land for agricultural purposes. Therefore, the potential of arable land reserve resources was 7.02 million hm^2, with arable marsh and reed land being excluded
		Zhang et al. (2010) concluded that the land potential for sweet sorghum cultivation in the category of unused land was only 786,000 hm^2, considering such factors as climate, soil, terrain, and land utilization. In this context, 3.0436 million hm^2 was taken as the minimum value of land suitable for agricultural use, which was based on a compromise between the above two research results of the Editorial Board of Sustainable Development of Energy Crops in the People's Republic of China (2009)
		The latest survey result released by the Ministry of Land and Resources in 2011 indicates that the total area of China's continuous arable land reserve resources amounted to 7.3439 hm^2.
Land suitable for forestation in the category of unused land	3,736.06[a] (3,736.06 – 10,000)	According to Lv (2008), roughly 100 million hm^2 of marginal land is unsuitable for agricultural use in China, including moderate and mildly saline-alkaline land, arid and semiarid sandy land, mine and oilfield reclamation land. The potential area in the category of unused land that can be exploited for growing oil-bearing plants amounts to about 24 million hm^2

		The lower value of the land area suitable for forestation in the category of unused land equals the area of usable but unused land (Wen and Tang 2005) minus the area of arable land reserves (Wen and Tang 2005) minus the area of land suitable for forestation in the category of forest land (State Forestry Administration 2009) = 8,874 − 734.4 − 4,403.54 = 3,736.06 (10,000 hm^2)
Land suitable for growing energy crops in the category of wooded land[b] (including oil-bearing plant forest and land suitable for forestation[c]; the potential of firewood forest and shrubbery is estimated based on residues)	4,824.14[a] (4,296.3–6,047)	Based on the 2005 Report on China's Forest Resources, the Research Group of China Renewable Energy Strategy (2008) concluded that the area of oil-bearing trees in China was about 3.43 million hm^2 and that the land area suitable for forestation was 57.04 million hm^2, which amounts to 60.47 million hm^2
		According to Xie et al. (2007), the land mainly consists of barren hills and wasteland with a slope of 15°–25°, cutover land (2.51 million hm^2), burnt land (600,000 hm^2), and sandy land suitable for planting xerophytic shrubbery energy crops (7.003 million hm^2), which amounts to a total of up to 42.963 million hm^2
		According to the Report on Forest Resources in China—Seventh Inventory of National Forest Resources released by the State Forestry Administration—the land suitable for forestation in the category of wooded land was 44.0354 hm^2. Zhang and Lv (2008) concluded that the area of woody oil-bearing forest was 4.206 hm^2. The total is 448.2414 hm^2
		Lv (2008) concluded that the normal land area for oil-bearing plants was about 36 million hm^2, among which 12 million hm^2 was in the category of land suitable for forestation
Land suitable for growing energy plants in the category of arable land (mainly including non-grain land with middle and low production in the category of nonbasic farmland)	2,000[a] (0–2,091)	Based on empirical estimates by the Research Group of China Renewable Energy Strategy (2008), 40 % of 50.24 million hm^2 (20 million hm^2) of low-production arable land in China could be used as marginal land to cultivate biomass feedstock
		Shi (2011) verified the above figures by referring to the Investigation and Assessment of Quality Classification of National Arable Land released by the Ministry of Land and Resources in 2009. That classified the national arable land into 15 levels, with the first level being the best and 15th level the poorest. The last three levels of arable land are naturally poor with low and unstable production; they cover an area of 20.91 hm^2

(continued)

Appendix 8.2 (continued)

Type of marginal land	Area	Overview of study results
Unclassified wasteland for energy crops	2,680	According to the Interim Regulations on the Subsidy for Bioenergy and biochemical Feedstock promulgated by the Ministry of Finance in 2007, feedstock bases had to meet the requirement of not occupying arable land or unused land that had been planned for agricultural purposes. From the perspective of conservation, growing energy plants on any arable land was likewise prohibited. The Ministry of Agriculture carried out a project called Investigation and Assessment of Marginal Land Resources Suitable for Planting Energy Crop Resources with the support of all levels of rural energy management departments. The investigation covered wasteland for energy crops and winter fallow land suitable for growing energy crops. Wasteland for energy crops refers to natural grassland, sparsely wooded land, shrubbery, and unused land suitable for growing energy crops. Results showed that the total area of marginal land suitable for growing energy crops was 34.2 million hm^2, including 26.8 million ha of wasteland and 7.4 million ha of winter fallow field (Kou et al. 2008). Shi (2011) also arrived at the figure of 26.8 million hm^2
Total	11,262.2[a] (8,336.72–20,274)	

Notes:
[a]This value is adopted in this chapter
[b]The definition of "wooded land" comes from the Regulations for the Implementation of Forestry Law of the Peoples' Republic of China, which consists of arbor forest with a crown density above 0.2, bamboo forest, shrubbery, and sparse woodland as well as land suitable for forestry planned by county-level or higher governments, i.e., barren hills and area suitable for forestry, sandy wasteland suitable for forestry, and others
[c]According to Major Technical Regulations on Planning, Design, and Inventory of Forest Resources issued by the State Forestry Administration in 2003, land suitable for forestation consists of barren hills and wasteland, sandy bare land, and others. Barren hills and wasteland suitable for forestation mainly includes barren hills, desolated beaches, and deserted ditches that have been listed in the category of wooded land by county-level or higher governments and that do not meet the standards of forest land, sparse wood land, shrubbery, and afforestation land. Sandy bare land suitable for forestation mainly includes fixed and migratory sandy land (sand dunes) and desertified land that can grow trees and has been listed in the category of wooded land by county-level or higher governments and that fails to meet the standards of forest land, sparse wood land, shrubbery, and afforestation land

Appendix 8.3 Relevant Policy and Measures During the Process Chain

Biofuels	Process	Policies	Source	Application
Sugar and starch energy crops for fuel ethanol	Feedstock	Subsidies for aged grain as feedstock	NDRC et al. Expanding Vehicle-Use Ethanol Gasoline Trial Promotion Schedule and Detailed Rules for Implementation (2004)	Designated projects only
		Subsidies of RMB 180 per *mu* on unused land	MOF. Interim measures on financial subsidies to production bases of feedstock for the bioenergy and biochemical industry (2007)	
	Conversion	Consumption tax is exempted	NDRC et al. (2004)	Designated projects only
		Value-added tax is refunded	NDRC et al. (2004)	Designated projects only
		Specific subsidies to offset extra cost	NDRC et al. (2004)	Designated projects only
		Entry regulation	NDRC. Announcement regarding greater management of bioethanol projects and promoting healthy development of the ethanol industry (2006)	
		Incentive funds for scaling up production of non-grain bioenergy and biochemical processes	MOF. Interim measures on incentive funds for the bioenergy and biochemistry industry (2007)	Has not been implemented
	Distribution	Standards	National standards of Denatured Fuel Ethanol (GB18350-2001)	
			Ethanol Gasoline For Motor Vehicles (GB 18351-2001)	
		Regional mandatory mixture of 10 % in pilot areas	NDRC et al. (2004)	Designated projects only
		Settlement prices of denatured fuel ethanol between producers and distributors guaranteed	NDRC et al. (2004)	Designated projects only

(continued)

Appendix 8.3 (continued)

Biofuels	Process	Policies	Source	Application
		Retail price of ethanol gasoline is the same as that of gasoline.	NDRC et al. (2004)	Designated projects only
Oil-bearing trees for biodiesel	Feedstock	Subsidies of RMB 200 per *mu* on unused land	MOF. Interim measures on financial subsidies to production bases of feedstock for the bioenergy and biochemical industry (2007)	
			SFA. Check Methods of Forest Energy Planting Base 2009	
		Anticipated ratio target for major commercial forest among total forest land	SC. National Plan for Conservation and Utilization of Forest Land (2010–2020) (2010)	
		Sustainable guidelines	SFA. Guideline of Sustainable Planting of Energy Forest and the Guideline of Sustainable Planting of *Jatropha* (2011)	
	Conversion	Incentive funds for scaling up production of non-grain bioenergy and biochemical processes	MOF. Interim measures on incentive funds for the bioenergy and biochemical industry (2007)	Has not been implemented
	Distribution	Standards	Biodiesel Blend Stock for Diesel Engine Fuels (BD100) (GB/T20828-2007)	
			Biodiesel fuel blend (B5) (GB/T 25199-2010)	
		Regional mandatory mixture of 5 % in pilot areas	HPPGO. Work program on biodiesel deployment in Hainan Province (2009)	Designated projects in Hainan Province only
Waste oil to biodiesel	Feedstock	Collection	MEP, MHURD, and NDRC. Notice on improving the collection of municipal waste (2010)	
	Conversion	Value-added tax refunded	MOF et al. Notes on value-added tax policies related to comprehensive utilization of resources (2008)	
		Income tax relief of 10 %	MOF. Notes on list of income tax relief preferential policies for comprehensive utilization of resources (2008)	

		Consumption tax exempted	MOF & SAT. Notes on consumption tax exemption for biodiesel using waste oil as feedstock (2010)	
	Distribution	Standards	Biodiesel Blend Stock for diesel engine fuels (BD100) (GB/T20828-2007)	
			Biodiesel fuel blend (B5) (GB/T 25199-2010)	
Second-generation biofuels	Feedstock	Utilization of agricultural residues and forestry residues encouraged	SC & NDRC. Adjustment list of industry structure 2011 (2011)	
	Conversion	Reward fund for scale up production of non-grain bioenergy and biochemistry	MOF. Interim measures on reward fund for bioenergy and biochemistry (2007)	Has not been implemented

Notes: *HPPGO* Hainan Provincial People's Government Office, *MEP* Ministry of Environment Protection, *MHURD* Ministry of Housing and Urban–rural Development, *MOF* Ministry of Finance, *NRDC* National Reform and Development Commission, *SAT* State Administration of Taxation, *SC* State Council, *SFA* State Forestry Administration

References

Bauen A, Berndes G, Junginger M et al (2009) Bioenergy – a sustainable and reliable energy source: a review of status and prospects. IEA Bioenergy. http://www.ieabioenergy.com/LibItem.aspx?id=6479. Accessed 12 Dec 2011

Bhattacharya SC, Salam AP, Hu RQ et al (2005) An assessment of the potential for non-plantation biomass resources in selected Asian countries for 2010. Biomass Bioenergy 29:153–166

Cai YQ, Qiu HG, Xu ZG (2011) Evaluation on potentials of energy utilization of crop residual resources in different regions of China. J Nat Resour 26(10):1637–1646 (in Chinese)

Campbell JE, Lobell DB, Genova RC, Field CB (2008) The global potential of bioenergy on abandoned agriculture lands. Environ Sci Technol 42:5791–5795

Chang SY, Guo QF, Zhang XL (2009) Report on biomass potential (forestry). China Automotive Energy Research Centre, Tsinghua University, Beijing (in Chinese)

China Renewable Energy Society (2011) Yearbook of new energy and renewable energy in China 2010. Guangzhou Institute of Energy Conversion, Chinese Academy of Science, Guangzhou (in Chinese)

IPCC (Intergovernmental Panel on Climate Change) (2011) Special report on renewable energy sources and climate change mitigation. http://srren.ipcc-wg3.de. Accessed 1 Jan 2012

de Wit MP, Faaij A (2010) European biomass resources potential and costs. Biomass Bioenergy 34:188–202

Department of Pricing of National Development and Reform Commission (2004) Compilation of national agricultural product cost and benefit information. http://www.npcs.gov.cn/web/Column.asp?ColumnId=46&ScrollActio=2 (in Chinese) Accessed 1 Jan 2010

Department of Pricing of National Development and Reform Commission (2010) Compilation of national agricultural product cost and benefit information. China Statistics Press, Beijing (in Chinese)

Department of Rural S&T, Ministry of Science and Technology (2009) Yearbook of China agricultural products processing industries. China Agriculture Press, Beijing

Editorial Board of Sustainable Development of Energy Crops in the People's Republic of China (2009) Sustainable development of energy crops in the People's Republic of China. China Agriculture Pressing, Beijing (in Chinese)

Erb KH, Haberl H, Krausmann F, Lauk C, Plutzar C, Steinberger JK, Muller C, Bondeau C, Waha K, Pollack G (2009) Eating the planet: feeding and fuelling the world sustainably, fairly and humanely – a scoping study institute of social ecology, Potsdam Institute of Climate Impact Research, Vienna

Fan XF, Hou XC, Zuo HT (2010) Effect of marginal land types and transplanting methods on the growth of switchgrass seedlings. Pratacultural Sci 27(1):97–102 (in Chinese)

Fang J, Pu WH, Zhang HJ (2010) The development status of cassava industry at home and abroad. Chin Agric Sci Bull 26(16):353–361 (in Chinese)

Field CB, Campbell JE, Lobell DB (2008) Biomass energy: the scale of the potential resource. Trends Ecol Evol 23:65–72

Food and Agriculture Organization of United Nations (2008) Bioenergy, food security and sustainability: towards an international framework. High-level conference on world food security: the challenges of climate change and bioenergy. Rome, http://www.fao.org/fileadmin/user_upload/foodclimate/HLCdocs/HLC08-inf-3-E.pdf. Accessed 1 Jan 2010

Haberl H, Beringer T, Bhattacharya SC, Erb K-H, Hoogwijk M (2010) The global technical potential of bio-energy in 2050 considering sustainability constraints. Curr Opin Environ Sustain 2:394–403

Hoogwijk M, Faaij A, Eickhout B, de Vries B, Turkenburg W (2005) Potential of biomass energy out to 2100, for four IPCC SRES land-use scenarios. Biomass Bioenergy 29:225–257

Hou XC, Fan XF, Wu JY, Zuo HT (2011) Evaluation for production potentials of bioenergy grasses grown in abandoned sandpits in Beijing suburbs. J Nat Resour 26(10):1768–1774 (in Chinese)

Huang J, Li KM, Ye JQ, Qin HL, Shao NF, Chen MY (2008) A summary review of dominant regions of cassava growing in China. Guangxi Agric Sci 39(1):104–108 (in Chinese)
IEA (2008) Energy technology perspectives 2008: scenarios & strategies to 2050. OECD/IEA, Paris
IEA (2010) Status of 2nd generation biofuels demonstration facilities in June 2010. http://www.ascension-publishing.com/BIZ/IEATask39-0610.pdf. Accessed 1 Jan 2010
IEA (2011) Technology roadmap: biofuels for transport. www.iea.org. Accessed 1 Jan 2012
Institute of Agricultural Economics and Development of CAAS (2007) China agricultural policy analysis and decision support study (II). Science Press, Beijing (in Chinese)
Japan Institute of Energy (2007) Manual of biomass and bioenergy. Chemical Industry Press, Beijing
Jin SY, Sun SF, Song AP, Zhang FQ et al (2011) Analysis of raw material resources for non-grain fuel ethanol in China. Sino-Global Energy 16:40–45 (in Chinese)
Kou JP, Bi YY, Zhao LX, Gao CY, Tian YS, Wei SY, Wang YJ (2008) Investigation and evaluation on wasteland for energy crops in China. Renew Energy Resour 26(6):3–9 (in Chinese)
Li SZ, Chan-Halbrendt C (2009) Ethanol production in China: potential and technologies. Appl Energy 86:S162–S169
Li JF, Hu RQ, Song YQ et al (2005) Assessment of sustainable energy potential of non-plantation biomass resources in China. Biomass Bioenergy 29:167–177
Liu Q, Zhai GW, Jiang BX et al (2011) Fuel. In: Liu TN (ed) Report on China's energy development for 2011. Economic Science Press, Beijing
Lv W (2008) Forest energy resource and potential analysis. In: Zhang XL, Lv W (eds) China forest energy. China Agriculture Press, Beijing (in Chinese)
Ministry of Agriculture (2007) Agricultural bioenergy industry development plan, 2007–2015. http://www.moa.gov.cn/zwllm/ghjh/200808/t20080826_1168529.htm. Accessed 17 Jan 2013
Mogensen J (2009) 2010.2015 Production cost of cellulosic ethanol in China. In Novozymes. Current status and projection of cellulosic ethanol in China. Novozymes, Danmark
National Bureau of Statistics of China (2011) China statistics abstract 2011. China Statistics Press, Beijing (in Chinese)
OFID/IIASA (2009) Biofuels and food security: implications of an accelerated biofuels production. http://www.ofid.org/PUBLICATIONS/SpecialPublications.asp. Accessed 17 Jan 2013
Ralph EHS, Mabee W, Saddler JN, Taylor M (2010) An overview of second generation biofuel technologies. Bioresour Technol 101:1570–1580
Research Group of Biomass Resource Strategy of Chinese Academy of Science (2009) China's science and technology roadmap of biomass resource to 2050. Science Press, Beijing (in Chinese)
Research Group of China Renewable Energy Strategy (2008) Comprehensive report on China's renewable energy development. China Electric Power Press, Beijing (in Chinese)
Ros M, Jeeninga H, Godfroij P (2006) Policy support for large scale demonstration projects for hydrogen use in transport. ECN, Petten
Shi YC (2011) China's resource of biomass feedstock. Eng Sci 13(2):16–23 (in Chinese)
Sims REH, Hastings A, Schlamadinger B, Taylor G, Smith P (2006) Energy crops: current status and future prospects. Glob Change Biol 12:2054–2076
Sims REH, Mabee W, Saddler JN et al (2010) An overview of second generation biofuel technologies. Bioresour Technol 101:1570–1580
Smeets EMW, Faaij APC, Lewandowski IM, Turkenburg WC (2007) A bottom-up assessment and review of global bio-energy potentials to 2050. Prog Energy Combust Sci 33:56–106
State Council (2008) National medium- and long-term planning framework of food security, 2008–2020. http://www.gov.cn/jrzg/2008-11/13/content_1148414.htm. Accessed 17 Jan 2013
State Council (2011) Notice of national audit opinion on annual deforestation limits during the twelfth Five-Year Plan (2011). http://www.chinaacc.com/new/63_73_201112/07ya1612616082.shtml. Accessed 17 Jan 2013
State Forestry Administration (2009) China forest resource report – seventh national forest resources inventory. China Forestry Publishing House, Beijing (in Chinese)

State Forestry Bureau (2009) Forestry action plan to address climate change. http://www.forestry.gov.cn//portal/xby/s/1300/content-127583.html. Accessed 17 Jan 2013
Tian YN, Lu SQ, Fu HT et al (2010) Development potential of cassava based bioenergy in China. J Anhui Agric Sci 38(28):15763–15766 (in Chinese)
van Vuuren DP, Van Vliet J, Stehfest E (2009) Future bio-energy potential under various natural constraints. Energy Policy 37:4220–4230
Wang XH (2011) Current status of biodiesel production in forestry sector. Seminar on key conversion technologies and industry development of biodiesel in China 2011, Beijing (in Chinese)
Wang GS, Lv W, Liu JL, Wang SS, Lv Y, Wang GT, Xu JQ, Wang Y (2006) Survey of China forest energy resource cultivation and development potential. China For Ind 1:12–21 (in Chinese)
Wang ZY, Ren DM, Gao H (2010a) The renewable energy industrial development report 2010. Chemical Industry Press, Beijing (in Chinese)
Wang ZY, Zhao YQ, Zhang ZM et al (2010b) Biofuel industry in China: strategy and policy. Chemical Industry Press, Beijing (in Chinese)
WBGU (2009) Future Bioenergy and Sustainable Land Use. Earthscan, London
Wen MJ, Tang CJ (2005) Potential arable land in China. China Land Press, Beijing (in Chinese)
Xiao MS, Yang JX (2006) Showcase project: ethanol production through solid fermentation of sweet sorghum stalks. Trans CSAE 22(Sup.1):207–210 (in Chinese)
Xie GH (2011) An overview on classification and utilizations of energy plants. J China Agric Univ 16(2):1–7 (in Chinese)
Xie GH, Guo XQ, Wang X, Ding RE, Hu L, Cheng X (2007) An overview and perspective of energy crop resources. Resour Sci 29(5):74–80 (in Chinese)
Xie GH, Han DQ, Wang XY, Lu RH (2011a) Harvest index and residue factor of cereal crops in China. J China Agric Univ 16(1):1–8 (in Chinese)
Xie GH, Wang XY, Han DQ, Xue S (2011b) Harvest index and residue factor of non-cereal crops in China. J China Agric Univ 16(1):9–17 (in Chinese)
Zhang XL, Lv W (2008) China forestry energy. China Agriculture Press, Beijing (in Chinese)
Zhang D, Zhang FZ, An LP, Liu LM (2004) Potential economic supply of uncultivated arable land in China. Resour Sci 26(5):46–52 (in Chinese)
Zhang CX, Xie GD, Li SM et al (2010) The productive potentials of sweet sorghum ethanol in China. Appl Energy 87:2360–2368
Zhao YQ, Shi JL, Gao H (2011) Current status, prospects and policy suggestions of renewable energy development in China. China Energy 33(4):5–9, in Chinese

Chapter 9
Electrical Energy for Vehicles

Zhu Guiping, Lu Zongxiang, and Wang Zanji

Abstract Current status of the power industry in China is presented, including the installed capacity and energy mix, power grid construction, and power consumption.

Based on historical data, the installed capacity and power consumption in China are forecasted up to 2050. It is evident that electricity demand for electric vehicles charging can certainly be supplied.

Thereafter, different charging modes and their corresponding power demand are analyzed and calculated. The power rush caused by rapid charging with DC voltage cannot be neglected; thus, orderly charging is required.

The interaction between electric vehicles and the power grid is examined in detail. Electric vehicles charging will inject harmonic currents into the grid; on the other hand, onboard batteries of electric vehicles can serve as a reserve capacity for power grid through vehicle-to-grid technology.

Brief introduction of national standards and policies related to electric vehicles is presented. Conclusions and comments are made about the development of electric vehicles in China.

Keywords Electric vehicle • Power grid • Charging mode • Power quality

Research and development of electric vehicles has emerged as an important trend in the automobile industry. As a new power load, electric vehicles exert considerable influence on the operation and development of power grid and result in a number of non-negligible technical and economic problems. The charging of electric vehicles has both a beneficial and a detrimental impact on the power grid (Chen et al. 2008; Lei and Liu 2000; Zou 2008). The beneficial impact is that charging the vehicle battery at the valley load can reduce the difference between the peak and valley

Zhu Guiping (✉) • Lu Zongxiang • Wang Zanji
Electrical Engineering Department, Tsinghua University, Beijing 100084, P.R. China
e-mail: gpzhu@tsinghua.edu.cn

CAERC, Tsinghua University, *Sustainable Automotive Energy System in China*,
DOI 10.1007/978-3-642-36847-9_9, © Springer-Verlag Berlin Heidelberg 2013

of the grid, improve utilization of the power system's facilities, and expand the terminal power consumption market. The detrimental impact is that as a form of nonlinear power load, the vehicle battery charger unit will inject harmonic current into power grid; this will result in a low power factor, which is disadvantageous to power quality. Charging in the peak-load period will increase the burden on power supply. Therefore, it is necessary to further examine the mutual effect between electric vehicles and the power system with respect to the development of these vehicles and assuring that the power system is properly planned and developed.

This chapter covers the present situation of the development of power system in China. It makes forecasts over a number of years and discusses a number of key factors that will influence this development; it also analyzes the basic factors that will limit the expansion of electric vehicles with respect to electric power supply. The possibility of electric energy as an alternative to fossil energy for vehicles was addressed and feasible technological routes were forecasted in this chapter.

9.1 Prediction for Power Development and Analysis of Supply and Demand of Electricity for Electric Vehicles

9.1.1 Analysis of the Current Situation and Development Trends for Power System in China

9.1.1.1 Power Supply

The installed capacity and power generation in China has ranked second in the world for the past 14 years. Table 9.1 presents the total installed capacity and power generation in China for 2007–2010.

By the end of 2010, the installed capacity in China was 962.19 million kW. Of that amount, 213.40 million kW came from hydropower (22.2 %), 706.63 million kW from thermal power (73.4 %), 10.82 million kW from nuclear power (1.1 %), and 31.07 million kW from wind power (3.2 %). Figure 9.1 shows the composition of the installed capacity in China for 2010. In that year, total power generation in China was 4,228 billion kW · h, which ranked the country second in the world.

Table 9.1 Total installed capacity, power generation, and rate of increase in China, 2007–2010

Year	2007	2008	2009	2010
Total installed capacity (MW)	713,290	792,530	873,948	962,190
Net increase in installed capacity (MW)	91,290	74,314	81,418	88,242
Increase over the previous year (%)	14.48	10.35	10.27	10.10
Total power generation (100 million kW · h)	32,559	34,334	36,639	42,280

Source: China Electricity Council (2011)

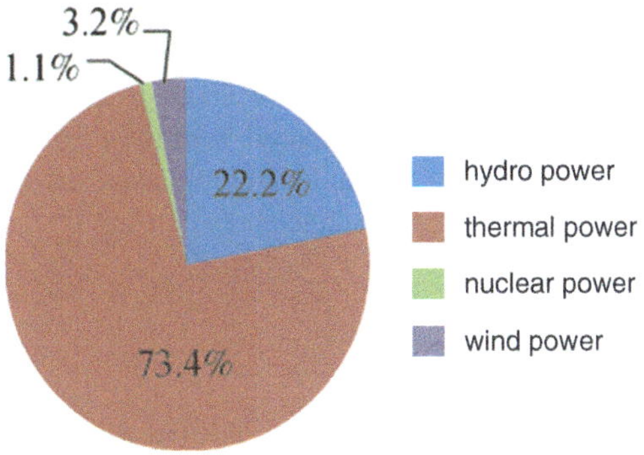

Fig. 9.1 Composition of the installed capacity in China in 2010

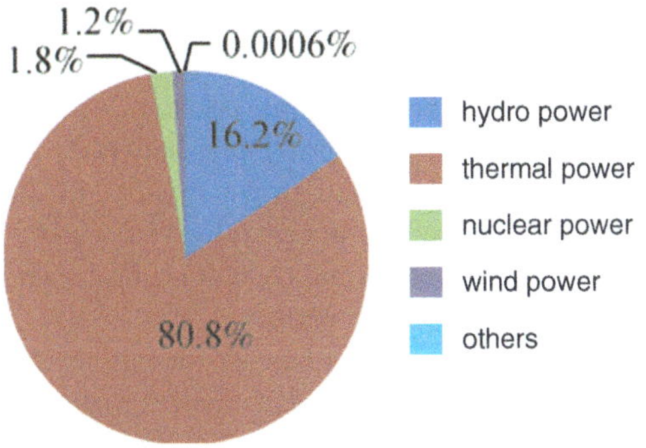

Fig. 9.2 Composition of power generation in China in 2010

Of that amount, 686.3 billion kW · h was generated by hydropower, 3,414.5 billion kW · h by thermal power, 76.8 billion kW · h by nuclear power, 50.1 billion kW · h by wind power, and 277 million kW · h by other sources (including photovoltaic energy). Figure 9.2 presents the composition of power generation in China for 2010.

At present, China gets its electricity mainly from coal-fired power plants, and followed by hydropower. There is a small proportion of other renewable energy sources as well as gas and oil. This complies with the natural resources endowment in China: the country is rich in coal, but poor in oil and gas. Electricity generation is also the best way to use coal.

In light of development trends, the energy mix for power generation in China is undergoing continual optimization, and the proportion of clean energy is constantly increasing. Thermal power generation is developing in the direction of large-capacity, high-efficiency, environment-friendly units. The installed proportion of hydropower is improving, and China has become the country with the largest hydropower capacity in the world. The installed capacity of non-fossil energy is increasing.

However, excessive proportions of thermal power and hydropower and high proportion of hydropower and wind power currently exist in some areas of China. Although the power supply has been sufficient following the global financial crisis in 2008, the imbalance among various resources in the energy mix in power generation is becoming acute. The difficulties in integrating wind power and its market utilization increase in winter in North China, Northeast China, and Northwest China. In the second half of 2009, hydropower output significantly decreased in southwestern areas compared with the previous year, and the installed capacity of thermal power was unable to compensate the gap in hydropower supply. Thus, there is an outstanding imbalance between supply and demand.

9.1.1.2 Power Grid

Regarding transmission and substation facilities, the length of transmission lines of 220 kV and above throughout the country was 399,400 km by the end of 2009—an 11.29 % increase over the previous year. In the same year, the capacity of transformer equipment of 220 kV and above was 1.762 billion kV·A—a 19.40 % increase over the previous year. In the power grid, the trans-regional, trans-provincial, and provincial backbone network of 500 kV and above AC/DC transmission lines has grown rapidly. In 2009, the loop length and transformer capacity showed an increase by 16.64 and 25.97 %, respectively, compared with 2008. At present, China's power grid is larger than that of the United States, thereby ranking it first in the world.

China's power grid has grown rapidly, and it is undergoing further expansion. Six regional power grids have been developed; hydropower transmission from the Three Gorges project, thc West-East electricity transmission project, and receiving-end networks throughout the country have been utilized as their backbone. Interconnection at different scales has been achieved among regional power grids throughout China. As the trans-regional power grids expand and the voltage level further improves, there is enhanced optimization of resources.

However, the transmission capacity of the power grid is relatively low, network power loss is somewhat high, and the distribution network also needs to be improved. At present, the transmission capacity of 500 kV long-distance transmission lines is limited by the transient stability limit: most of them are 0.6–1 million kW, which is 0.4–0.8 million kW lower than the advanced levels abroad. The network power loss rate in 2010 was 6.49 %, which is 1–2 % higher than the advanced levels in the world; that is equivalent to a yearly loss of 37–65 billion kW·h. With respect to distribution network, the structure of some receiving power grids is not perfect, and power supply distance of the distribution network in receiving areas is too long. In addition, there exist many transformers with high power loss and the capacity and regulating ability of reactive compensators is low; this affects the power supply capacity and voltage quality, resulting in a high power loss in transmission lines. Distribution network construction has been slower than power grid for transmission for many years. Thus, weak distribution network is the bottleneck in power grid.

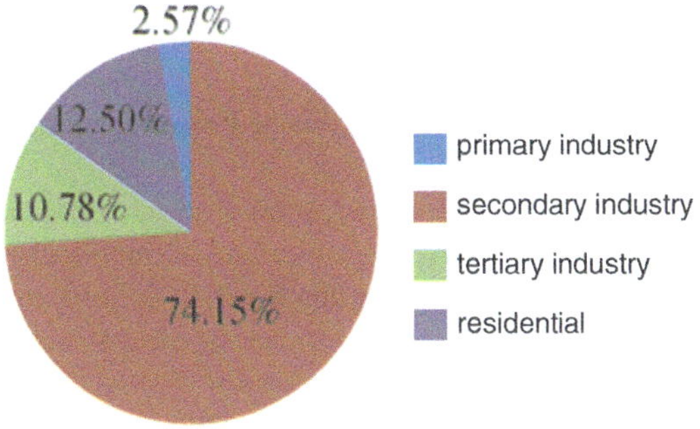

Fig. 9.3 Proportion of total electricity consumption in China in 2009 (*Source*: China Electricity Council 2010)

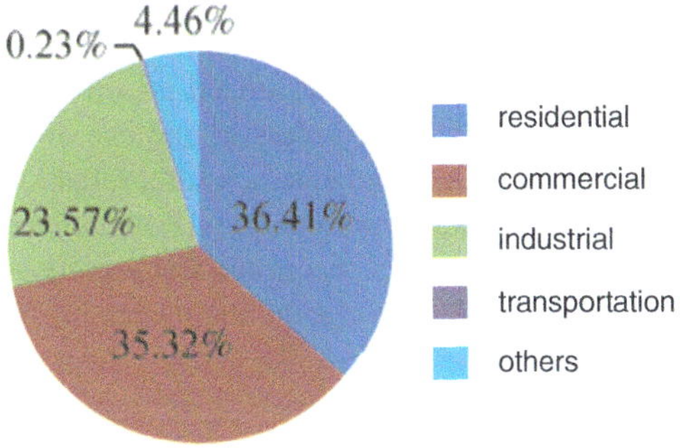

Fig. 9.4 Proportion of total electricity consumption in the United States in 2009 (*Source*: U.S. Energy Information Administration 2009)

9.1.1.3 Electricity Consumption

In 2009, China's total electricity consumption was 3.6596 trillion kW · h—a 6.44 % increase in amount and 0.95 % increase in growth speed over 2008. The electricity consumption by China's primary, secondary, and tertiary industries and individual consumers in 2009 was 94 billion kW · h, 2.7137 trillion kW · h, 394.4 billion kW · h, and 457.5 billion kW · h, respectively; this was a respective increase of 6.90, 4.69, 12.74, and 12.08 % over 2008 levels and accounted for 2.57, 74.15, 10.78, and 12.50 % of total consumption. Those proportions are presented in Fig. 9.3. In 2009, China's industrial electricity consumption was 2.6755 trillion kW · h. That was a 4.60 % increase over 2008 levels and accounted for 73.11 % of total consumption; the latter figure was a 1.29 % reduction from the 2008 figure. In 2009, heavy industry accounted for 82.67 % of industrial electricity consumption—an increase of 0.55 % from 2008. By way of comparison, Fig. 9.4 presents details of consumption in the United States in 2009.

From Figs. 9.3 and 9.4, it is evident that the structure of China's electricity consumption was quite different from that of the United States in 2009. In China, industrial electricity consumption accounted for approximately 75 %, whereas in American industrial, commercial, and residential consumption, each accounted for about one-third of electricity consumption.

9.1.2 *Prediction of Power Development in China*

9.1.2.1 Situation Analysis of Power System Development

With the introduction and rapid promotion of smart power grids, building a cleaner, more efficient, safe, reliable, and economical new generation of power grid has become a global technology trend. In Europe and North America, smart power grid strategies have been developed. China has also established the goal of building a smart, efficient, and reliable power grid system covering both urban and rural areas. A smart grid is an important platform for implementing a new energy strategy and optimizing the allocation of energy resources. This type of grid embraces power generation, transmission, transformation, distribution, consumption, and scheduling; it employs advanced information technology and material technology to achieve large-scale access to and use of clean energy, improve the efficiency of energy utilization, and ensure a safe, reliable, quality power supply.

Climate change and energy crises demand improved use of clean energy. With the development of its economy and society as well as its rapid industrialization and urbanization, China is facing increasing pressure to bring about energy conservation and emission reductions. To actively respond to climate change, ensure a reliable energy supply, achieve sustainable development of its power industry and environment, and meet the power demands of a rapidly developing economic society, China has to expand its use of clean energy. In addition, the following are necessary steps: improving transmission capacity of power lines and flexibility of power grid operation control; improving the ability of resource optimization and against failure to various serious problems with such optimization; meeting requirements for convenient access and quit for all power sources and users; and achieving a harmonious interaction among the power supply, the power grid, and user resources. The rapid development of new energy requires the following: building up the power grid infrastructure; improving the adaptability and acceptability of the grid; eliminating intermittence and uncertainty with large-scale power generation by wind, solar, and other renewable energy resources; and ensuring the large-scale access and long-distance transmission of the power.

Because of the imbalanced distribution of energy resources, optimal allocation in wide area and intensive development of energy base need to be achieved. Reverse distribution has long been a problem with China's primary energy with respect to the place of production and areas of main consumption. With the country's rapid economic development, there is a great distance between the place of energy production and that of consumption, which has resulted in the large-scale movement of energy across regions. This has necessitated the building of large-capacity transmission lines and transmitting power over long distances from western and northern regions of China to central, eastern, and southern regions. Thus, there has been continuous demand for long-distance, large-capacity power transmission as part of China's energy development.

Increasingly, limited land resources have added to the difficulties of developing the power grid. As China's economy continues to expand and land resources become ever more restricted, the corridor for transmission lines will narrow and engineering costs will rise sharply. At the same time, social development will call for stricter requirements regarding environment protection. To preserve precious land resources, narrow the corridor for transmission lines, and reduce transmission power losses and costs, it will be necessary to build strong super- or ultra-high-voltage grids and develop large-capacity AC/DC transmission technology. Achieving sustainable development of China's power industry will also demand the increased use of large cross-section cables, compact transmission lines, multiple circuits in a single tower, large-capacity transformers, and series capacitance compensation, in addition to other advanced technologies.

9.1.2.2 Medium- and Long-Term Forecasts

There are three stages in the development of a power system. The first is the building and initial development stage, which is characteristic of small-scale power grids, largely isolated, urban power grids, and small generator sets. The second stage is that of modern industrialization, which is characterized by large units, large power plants, large systems, high voltage, intensive capital investment and technology, and an advanced information technology. The third stage is that of sustainable development, and its goal is harmonious, sustainable development and cleanliness. Conventional power generation requires cleanliness and efficiency, and new modes of power generation use high technology to achieve overall efficiency of power generation, environmental protection, and economic production. Establishing a safe, economic, efficient, clean, low-carbon, flexible, smart, and sustainable power grid is an important objective at this third stage.

According to the General Report on Strategy Research and Consulting Project of Medium- and Long-Term (2030, 2050) Energy Development in China (China Energy Medium- to Long-Term Development Strategy Research Project Group 2011), the country is going to be facing an energy shortage in the future. Controlling energy consumption and implementing an ecological, green, and low-carbon energy strategy is an inevitable option. Social and environmental pressure will obstruct the development of power technology. Therefore, to attain social and economic development goals at the national level, it will be necessary to bring about leapfrog development of the power industry by using high technology to improve efficiency and slow down the increase in the installed capacity. Using model prediction, medium- and long-term prediction results for the installed capacity in China were obtained (Table 9.2). The prediction results for power consumption by different sectors are presented in Table 9.3.

Table 9.2 Medium- and long-term prediction results of the installed capacity in China (unit: 100 million kW)

Year	2020	2030	2040	2050
Coal	7.23–7.62	6.34–6.70	6.01–7.11	6.09–6.80
Oil	0.13	0.11–0.12	0.10–0.11	0.09–0.10
Gas	0.54–0.57	1.13–1.30	1.61–1.67	1.76–1.92
Hydropower	2.81–2.95	4.25–4.51	4.75–4.85	4.80–4.89
Nuclear	0.83–0.87	2.14–2.25	3.33–3.55	4.29–4.63
Wind	1.46–1.54	2.96–3.16	3.74–4.17	4.10–4.88
Solar	0.24–0.25	0.78–1.07	2.18–2.23	4.32–5.06
Biomass	0.18	0.21–0.23	0.21–0.23	0.20–0.23
Total	13.41–14.11	17.93–19.34	21.99–23.85	25.82–28.35

Source: China Energy Medium- and Long-Term Development Strategy Seminar (2011)

Table 9.3 Prediction results of electricity power consumption by sector (unit: 1 billion kW)

Year	2020	2030	2040	2050
Agriculture	144–157	176–208	194–250	208–280
Industry	3,278–3,386	3,813–4,137	4,394–4,891	5,034–5,499
Building	52	52	54	55
Services	808–909	1,244–1,257	1,618–1,634	1,860–1,880
Urban	462–508	740–846	903–1,066	1,047–1,257
Rural	254	326	369–379	391–416
Transportation	59	186	279	344
Power	571–605	752–808	908–993	1,048–1,135
Total	5,627–5,929	7,304–7,809	8,744–9,519	10,033–10,823

Source: China Energy Medium- and Long-Term Development Strategy Seminar (2011)

Table 9.4 Medium- and long-term forecasts of electricity consumption by electric vehicles (in 100 million kW · h)

Year	2020	2030	2040	2050
Reference scenario	0.215	0.361	0.515	0.644
Comprehensive alternative policy scenario	0.535	1.240	2.220	2.926

9.1.3 Analysis of Electricity Supply and Demand for Electric Vehicles

The total amount of energy consumed by electric vehicles in the reference scenario and comprehensive alternative policy scenario was estimated, and the results appear in Table 9.4. This was carried out according to the predicted installed capacity of power generation in China given in Table 9.2 and the electric vehicle ownership in the corresponding scenarios in Chap. 12.

Comparing the data in Tables 9.3 and 9.4, it is evident that electricity consumption by electric vehicles is very small, and it has no significant impact on the balance of power supply and demand in either scenario.

9.2 Comparison of Charging Modes for Electric Vehicles and Their Respective Power Demand

9.2.1 Three Typical Modes of Electric Replenishment

A typical electric vehicle battery charging circuit is shown in Fig. 9.5.

In Fig. 9.5, u_a, u_b, and u_c signify three-phase voltages; i_a is the charging power of phase A; u_d and i_d are the DC voltage and current after the three-phase rectifier; R_f, L_f, and C_f signify inductance and capacitance parameters of the filter; I_I is the input current of the power converter after filtering; U_B is the input DC voltage of the power converter; and U_0 and I_0 are, respectively, the actual output charging voltage and charging current. Regarding present development of power electronics technology, it is not difficult to achieve such a charging circuit (Zhang et al. 2011). Thus, decisive factors of electric vehicles charging modes are reasons other than the charging equipment itself.

At present, three main modes of replenishment for electric vehicles are commonly used around the world. They are normal charging with a low current, rapid charging with a high current, and battery replacement (also known as mechanical charging) (Yin and Wang 2010).

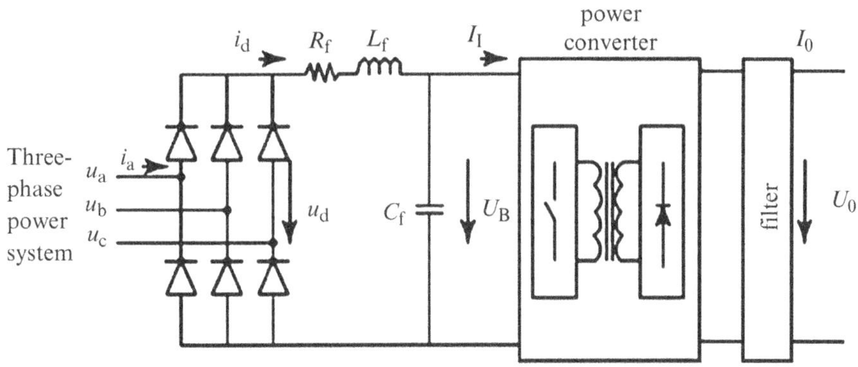

Fig. 9.5 Vehicle battery charging circuit

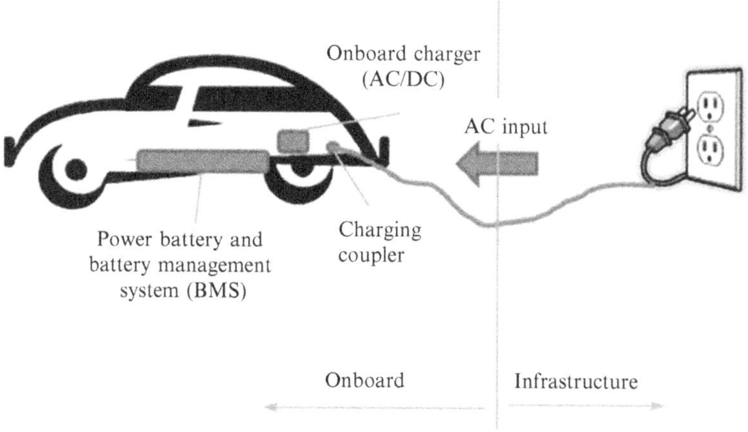

Fig. 9.6 Normal charging by an onboard charger

9.2.1.1 Regular Charging

The current used for normal charging is generally low, about 0.1 C, and multistage (constant-voltage, constant-current) charging is usually adopted. With this mode, the charging time is in the range of 5–8 to 10–20 h depending on battery size.

Chargers used in normal charging are usually an onboard type or they are connected to an external infrastructure. If onboard chargers are adopted, as depicted in Fig. 9.6, the construction costs for infrastructure are relatively low since it is necessary to provide only a normal AC plug. The automaker provides the AC/DC conversion device used to charge the batteries with the car. Normal charging is also called AC charging. The onboard charger power is usually small, not exceeding 3–5 kW, and it is generally applicable to small electric vehicles with a low battery capacity. When the AC/DC conversion device is installed on charging posts or at charging stations that provide the DC source directly to the vehicle battery, this is called off-board charging (Fig. 9.7); it is also known as DC charging. Off-board charging devices may have larger power and are usually used for medium-sized electric vehicles with larger battery capacities (Fig. 9.8).

Normal charging is characterized by a low charging current and low charging power. This charging has little impact on the power grid and is beneficial in terms of extending battery life cycles. The use of normal charging during load valleys can also effectively reduce charging costs. The main disadvantage with normal charging is the long charging time and the necessity of a parking space when charging; this mode is also unsuitable for emergency charging.

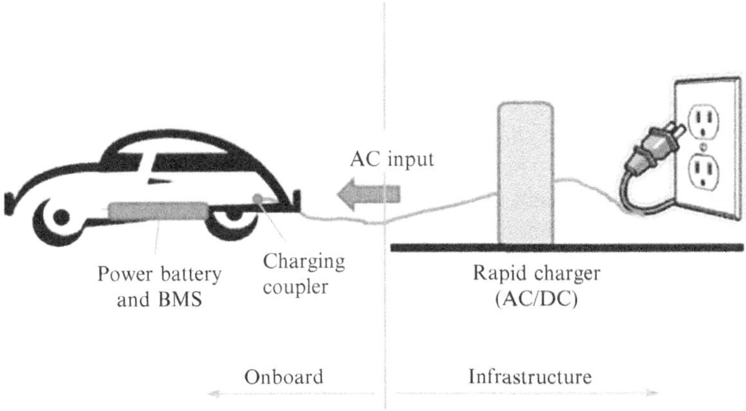

Fig. 9.7 Normal charging at a charging station with an off-board charger

Fig. 9.8 Electric buses being charged with off-board chargers in Beijing

9.2.1.2 Rapid Charging

The long charging time with normal charging has largely restricted the widespread use of electric vehicles. Rapid battery-charging technology seeks to solve this problem by employing off-board chargers at charging stations, which provide strong support for the commercial application of pure electric vehicles. The setup for rapid charging is the same as that shown in Fig. 9.7. However, with rapid charging, electric vehicles are charged with large currents of about 150–400 A, and batteries can be filled to 60–80 % of their capacity in 20 min to 2 h. For example, medium-sized electric vans in Bordeaux, France, can be charged to 90 % of their battery capacity within 25 min by rapid charging, and this has led to the widespread use of these vehicles in that city.

A pulse-type charging current is commonly adopted in rapid charging for depolarization during the charging. It should be noted that the various advantages of pulse-type charging are heavily dependent on the pulse parameters. However, it is usually difficult to obtain the most appropriate charging pulse. This may result in poor control of the charging current pulse and lead to compromised charging effects—sometimes even to such negative effects as battery damage.

At present, the main difficulties in promoting the use of rapid charging are as follows:

1. Significant impact on the battery life cycle.
2. Poor interoperability of rapid charging devices with different types of batteries owing to a lack of standards.
3. The higher cost of rapid charging equipment compared with normal low-current charging as well as the costs of operation and maintenance.
4. The output current with rapid charging changes dramatically in the working process, with a peak current of up to hundreds of amps. Thus, a power rush is usually generated with rapid charging. This is particularly the case when a large number of electric vehicles are being rapidly charged simultaneously, and this presents a great challenge to the stability and bearing capacity of the power grid.

The rapid charging mode is thus indispensable for the popularization of electric vehicles, though orderly charging is essential to avoid a catastrophic power rush imposed on the grid. This charging mode should be positioned as a necessary supplement to normal charging in cases of emergency.

9.2.1.3 Rapid Replacement of Battery Packs

The rapid replacement of battery packs—also known as battery swapping or mechanical charging—is done by replacing exhausted battery packs in vehicles with newly charged battery packs, thereby "filling" the cars. Such battery packs need to be replaced quickly and serviced by special personnel with appropriate tools considering the heavy weights and advanced technologies used in the batteries. The electric buses operated during the 2008 Beijing Olympic Games employed such a complementary power approach, as shown in Fig. 9.9.

Battery swapping offers a number of benefits as follows. Owners of electric cars may rent removable batteries to reduce the initial purchase cost. The electric supplement is achieved rapidly. The replaced batteries may be charged during off-peak hours of the grid to reduce charging costs. Quick professional charging and maintenance are conducive to promoting battery life. With a great many battery packs, battery swap stations could also be used as energy-storage stations and participate in grid frequency regulation, peak shaving, and smoothing power fluctuations generated by renewable energy.

Fig. 9.9 Workshop for changing batteries at a charging station for buses during the Beijing Olympic Games

Several problems remain to be solved in promoting the above charging modes. These include the following: standardization and serialization of car batteries; improving the performance of batteries designed for electric vehicles; setting up charging stations and management of the network; managing the distribution of standardized batteries; accurate control of battery energy; quantitative evaluation of battery quality; and appropriate commercial modes of operation.

9.2.1.4 Comparison of Modes of Energy Supplement for Electric Vehicles

The modes of energy supplement largely depend on the future marketing position of electric vehicles. Appropriate modes need to be designed and selected, taking into account the operating modes and technical features of the electric vehicles.

Vehicles that are able to run a whole day on a single charge (such as urban sanitary vehicles and engineering vehicles) could be filled by means of normal charging in the night hours. Other vehicles (such as buses, sightseeing vehicles, and taxis) should be supplemented by either rapid charging or battery swapping to meet their energy demands. Although the travel routes of private, government, and corporate vehicles vary greatly, these cars could also be charged during off-peak hours of the grid at charging stations set up in residential areas or parking lots since these vehicles are usually stationary at night.

The needs at different stages of electric vehicle development also have to be taken into consideration when selecting charging modes. In the early stages of development, unmanned regular charging devices (chargers and charging posts) or rapid charging stations could be installed in public parking lots (parking spaces) or in residential parking areas. Drivers would be able to charge their cars quickly or during certain appointed periods paid by debit cards according to their travel

Table 9.5 Ratings of power supply units under different charging modes

Charging mode	Rated voltage (V)	Rated current (A)	Remarks
1	250/440 (AC)	16	The AC network is connected by standard plugs and sockets with a rated current not less than 16 A
2		32	The AC power for EVs is supplied by a specific power supply unit, which has the control guided circuit
3	400/750 (DC)	125 250 400	EVs charged by means of off-board chargers

needs. With the promotion of the electric vehicle (EV) industry and standardization of EV batteries, large-scale rapid charging stations integrated with rapid charging and speedy battery swap services could be brought into operation to satisfy various user needs.

The above-mentioned three modes of energy supplement have their various advantages and disadvantages, and these will play a role in their future utilization (Xia 2010; Zhang 2010). From a comparative viewpoint, battery swapping would appear to be the optimal means for EVs in terms of both battery maintenance and impact on the power grid. However, since this mode requires restructuring the entire production process of the auto industry, the start-up costs associated with its adoption would appear to present the most difficulties at present.

9.2.2 Power Demand with Different Charging Modes

The instantaneous power demand by different charging modes varies significantly. A national standard for power requirements with different EV charging modes in China was specified in the GB/T 20234.2-2011 "Conductive Charging Coupler for Electric Vehicles," which was released in May 2012. The details are shown in Table 9.5.

The power demand with different charging modes can be calculated. From the predicted installed capacity of power generation in China given in Table 9.2 and the EV ownership predicted under the reference scenario and comprehensive alternative policy scenario detailed in Chap. 12, it is easy to determine the extent of the power surge that these charging modes would cause to the grid. The results are presented in Table 9.6. It is assumed that factor of simultaneous charging (FSC) is 80 % for AC charging (slow charging) and 20 % for DC (rapid) charging.

As seen in Table 9.6, even if EVs are underused in the reference scenario, the maximum power demand of these vehicles under DC rapid charging will account for 16.4 % of the installed capacity of the whole electric power system in China up

Table 9.6 Predicted power demand with different charging modes

Year					2010	2015	2020	2030	2040	2050
National installed capacity (100 million kW)					9.5	13.5	14.1	19.4	23.9	28.4
Reference scenario	EV ownership (10,000 units)				0.0	370.0	850.0	1,102.0	1,356.0	1,504.0
	Power demand percentage of the installed capacity over the same period under different charging modes	1	220 V AC	16 A	0.0	0.8	1.7	1.6	1.6	1.5
		2	220 V AC	32 A	0.0	1.5	3.4	3.2	3.2	3.0
		3	400 V DC	125 A	0.0	2.7	6.0	5.7	5.7	5.3
				250 A	0.0	5.5	12.1	11.4	11.3	10.6
				400 A	0.0	8.8	19.3	18.2	18.2	16.9
			750 V (DC)	125 A	0.0	1.2	11.3	10.7	10.6	9.9
				250 A	0.0	10.3	22.6	21.3	21.3	19.9
				400 A	0.0	16.4	36.2	34.1	34.0	31.8
Comprehensive alternative policy scenario	EV ownership (10,000 units)				0.0	258.0	515.0	5,116.0	15,658.0	27,364.0
	Differential ratio (%) of power demand of the installed capacity over the same period under different charging modes	1	220 V AC	16 A	0.0	0.5	1.0	7.4	18.4	27.1
		2	220 V AC	32 A	0.0	1.1	2.1	14.9	36.9	54.3
		3	400 V DC	125 A	0.0	1.9	3.7	26.4	65.5	96.4
				250 A	0.0	3.8	7.3	52.7	131.0	192.7
				400 A	0.0	6.1	11.7	84.4	209.6	308.3
			750 V DC	125 A	0.0	3.6	6.8	49.4	122.8	180.7
				250 A	0.0	7.2	13.7	98.9	245.7	361.3
				400 A	0.0	11.5	21.9	158.2	393.1	578.1

Note: The charging power under these modes is calculated by the following formula: charging power $P = VI \times \text{FSC} \times \text{EV ownership}$, whereby V is the charging voltage and I the charging current to 2015. Under the reference scenario, the fossil energy required for vehicles is not renewable if demand continues to increase. Thus, the widespread application of EVs is indispensable to China for the sake of both energy and environmental security, and increased EV ownership is predicted under the comprehensive alternative policy scenario. It is evident under that scenario that the charging power demand will account for 7.4 % of the whole installed capacity up to 2030 even with normal charging using the lowest charging current. Therefore, it is necessary to develop appropriate technologies and formulate policies to regulate the charging mode and time for EVs to achieve orderly charging within that sector.

9.3 Interaction Between EVs and the Power Grid and Their Coordinated Development

9.3.1 Power Quality Issues Arising from EV Charging

The general circuit used for a large power-charging system for pure EVs is depicted in Fig. 9.5. The three-phase power system supplies the AC source, which is then rectified to a DC source by a diode-based three-phase rectifier. After being filtered, the DC source is then supplied to a power converter, from which the desired charging power is produced; it is then filtered once again before being supplied to the battery of the pure EV.

Since the charger is naturally a rectifier load, the most prominent power quality problem may concern harmonics (Lu and Zhang 2006; Huang 2008; Chen et al. 2008). The considerable amount of current harmonics produced by EV charging stations (or chargers) inserted into the power grid can result in many hazards:

1. To the electricity metering system: the harmonic current is metered as active current, resulting in increased user cost.
2. To precise electronic equipment, such as computers: more harmonics could lead to control equipment malfunction and cause production outages.
3. To transformers: harmonic currents could lead to increased copper loss and stray loss; harmonic voltage could increase iron consumption and exacerbate the heating of transformers, and it could also cause an increase in the operation noise of transformers.
4. To power cables: more harmonics will increase cable loss.
5. To engines and motors: mechanical vibrations will be affected by the harmonic current and fundamental frequency magnetic field; if the mechanical resonant frequency coincides with the electrical excitation frequency, resonance will be induced to produce high mechanical stress, leading to the risk of damage to machinery.
6. To control and communications equipment: voltage harmonic distortion could lead to incorrect judgment of the zero-crossing time of the voltage and location

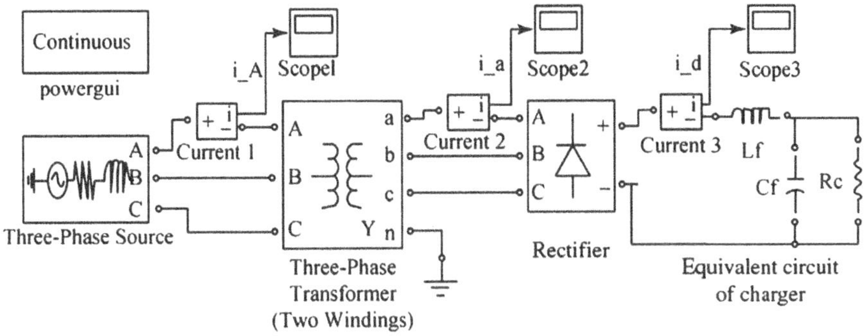

Fig. 9.10 MATLAB model of a charging station, including a single charger

of the voltage point by the control system, resulting in faulty control systems; inductive or capacitive coupling between power and communication lines may also cause interference with communications equipment.

7. To switches and relays: abnormal trips may occur in solid-state trip devices of electronically protected low-voltage circuit breakers. General harmonics in the grid could interfere with the protective automatic devices of the start-up elements of sequence component filters.
8. To power factor compensation capacitors: the heating caused by harmonics and the increased voltage may reduce the capacitors' life cycles. Local parallel resonance and series resonance produced by harmonics in the grid could result in higher harmonic voltage and current than when resonance is absent.

In light of the above hazards with harmonics, appropriate measures need to be taken in charging stations to suppress harmonics. Harmonic suppression effects need to meet national standard GB/T14549-93: Quality of Electric Energy Supply—Harmonics in Public Power Systems.

The charging process of single chargers and multiple chargers can be simulated using MATLAB/Simulink® software. A simulation model with a single charger is shown in Fig. 9.10. The batteries are charged by phased normal charging (initially, constant-current charging; then, constant-voltage charging). The parameters of the power supply, transformers, rectifiers, and other modules in the simulation model are based on the device parameters used in electric bus charging stations at the Beijing Olympics.

9.3.1.1 Simulations and Analysis of a Single Charger

The following conclusions can be reached based on an analysis of simulation results with a single-charger model:

1. The orders of harmonics on the grid side are $6k \pm 1$, $k = 1, 2, 3, \ldots$, i.e., 5, 7, 11, 13, ...; the higher the harmonic frequency, the smaller is the harmonic amplitude.
2. The orders of harmonics on the load side are $6k$, $k = 1, 2, \ldots$, i.e., 6, 12, ...; the higher the harmonic frequency, the smaller is the harmonic amplitude.
3. The contents of the fifth and seventh harmonics on the system side have roughly the same distribution throughout the charging process. The highest percentage of the fifth and seventh harmonics appears when converting from constant-current charging to constant-voltage charging.
4. Throughout the charging process, the highest percentage of the 11th harmonic appears in the period of constant-voltage charging; however, the percentage of the 13th harmonic has no significant change, and it is higher in the constant-current charging stage.
5. When charging, the voltage and load equivalent resistance are determined; greater filter inductance contributes more to reducing the harmonic currents and total harmonic distortion. This is because series inductance can effectively suppress the impact of the current, thereby inhibiting the distortion of the AC current.
6. When the transformer's wiring system changes from $\Delta{-}Y$ to $\Delta{-}\Delta$, the harmonic content is slightly increased in the system. This is because the links of $\Delta{-}Y$ or $Y{-}\Delta$ may reduce high-order harmonics of a multiple of three.

Throughout the charging process, there is a change in the percentage of each order of harmonic compared with the fundamental wave. The smaller the load equivalent resistance, the greater are the harmonic currents. Variations in each order of harmonic are not exactly the same. In the constant-current charging stage, the harmonic content is essentially unchanged. At the early stage of constant-voltage charging, the low-order harmonic content is significantly increased; subsequently, with the decrease in charge current, the harmonic content is also reduced.

9.3.1.2 Simulations and Analysis of Multiple Chargers

Multiple chargers working simultaneously at a charging station may be simulated with a single-charger model and randomly setting the equivalent nonlinear resistance of the chargers. The following conclusions can be drawn:

1. The orders of harmonics on the grid side are still mainly the 5th, 7th, 11th, and 13th when multiple chargers are operating simultaneously. The harmonic amplitudes decrease with increasing frequency.
2. Parts of the harmonic currents are offset in bus lines of charging station when multiple chargers are operating simultaneously. Therefore, although each harmonic current increases, the total is less than the algebraic sums generated by each charger. Consequently, with an increase in the number of chargers, the ratio of each harmonic current compared with the fundamental wave shows a downward trend.

Taking into account the results of the above simulations and from a comparison of the harmonics produced in the working processes of single and multiple chargers, the following suggestions are made for charging stations when selecting appropriate charging equipment:

1. If other conditions are unchanged, the input transformer in charging stations should be connected at a lower voltage level, and it should be able to supply power to additional chargers provided that national harmonic standards are met.
2. When multiple chargers are operating at the same time, although each harmonic current increases, the total harmonic current is still lower than the algebraic sums generated by each charger. It is therefore recommended that a harmonic suppression device be installed at the main output terminal of the whole charging circuit.
3. If economic conditions and power capacity permit, it is recommended that a single charger with larger power capacity be chosen.
4. Charging stations should use a higher-level power supply that is located closely. The use of the DY_n-11 wiring system for distribution transformers in charging stations also helps to suppress the third harmonic.

9.3.2 Bidirectional Energy Conversion Technology of EVs

At present, enlarged peak-valley difference and lack of peak regulation capacity have become the basic situation in power grid operations, with the average peak-valley difference in major grids having reached 60 %. With the development of regional economies, the grid peak-valley difference will continue to increase. In addition to conventional means of peak shaving, new technological trends and research are focused on how to incorporate users in peak shaving of grid operations by taking advantage of demand-side management. The development of EVs offers a new possibility for grid "valley filling."

As long as appropriate technologies and control methods are used to connect numerous EVs with the grid, it will be fully possible in the future to make EVs a group of distributed power supplies to the grid in emergency cases, thereby improving grid stability (Yang Jian et al. 2010). This possibility could be achieved using vehicle-to-grid (V2G) technology, which was first proposed by Willet Kempton and Steven E. Latendre, professors at the University of Delaware in the United States in 1997. V2G describes the relationship between EVs and the grid. With this technology, the electricity of onboard batteries in EVs is sold to the grid when the vehicles are not in use; power flows from the grid to the vehicles when they need to be charged. Thus, a bidirectional flow of energy between the grid and EVs is achieved with V2G. Bidirectional information communication is necessary to achieve good control of the energy exchange. V2G is a powerful force for both the EV industry and smart grid development (Sidebar 9.1).

Sidebar 9.1: Three Modes of Energy Obtaining for EVs

1. Unidirectional Unordered Electric-Energy Supply: V0G mode
   With the vehicles plug in without logic/control (V0G) mode, EVs are charged as soon as they are connected to the grid. V0G is currently the most common charging mode for EVs. With this mode, such EVs as electric buses, golf carts, and airport shuttle buses are treated as normal loads. The charging equipment in this mode mainly uses a unidirectional converter, whose technology is relatively mature both at home and abroad. Some public charging facilities for EV charging stations, such as those used at the Beijing Olympic Games, have already been built. The problem with this mode is that the unrestricted use of EVs as high-power electrical loads may increase the difficulty of peaking regulation.
2. Unidirectional Ordered Electric-Energy Supply: TC and V1G mode
   With the timed charging (TC) mode, EVs start charging during a given time. The TC mode takes into account the impact of EVs on the power grid during its peak-load period. Users may share the economic benefits from charging in off-peak periods by controlling the starting time of charging. But the mode is too simple to achieve the flexible control of the charging process according to the real-time price of electricity or the grid peak-valley state. The charging equipment with this mode is a unidirectional converter, but it is unable to achieve real-time communication with the grid. This technology is mature, and it has been introduced in pilot demonstrations.

   With the vehicles plug in with logic/control regulated charge (V1G) mode, charging of EVs is controlled by the power grid. EVs communicate with the grid in real time and start charging during any allowed period. This may optimize the charging arrangement and improve grid efficiency, but it does not provide feedback power from the vehicle to the grid.
3. Bidirectional Ordered Electric-Energy Supply: V2G mode
   With the vehicles plug in with logic/control regulated charge/discharge (V2G) mode, EVs communicate with the energy-management system of the grid; subject to its control, energy exchange (charging and discharging) is achieved between the vehicles and the grid. In this mode, EVs can be used as an electrical energy-storage device and backup power supply unit.

   At present, the V2G mode is in the pilot demonstration stage. Bidirectional converters and communications equipment with low cost and high reliability are still under development to meet the requirement of V2G commercialized operation. The intelligent levels of the power grid and electricity market environment in China currently fail to meet the requirements of the V2G mode. On the one hand, this mode needs the support of advanced technologies of grid communication, dispatching,

(continued)

(continued)

control, and protection. On the other hand, it requires support from peak-valley pricing policy and other policies related to paid services for EVs being connected to the grid to provide peak regulation, load adjustment, and demand response.

V2G technology is proposed if the following technical background applies:

1. The power system is unable to achieve effective large-scale energy storage, and generators and grids have to be regulated in real time with fluctuation loads.
2. The vehicle power system reserves considerable capacity for adapting to variability in traffic conditions, with the average capacity factor being about 57 %.
3. The service time of household vehicles is very low—generally only 5–10 % (i.e., running about 2 h a day); about 90 % of the day the vehicle is idle.

Therefore, when EVs are developed on an appropriate scale, they will have adequate capacity and idle time so that they can be used as a new type of energy-storage device in power system operation. As the local supplemental power supply to the load center, this "part-time" energy-storage device has no transmission losses. It is conducive to energy saving and environment protection, can be used to achieve load shaving and filling, and will improve the effective use of EV batteries. While taking into account the car's main function as a means of transportation, V2G-equipped EVs involved in power marketing will be able to supply power at peak-load hours; this will be a high-value power supply over a short period, rather than being for the base load.

The typical mode of operation of V2G is shown in Fig. 9.11. First, it is necessary to locate EVs using GPS to determine their distribution in the grid. Then, grid operation personnel send power commands to EVs via the Internet or wireless communications. EVs respond to these commands in real time. They communicate with grid operation personnel, aggregator, or power companies by various means of communications. The "$" symbol in Fig. 9.11 indicates that under this mode of operation, the cost information of electricity charging or provision can be settled in real time via the network.

Several key issues concerning V2G technology are as follows:

1. Different types of EVs play different roles in the power grid. In general, an EV purely driven by batteries can be considered only as a pure load; this is because the energy it delivers can never be greater than what has been filled in. However, hybrid or fuel-cell-powered EVs can be used as small engines to provide energy for the grid if they have the capability of transforming other forms of energy into electric energy. The excess renewable energy that is unable to be accepted by the grid will be absorbed by EVs as long as they are the plug-in type.

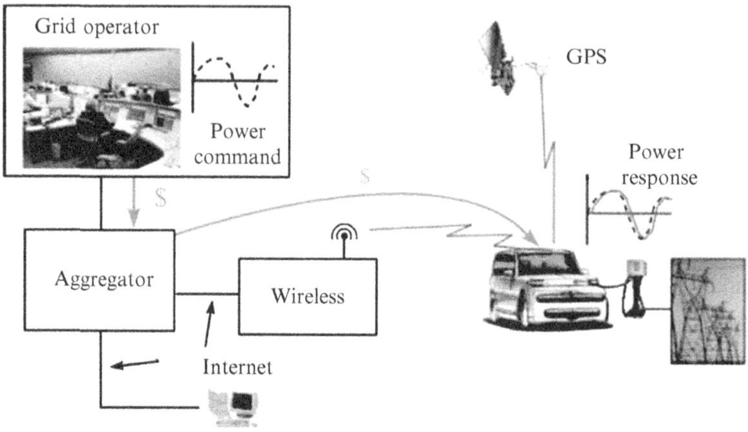

Fig. 9.11 Typical mode of operation with V2G

2. The power grid buys electricity from EV users only when the unit cost of EV batteries is much lower than that of energy storage (such as pumped storage) used in the power system. However, considering such factors as battery life, users would be willing to sell electricity only when the grid agrees to offer a higher price.
3. EV batteries will be sufficient for providing support for regional power distribution grids only when their total capacity reaches a certain scale.
4. It is necessary to adopt advanced energy metering technology (advanced measurement interface). Unlike with common loads, the energy flow between EVs connected to the grid by V2G technology and the grid is bidirectional; thus, it is necessary to develop advanced, convenient energy-measurement technology. Figure 9.12 shows an energy metering interface proposed by Kempton and Latendre (1997).

At present, the major technical difficulties that V2G faces are mainly with the following aspects:

1. Limited battery life cycle—thus, the cost of selling electricity to the grid is very high.
2. Lack of infrastructure support, such as charging stations and posts.
3. The batteries of EVs are DC sources, so converters and transformers are necessary when supplying power to the grid; thus, the costs of infrastructure construction are high.
4. There are many EVs and their use is widespread: the difficulty is how to coordinate and manage such a large-scale power source with small capacity.
5. Harmonic waves and other power quality problems.

From data issued by the China Electricity Council, by the end of 2009 the gross installed capacity of all kinds of power generation in China was 0.894 TW. At the same time, the vehicle population in the country was about 76.19 million. Provided

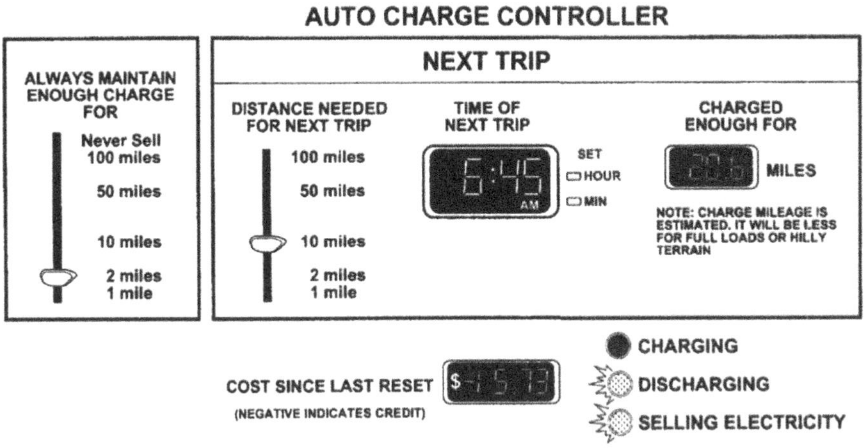

Fig. 9.12 Electric power metering screen on EVs equipped with V2C technology. *Notes*: The driving distance is an estimate: it will be lower in the case of a full load or driving on mountain roads (Reprinted from Kempton and Latendre (1997), Copyright 1997, with permission from Elsevier)

that the engine power of each vehicle is 100 hp, that amounts to 735 kW (Kempton and Latendre (1997) adopt the figure of 125 hp [93 kW] as the mean engine power of American vehicles); then, the engine power of all vehicles amounts to about 5.6 TW, which is over six times the installed capacity of that time. From this, it is evident that even if 1 % of vehicles were electrified and connected to the grid using V2G technology, this would have a significant effect on the power grid.

9.4 Infrastructure and Technical Standards of the Power Grid Supporting EV Development

9.4.1 Infrastructure Construction and Upgrading the Electrical Power System

Similar to the situation in Japan and France, the charging infrastructure for EVs in China is mainly executed by leading power operators—the State Grid Corporation of China and China Southern Power Grid Co., Ltd. The State Grid Corporation of China (SGCC) established 75 charging stations and 6,209 charging posts in 27 cities across China in 2010. By the end of 2010, SGCC had established and operated 87 charging/battery swap stations, 5,179 charging machines, and 7,031 AC charging posts. Thus, the charging/battery swap service network has initially taken shape. At the end of 2010, China ranked first in the world in terms of the number of commissioned pieces of charging/battery swap equipment. The State Grid Corporation of China determined that the best business mode of charging/battery swap stations for EVs was to swap batteries instead of charging them, charge batteries together, and distribute them uniformly. This state corporation also improved its research efforts into charging/battery swap equipment standards and technologies. It has undertaken or participated in the preparation of 12 national standards and 13 professional standards; it has set nine enterprise standards and performed research and development into four kinds of battery swap equipment as well as applying for 38 invention patents, 31 utility model patents, and 11 design patents.

The State Grid Corporation of China established the first intelligent charging/battery swap network for EVs—in Zhejiang Province in the beginning of 2011. This was also the first charging/battery swap service network with an intercity connection in the world. With this trial project, the power supply mode was mainly battery swapping and secondarily filling and charging. With integration of intelligent power grid technologies and use of the Internet with modern logistics technology, this demonstration project in Zhejiang Province has successfully connected 14 charging/battery swap stations in Hangzhou, Jinhua, Ningbo, and some other cities together with 500 AC charging posts in 11 regions of the province. Thus, an intelligent charging/battery swap service network has been established whose operation mode is characterized by joint charging, uniform distribution, flexible swapping, wide coverage, billing by mileage, and professional operations.

This power supply method has the following advantages:

1. Batteries are provided by operators, such as grid companies. This reduces the cost of purchasing EVs, improves their competitiveness in the market, and saves other costs, such as battery maintenance for consumers.
2. Battery swapping can be completely implemented by mechanized, automatic operation. Thus, it requires less time and the power-charging station needs only a small amount of space, allowing for flexible arrangements in big cities with limited land resources.
3. Operators are responsible for releasing batteries, and they can intensively charge and professionally maintain batteries in power-charging stations, which will effectively warranty their service life.
4. By charging during valley-power hours, a good EV charging service can be provided; additional peak-load pressure for the power grid through large-scale EV charging is eliminated. This ensures safe operation of the power grid and improves the social resource utilization rate.
5. Energy-storage stations are established for charging during off-peak hours (at night) using retired power battery packs and delivering power to the grid during peak hours (daytime) to achieve a cascaded utilization of power battery.

At the end of 2010, the National Development and Reform Commission issued three national standard drafts related to the charging interface of EVs on its official Web site. As long as the standards for batteries can be unified in the future, an economical and convenient power swap method will stimulate rapid development of EVs. Nevertheless, there are still many problems with this method at present, as noted in Sect. 9.2.1.3.

China Southern Power Grid Co., Ltd. released plans at the beginning of 2011: by 2012, it aimed to establish 89 power-charging stations and 29,500 power-charging posts in Shenzhen. It is predicted that the gross investment by this company will be over RMB one billion and the charging service network of mainly slow-charging facilities supplemented by quick charging will cover main roads, residential areas, and public parking lots throughout the city.

At the end of 2010, the first automatic power battery swap system in China, which was researched and designed by China Putian Information Industry Group Corp., Ltd., came into use. With this system, power batteries of different types of EVs can be swapped in just 3 min, which is almost as quick as with fuel and gas filling. As well as providing the power battery swap service, China Putian purchases, manages, maintains, and realizes cascade utilization of power batteries; comprehensive and intelligent management extend the life cycle of power batteries and also solve potential pollution problems resulting from recycling batteries.

In addition, charging stations for EVs are being established in Shanghai, Chongqing, Wuhan, Guangxi, and other cities (Li 2010). A complete infrastructure is an important prerequisite for the large-scale use of EVs, but the investment for infrastructure is huge. Even excluding the land cost, the total investment in battery-leasing stations (excluding batteries) for a battery swap service for 50 electric buses is as high as RMB 30–40 million. The total investment for power swap stations for 30 electric buses established on existing parking lots is RMB 10 million. The investment for commercial power-charging stations with nine charging posts for cars is at least RMB five million. The investment for roadside charging posts is at least RMB 250,000 (Wang 2010). By the end of 2011, 243 power-charging stations and 13,283 AC charging posts had been installed and put into operation in China. Because the number of EVs is still small, it is difficult to make profits from these charging stations and posts in a long period in the future. Thus, undoubtedly investment risk exists on the part of governments and enterprises, and investors' interest will be reduced if business mode of charging facilities is unclear. This is obviously not good for the long-term development of EV industry.

In addition to construction of the infrastructure starting from zero, it is also necessary to upgrade related electrical facilities. Normal charging exerts a small impact on the power grid. Thus, the impact can be ignored if charging posts are not widespread, which is the case at the initial stage of development of the EV industry. However, considerable power is required for quick-charging and high-power-charging stations; therefore, specialized substations and a power distribution network need to be established. If the number of charging posts increases, the existing distribution capacity of parking lots and residential areas cannot meet the power demands of all the expanding number of charging posts. Investment in upgrading the distribution system is about the same as for constructing the power-charging infrastructure, which is mainly undertaken by power grid companies (State Grid Corporation of China and Southern Power Grid Corporation). From the perspective of long-term development of the EV industry, it is clear that power grid companies cannot take on such a huge investment alone. Thus, adequate investment sources and proper modes of operation are necessary to guarantee upgrading of the power system to satisfy the charging needs for EVs.

9.4.2 Technical Standards for EVs

The preparation of technical standards for EVs in China began in 1998. On May 29, 2010, the Ministry of Industry and Information Technology issued the first batch of newly formulated industrial standards and revisions to older standards. That batch included 61 standards related to the vehicle industry (all were recommended industrial standards): 38 were newly formulated standards and 23 were revised standards. There were also revisions and 11 new standards pertaining to EVs (one of these was for hybrid EVs) (Xiao 2010). The details appear in Table 9.7.

By the end of 2010, there were 46 national and industry standards for EVs in China, including 15 that had been revised and submitted for approval. There were 50 standard items under formulation and revision; following industry recommendation and expert review, over 200 standards for EVs will require formulation or revision by the end of 2014. This means that China will possess the most extensive EV standard system with the largest number of standards in the world.

The classifications of the issued standards are as follows:

1. Thirty-one recommended national standards, two guiding national standards, and 13 industrial standards
2. Twelve pure EV standards, seven hybrid EV standards, four fuel-cell-battery vehicle standards, seven power battery standards, three motor standards, seven relevant charging standards, and six electric motorcycle standards
3. Four standards of equal or equivalent adoption of international standards, 16 reformulations by modifying, adopting, or referencing to international standards, and 26 completely autonomous formulation standards

The publication and implementation of these standards will play an important role in the development of EV technology and industrialization in China.

9.5 Conclusions and Recommendations

From the perspective of the electrical power system, the main factors at present constraining the development of EVs in China are listed below:

1. There are major regional differences. Regional economic development in China is very imbalanced. Thus, there are very significant differences in the power supply of regional power grids for the development of EVs. If priority is given to large- and medium-sized cities for EV development, the imbalance between the power supply and demand will intensify; however, the infrastructure is relatively

Table 9.7 EV industry standard formulations and revised plans in 2010

Serial no.	Standard no.	Standard name	Nature	Version	Completion time
1	2010-1843T-QC	Low-speed EV technical request	Recommendation	Formulation	2011
2	2010-1844T-QC	Technical specification of onboard charger for EVs	Recommendation	Formulation	2011
3	2010-1845T-QC	General technical requirements of charging station supervisor management	Recommendation	Formulation	2011
4	2010-1846T-QC	Technical specification for EV charging posts	Recommendation	Formulation	2011
5	2010-1847T-QC	EV conductive charge coupler	Recommendation	Formulation	2011
6	2010-1848T-QC	Communication protocols between battery management system and off-board chargers for EVs	Recommendation	Formulation	2011
7	2010-1849T-QC	Technical specification of battery management system for EVs	Recommendation	Formulation	2011
8	2010-1850T-QC	Technical specification of off-board charger for EVs	Recommendation	Formulation	2011
9	2010-1851T-QC	Specifications and dimensions of traction battery for EVs	Recommendation	Formulation	2011
10	2010-1853T-QC	Cycle performance of power batteries for EVs	Recommendation	Formulation	2011
11	2010-1854T-QC	Hybrid EV power assemblies system performance test methods	Recommendation	Formulation	2011

good, and upgrading the cost of the power grid is low. If priority is given to less-developed or rural areas for EV development, the imbalance between the power supply and demand will not be so great; however, the infrastructure is poor, and so the cost of upgrading the power grid for EV development is high. Because of such regional differences, there are a number of choices in development plans for EVs. In major cities, middle- and high-grade passenger cars should be the priority for EV development; in less-developed areas, middle- and low-grade mini cars should be given priority for EV development.

2. There is a lack of technical standards. Technical standards for charging facilities are particularly deficient, which will seriously impact the availability and profitability of charging facilities in the future.

3. The profit model is not clear. This will have a serious impact on energy providers' interest in investing in the charging infrastructure (including the State Grid Corporation of China) and delay the speed of construction in the charging network.

On the basis of the analysis and predictions made in Sect. 9.2, from the perspective of electric power systems, the following conclusions and recommendations are made for EV development in China:

1. Irrespective of the power source or state of the grid, the Chinese electric power system will continue to show rapid development up to 2030, and the total electricity consumption by EVs can be guaranteed. However, in terms of typical daily load curves, some technologies can be adopted to control the charging time of EVs to ensure that the power valley is filled (e.g., embedded chips and wireless communication can be used to charge at an appointed time). Ensuring orderly charging for EVs is the key to achieving an effective interaction between vehicles and the power grid.
2. At present, power batteries have become the bottleneck that is constraining EV development. It will be difficult to overcome such problems as low battery energy density, expensive battery packs, and limited life cycles.
3. Compared with normal and rapid charging, the battery swap method can effectively extend the battery service life by professional maintenance, thereby reducing the negative impact of large-scale charging on the power grid and improving the use value of batteries by cascaded utilization (i.e., retired batteries that are unsuitable for vehicles can be used as static energy-storage devices for the grid). The difficulty in extensively popularizing the battery swap method is in extending and adopting the standard charging coupler, unified battery dimensions, and reasonable operation modes.
4. Investment in the charging infrastructure for convenient use of EVs may be greater than that of the vehicles themselves. Thus, identification of appropriate investors and modes of operation is important to improving the charging infrastructure and healthy development of the EV industry.
5. EV charging will clearly impact the quality of the electric power system, which can be solved by power technology. If EVs develop on a certain scale and a major breakthrough is made with battery performance, batteries can be used as distributed power sources to help peak or frequency regulation by V2G. This will not only reduce the spinning reserve capacity of the electric power system but also increase gains by EV owners, thereby achieving a win-win situation.
6. It is necessary for battery manufacturers, EV makers, and charging equipment manufacturers to work together. They need to quickly formulate relevant technical standards for charging equipment, taking into consideration the requirements of the electric power system, including coupler types (AC or DC), rated voltage and current, and communication protocols. The compatibility and utilization rate of the charging facilities need to be improved, and it is necessary to promote the rapid and healthy development of the EV industry.

References

Chen X, Li P, Hu W et al (2008) Analysis of the impact of electric vehicle chargers on power grid harmonics. China Power 41(9):31–36

China Electricity Council (2010) Report of data of National Electric Power Industry 2009, Issued by the China Electricity Council. http:www.cec.org.cn/nengyuanyudianlitongji/hangyetongji/2010-11-27/29958.html. 15 Jan 2011

China Electricity Council Electric Power Reliability Management Center (2011) China electric power reliability management briefing (Power transmission and transformation part): 2001–2010

China Energy Medium- to Long-Term Development Strategy Research Project Group (2011) China Energy Medium- to Long-Term (2030, 2050) Development Strategy Research. Science Press, Beijing

Huang S (2008) Research on harmonics of electric vehicle chargers. Beijing Jiaotong University, Beijing

Jun Z, Xiaochun L et al (2007) Progress in international energy strategy and innovative technology. Science Press, Beijing

Kempton W, Latendre SE (1997) Electric vehicles as a new power source for electric utilities. Transp Res D Transp Environ 2(3):157–175

Lei L, Liu Q (2000) Exploration of impact of electric vehicles on the power grid load curve. Electr Mot Technol 2000(1):37–39

Li L (2010) Present status and development trend of batteries for electric vehicles. Commercial Car News, p 7, 13 Sept 2010

Lu Y, Zhang X (2006) Harmonic study of electric vehicle chargers. Proc CSU-EPSA 18(3):51–54

U.S. Energy Information Administration (2009) Annual energy review 2009 – released on August 19, 2010, pp 226. http://www.eia.gov/totalenergy/data/annual/archive/038409.pdf

Wang Q (2010) Status survey of China's electric vehicles. China Business News, CO_2 edition, 8 Nov 2010

Xia D (2010) Overview on research in electric vehicles. Energy Technol Econ 22(7):49–55

Xiao X (2010) The first batch auto industry and vehicle industry standards and national standard revision plan in 2010. Commer Vehicle 2010(S6):9

Yang J, Mei W, Yi Z et al (2010) Applying power battery of electric vehicles for regulating the peak in the grid. East China Electr Power 38(11):1685–1687

Yin H, Wang K (2010) If charging stations replace gas stations. Mod Veh [J], Period 2 in 2010: 68–75

Zhang S (2010) Prospect of electric vehicles in china from the perspective of technology dynamics. Energy Technol Econ [J] 22(8):37–41

Zhang Y, Jiazheng L, Bo L (2011) New electric vehicle AC charge posts using active power filter. High Volt Technol [J] 37(1):150–156

Zou Y (2008) China's Power industry development level and structure analysis. Electr Indus Policy Res [J] 2008(9):16–21

Chapter 10
Hydrogen and Fuel-Cell Vehicle Technology

Wang Hewu, Huang Haiyan, Deng Xue, and Ouyang Minggao

Abstract The hydrogen production and utilization and thus the flow chart are analyzed, and the fuel-cell vehicle technology and its demonstration activities are summarized; based on the analysis of government policies and vehicle development scenario, the hydrogen fuel-cell vehicle development pathway is suggested and recommended.

There is about 12 million tons of commercial hydrogen in China produced from fossil fuels and 6 million tons hydrogen as by-products, and these tons of hydrogen are consumed in large synthetic ammonia, oil refining, and methanol producing. The hydrogen directly used in fuel-cell vehicle demonstration activities is tiny.

The efficiency has improved from 2005 to 2010 by 49 and 62 % for pressured and normal fuel-cell systems, respectively; meanwhile, the specific power of the pressurized system has increased by 1.5 times. The fuel-cell buses and passenger cars have completed demonstrations in Beijing, Shanghai, and abroad during the last 5 years and achieved as low as 9.56 kg hydrogen consumption per 100 km for city bus.

China encourages the research and development of hydrogen production and fuel-cell vehicle technologies; the 6 million tons industrial by-product hydrogen capacity could satisfy the hydrogen supply demand for the fuel-cell vehicle development in recent 20 years.

Keywords Hydrogen production • Efficiency • Fuel-cell electric vehicle • By-production hydrogen • Demonstration

Wang Hewu (✉) • Huang Haiyan • Deng Xue • Ouyang Minggao
China Automotive Energy Research Center, State Key Laboratory of Automotive Safety and Energy, Tsinghua University, Beijing 100084, P.R. China
e-mail: wanghw@tsinghua.edu.cn

CAERC, Tsinghua University, *Sustainable Automotive Energy System in China*,
DOI 10.1007/978-3-642-36847-9_10, © Springer-Verlag Berlin Heidelberg 2013

10.1 Production of Hydrogen in China

10.1.1 Hydrogen from Coal

There are two main ways of producing hydrogen with coal as raw material—coal coking and coal gasification. Coal coking refers to the process whereby coal is heated to 900–1,000 °C in the absence of air to produce coke, with coke oven gas as the by-product. Generally, 300–350 Nm^3 of coke oven gas can be produced from every ton of coal. The content of hydrogen in the coke oven gas can reach as high as 60 %. Coke oven gas also contains a small amount of CO (5–8 %), unsaturated hydrocarbons with 2–3 carbon-carbon bonds (2–4 %), CO_2 (1.5–3 %), O_2 (0.3–0.8 %), and N_2 (3–7 %). The heat value of coke oven gas is 17–19 MJ/Nm^3. Therefore, it is suitable for use as a fuel for high-temperature industrial furnaces or city gas.

Coal gasification refers to the process by which coal reacts with vapor or oxygen (air) under high temperature and constant pressure or under elevated pressure and is then converted into gaseous products. The gaseous products formed contain such ingredients as hydrogen, whose content varies with different gasification methods. Coal gasification is used to produce chemical raw materials or city gas, and it is the prime means of producing hydrogen from coal. Following many years of development, coal gasification technology has become one of the most important technologies for coal deep processing (Xie Jidong et al. 2007).

Most of China's hydrogen from coal is used as a raw material in the production of ammonia and methanol (Fig. 10.1). In 2007, about 7.13 million tons of hydrogen was produced from coal in China; among that, about 5.6 million tons was used to produce ammonia, and about 1 million tons was used in the production of methanol (Deng Xue et al. 2010). In the making of these two products, a small quantity of hydrogen is emitted to the atmosphere along with the purge gas.

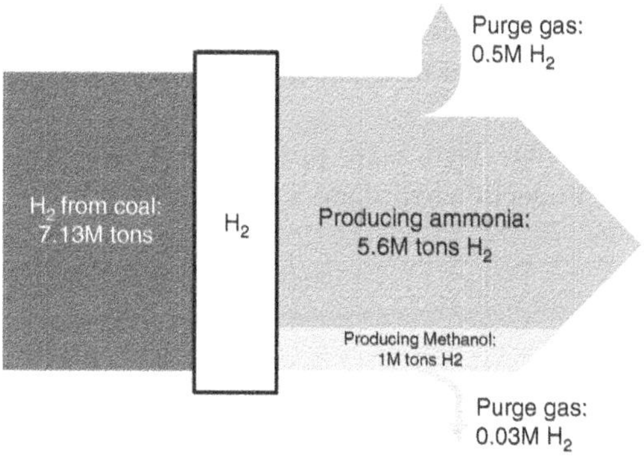

Fig. 10.1 Hydrogen production from coal and its use in China (2007)

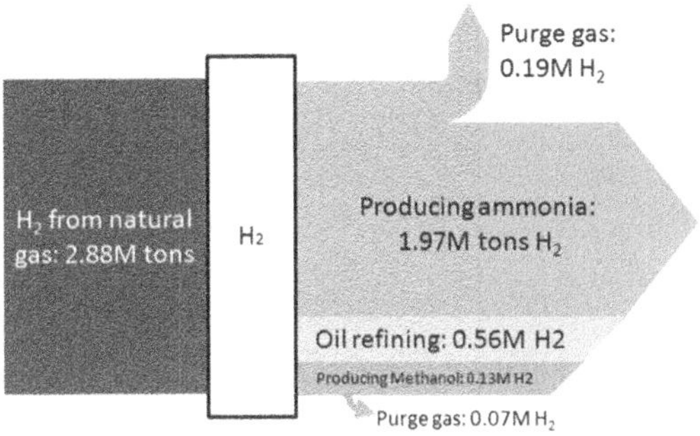

Fig. 10.2 Hydrogen production from natural gas and its use in China (2010)

10.1.2 Hydrogen from Natural Gas

Since the technology for producing hydrogen from natural gas is mature and the efficiency is high, it has been employed by small and medium-sized chemical plants. The investment levels and the cost of equipment used to produce hydrogen from natural gas are low, and so almost 60 % of the world's hydrogen is produced from natural gas. However, owing to the shortage of natural gas resources in China, the scale of hydrogen production from that source is limited.

The technologies used to produce hydrogen from natural gas include natural gas reformation, catalytic partial oxidation of natural gas, and natural gas pyrolysis. About 44 % of the synthesis gas generated in natural gas reformation is hydrogen. Theoretically, in the production of hydrogen by natural gas reformation, four molecules of hydrogen can be produced from one molecule of natural gas. However, in actual production, since partial natural gas or synthesis gas needs to be burned to provide heat for the production process, the hydrogen-producing capacity is significantly reduced. Currently, natural gas steam reforming equipment running under a full load in ammonia and methanol plants can generate one molecule of hydrogen from every two molecules of natural gas; the energy-conversion efficiency is around 60 %. In 2010, about 2.87 million tons of hydrogen was produced from natural gas in China (Fig. 10.2). Of that amount, about 1.974 million tons was used to produce ammonia, around 128.3 thousand tons went into making methanol, and around 561 thousand tons was used to meet the need for hydrogen in crude oil processing.

10.1.3 Hydrogen from Oil Refining

During the process of oil refining, a great deal of oxygen is consumed in improving the output and quality of the oil products. Usually, hydrogen is produced after naphtha or heavy oil in the oil refinery has been partially oxidized. Heavy oil raw

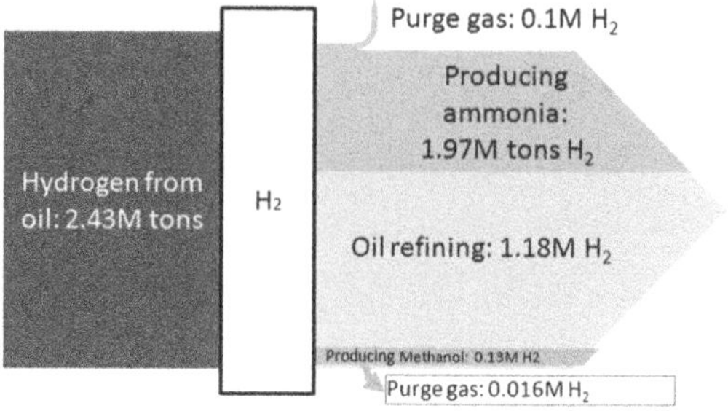

Fig. 10.3 Hydrogen production from oil and its use in China (2010)

materials include oil residuals produced during the oil refining process at constant pressure and reduced pressure as well as fuel oil produced after deep processing of oil. Gaseous products containing hydrogen are produced after naphtha or heavy oil reacts with water vapor and oxygen. The composition by volume of the gaseous products is roughly as follows: H_2, 46 %; CO, 46 %; and CO_2, 6 %. During the production of oxygen, some heavy oil needs to be burned to provide the heat required for conversion reactions and maintain temperatures. Currently, 18 % of the world's hydrogen is produced from naphtha and heavy oil (Tang Lin and Chen Baosheng 2002).

At present, China is equipped with large industrialized units for the production of hydrogen from naphtha and heavy oil after partial oxidation. The hydrogen produced by such units is used to supply oil refineries as well as being employed as a raw material for making ammonia and methanol. In 2010, about 2.44 million tons of hydrogen was produced from oil in China (Fig. 10.3). Of that amount, about 1.18 million tons was used to produce oil, about 1.03 million tons was used to make ammonia, and about 0.11 million tons was used to produce methanol. In addition, a small quantity of the hydrogen is discharged with the purge gas.

10.1.4 Industrial By-Product Hydrogen

Tail gases from chlor-alkali, coking, and steel plants in China are rich in hydrogen. Such hydrogen is usually known as industrial by-product hydrogen, and it can be converted to high-purity hydrogen by means of pressure swing adsorption technology.

10.1.4.1 By-Product Hydrogen from the Chlor-Alkali Industry

Industrial production of caustic soda commonly uses brine electrolysis, with hydrogen being simultaneously generated. In China, by-product hydrogen that appears in the production of caustic soda is normally used to manufacture such chemical products as hydrochloric acid. These products are then sold to reduce

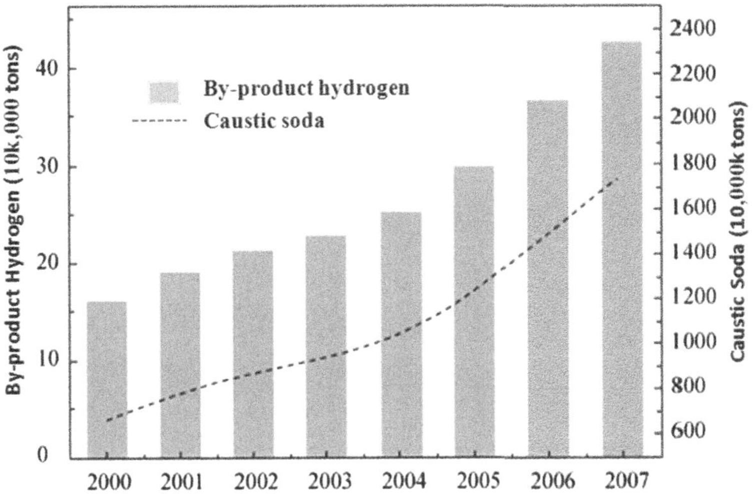

Fig. 10.4 Output of caustic soda and by-product hydrogen in China from 2000 to 2007

production costs. Typically, 270 m^3 of hydrogen is generated in the production of 1 ton of caustic soda (Chen Linxin 2002). Figure 10.4 shows the estimated volume of by-product hydrogen produced in China from 2000 to 2007 based on the output of caustic soda. It is evident in that figure that China's output of caustic soda in 2007 amounted to 17,592,900 tons and that of by-product hydrogen was about 415,700 tons.

10.1.4.2 By-Product Hydrogen from Coke Oven Gas

With the development of the steel industry, the output of coke—a raw material necessary for the steel industry—has increased considerably. China's output of coke in 2007 amounted to 335,000,000 tons. Coke oven gas is an important type of by-product gas that is generated during the coking process in the coal and steel industry. Normally, 425.6 Nm^3 of by-product coke oven gas is generated in the production of 1 ton of coke. Using pressure swing absorption technology, about 0.44 m^3 of hydrogen can usually be produced from 1 m^3 of coke oven gas (Shangguan Fanqin 2009). Figure 10.5 shows the estimated volume of by-product hydrogen based on the output of coke in China from 2000 to 2007. It is evident in the figure that China's coke output in 2007 amounted to 335,000,000 tons. In theory, the output of by-product hydrogen should have been 5,638,600 tons. According to surveys, some of the by-product gas was used as a fuel by enterprises or as domestic gas by residents. However, almost half of the by-product gas was discharged directly into the atmosphere rather than being efficiently and rationally used (Zhang Xuelei et al. 2006).

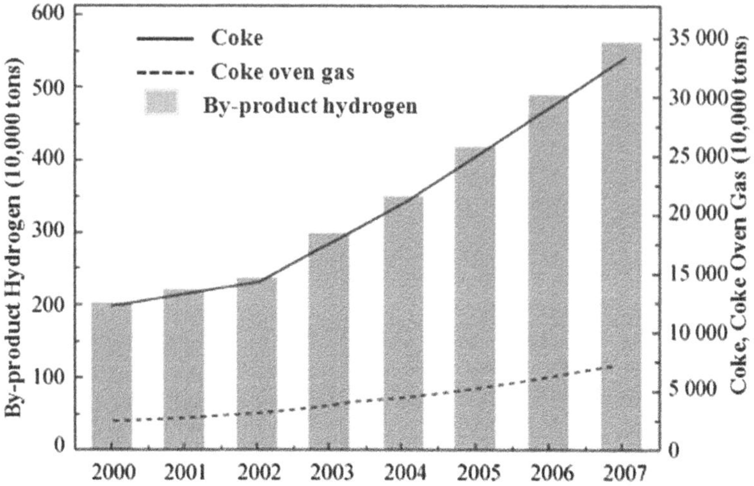

Fig. 10.5 Output of coke, coke oven gas, and by-product hydrogen in China from 2000 to 2007

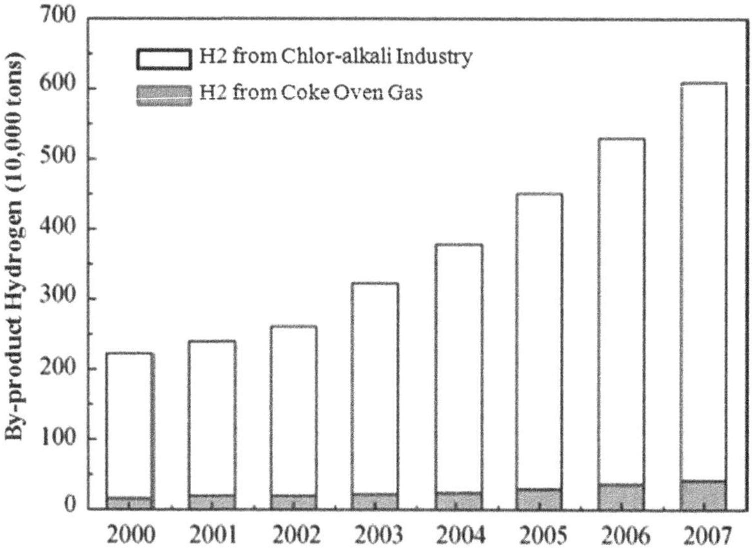

Fig. 10.6 Total output of by-product hydrogen in China from 2000 to 2007

In summary, by-product hydrogen generated during the chlor-alkali and coking processes is the main source of by-product hydrogen in China. Figure 10.6 shows the total output of by-product hydrogen in China in recent years. As an example, the total output of by-product hydrogen in 2007 exceeded 6,050,000 tons.

10.2 Hydrogen Utilization in China

Hydrogen in China has been mainly used as a raw material for semifinished products in chemical synthesis. Most hydrogen is produced and consumed in the large-scale manufacture of ammonia, oil, and refined methanol and in the chlor-alkali process. In addition, steel plants have a certain need for hydrogen. Most enterprises produce the hydrogen they need through coal coking, natural gas reformation, and partial oxidation of heavy oil with purchased equipment.

10.2.1 Hydrogen Used in Ammonia Production

The industrial production of ammonia usually employs a direct synthesis technique. In this process, hydrogen and nitrogen are synthesized at a certain temperature and pressure with the help of a catalyst. Hydrogen and nitrogen are the only raw materials used in synthesizing ammonia. In theory, 176.46 kg of hydrogen is required to produce 1 ton of ammonia. The synthesis of ammonia is a reversible reaction; to improve the synthesis rate of ammonia, purge gas containing hydrogen has to be released from the intermediate ammonia tank. Thus, the utilization rate of hydrogen is about 91.2 % after the loss of hydrogen in the purge gas is taken into account. The quantity of hydrogen consumed in the production of every ton of ammonia therefore has to be increased by 10 %, a little lower than that in 1997, which is 185 kg. In 1997, 27,000,000 tons of ammonia was produced in China, and 5,000,000 tons of hydrogen was consumed in the process (Zhu Qiming et al. 2011).

In 2008, China produced 51,589,000 tons of ammonia, which required almost 9,411,900 tons of hydrogen. Of that consumed hydrogen, about 828,200 tons was discharged as purge gas, some of which was recovered by enterprises that possessed tail gas recovery equipment. The main raw material used to produce ammonia in China is coal. Coal-derived hydrogen used to make ammonia accounts for 62–65 % of all the hydrogen consumed in the country. Oil-derived hydrogen used in producing ammonia accounts for 12–16 % of all the hydrogen consumed. Natural gas-derived hydrogen used to make ammonia accounts for 18–23 % of all the hydrogen consumed (Mao Zongqiang 2006). Figure 10.7 shows the consumption of hydrogen by China's synthetic ammonia industry in 2007 adopting the data of Deng Xue et al. (2010).

10.2.2 Hydrogen Used in Producing Methanol

Methanol is one of the most important chemical materials, and China is a large producer, accounting for over 20 % of the world's supply. The synthesis of 1 ton of

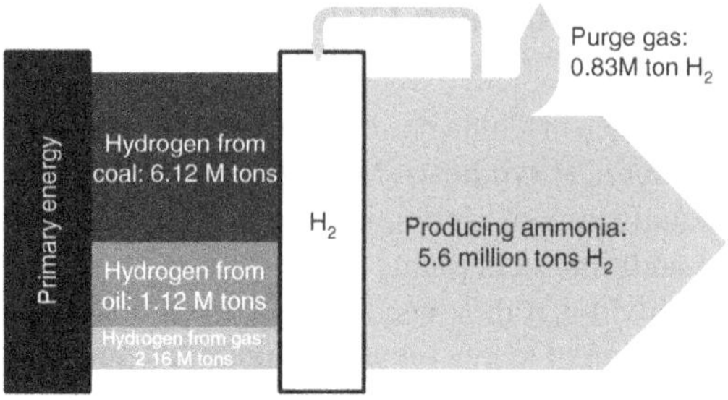

Fig. 10.7 Hydrogen consumed in ammonia production in China (2007)

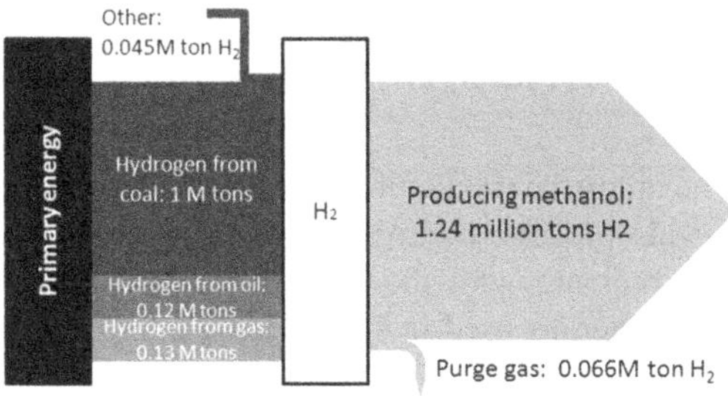

Fig. 10.8 Hydrogen producing methanol in China

methanol requires 125 kg of hydrogen. During the synthesis, purge gas containing such components as inert gases and hydrogen has to be discharged continuously from the system to prevent the reaction from slowing down or even stopping owing to the accumulation of inert gases. Hydrogen consumption is thus increased accordingly. After the loss of hydrogen in the purge gas is taken into consideration, the utilization rate of hydrogen varies with the size of the methanol plant, but it is in the range of 87.87–95 %. Zhu Qiming et al. (2011) estimated that in 1995, 200,000 tons of hydrogen was used in producing 1,500,000 tons of methanol in China; that is, about 133 kg of hydrogen was consumed to make every ton of methanol.

China's output of methanol in 2008 amounted to 10,367,700 tons. The production of this methanol required over 1,310,000 tons of hydrogen; of that, 65,500 tons was used in the purge gas. Coal is the main raw material used to produce methanol in China. Coal accounted for 77.0 % of the total production capacity, natural gas 10.3 %, acetylene 3.4 %, and heavy oil 9.3 % (Zhao Xiang et al. 2004). From this, the consumption of hydrogen in China's synthetic methanol industry in 2008 can be calculated, as indicated in Fig. 10.8.

10.2.3 Hydrogen Used in Making Oil Products

In the oil refining industry, hydrogen is mainly used for desulfuration and dedusting, improving the quality of oil products, improving the oil refining process, and protecting refinery equipment. Owing to insufficient output of crude oil in China, there has been a growing imbalance between supply and demand: the volume of

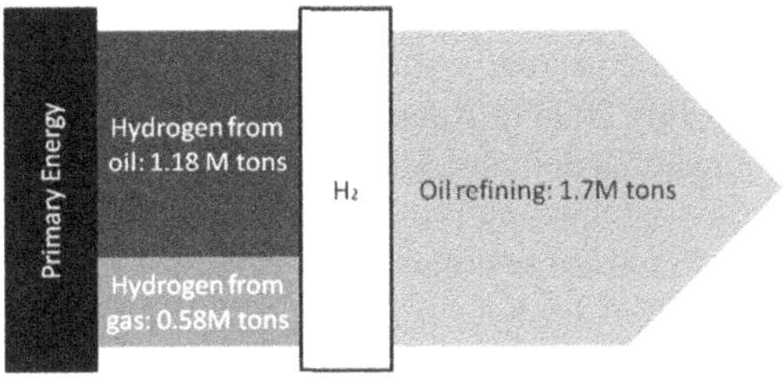

Fig. 10.9 Hydrogen consumed in processing crude oil in China

imported crude oil has increased rapidly. To lower costs, a considerable amount of high-sulfide crude oil has been purchased. With the raised level of sulfur crude oil being processed and higher requirements for oil demanded by national regulations, hydrogen has been increasingly consumed in refining processes.

The hydrogenation of oil consumes hydrogen. In this process, fresh hydrogen is mainly used in four areas—chemical reactions, dissolution loss, equipment leakage, and exhaust emission loss. Most of the hydrogen is consumed in chemical reactions, such as in the removal of sulfur, nitrogen, and oxygen in the incoming materials; olefin and aromatics saturation; hydrocracking; and open-loop systems. Typically, hydrogen consumption in refineries with crude oil hydroforming accounts for 0.8–2.7 % of the crude processing volume (Ren Hongli 2008). Among the various refinery processes, two main areas demand the use of hydrogen—hydrofining and hydrocracking. The amount of hydrogen consumed per ton of crude oil in hydrofining is 0.974 % whereas that per ton of crude oil in hydrocracking is 1.888 % (Hou Zhen 2004). The production capacity with hydroforming equipment in China is almost 30 % that of the crude processing capacity [11]. This proportion will increase as higher requirements are placed on the quality of oil products and the country's demand for crude oil and high-sulfur crude oil rises.

China's crude processing volume in 2007 amounted to 378,326,000 tons. In processing such a quantity of crude oil, nearly 1,700,000 tons of hydrogen was consumed. The amount of hydrogen produced from oil accounted for around 67 % of the hydrogen consumed in oil processing. Since the amount of hydrogen is insufficient, it is supplemented by simple hydrogen-production units, usually through the installation of natural gas-conversion devices. Figure 10.9 shows the consumption of hydrogen during crude oil processing in China.

10.2.4 Commercial Hydrogen

Currently, some hydrogen is also consumed in other industrial areas, such as in making semiconductors, metallurgy, food processing, float glass, fine organic synthesis, aeronautics and astronautics, and demonstrating operations of alternative-energy vehicles. Such commercial usage of hydrogen accounts for a small proportion of all the hydrogen consumed in China, and it does not have a great effect on the total consumption level. However, it is clearly increasing. The amount of hydrogen used directly as transport fuel is still very limited. At present, fuel-cell vehicles consume a small amount of hydrogen in demonstrations. For example, from July 2008 to August 2009, city buses driven by hydrogen fuel cells went into commercial demonstration operation in Beijing's public transport system. Three of these passenger-carrying vehicles traveled a total of 53,800 km, consuming a total of 5.33 tons of hydrogen.

10.3 Circulation of Hydrogen in China

It is evident from current production and consumption in China that industrial hydrogen is mainly used as a raw material in chemical synthesis, chiefly in the large-scale manufacture of ammonia, oil, and methanol. Hydrogen consumed by these enterprises accounts for over 95 % of all the hydrogen consumed. As is shown in Fig. 10.10, hydrogen consumption in China increased from about 7 million tons in 2000 to 12 million tons in 2007, with an annual average growth rate of over 9 %. In recent years, the output of ammonia has maintained an annual average growth rate of 8.8 %, accounting for around 75 % of the hydrogen consumed. Oil consumption has risen rapidly, with an average growth rate of 7.17 % per year. With increasing demand for oil, the import volume of high-sulfur crude oil has risen; as increased requirements are made for the quality of oil products, the amount of hydrogen consumed by oil refineries will continue rising. The production of methanol in different parts of China has developed rapidly, with an annual average growth rate of 38.7 %. Accordingly, hydrogen consumption has also risen.

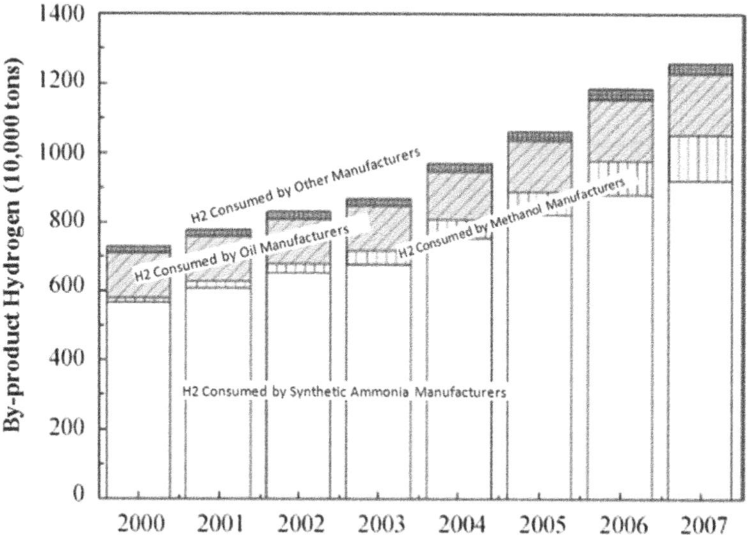

Fig. 10.10 Hydrogen consumption in China's industries from 2000 to 2007

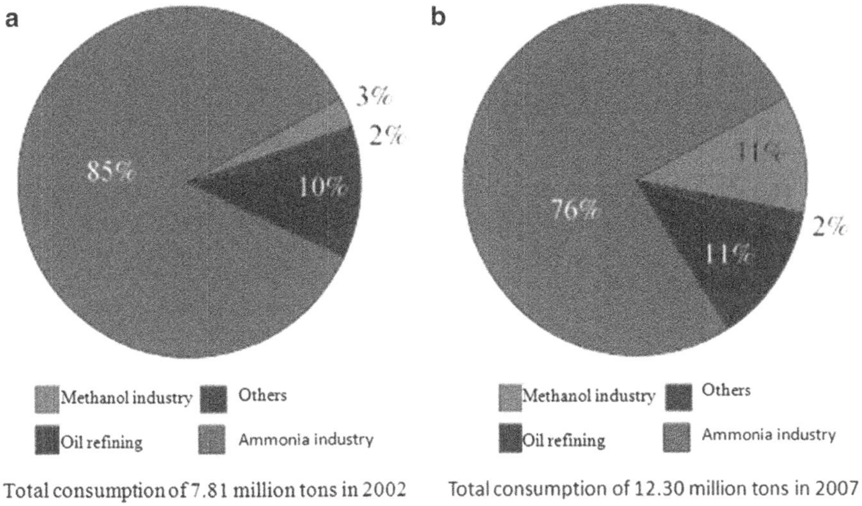

Fig. 10.11 Proportion of hydrogen consumed by different industries (for 2002 and 2007)

Owing to the uneven development of various industries, there is a considerable difference in the proportion of hydrogen that each consumes. As seen in Fig. 10.11, although synthetic ammonia still requires the greatest amount of hydrogen, its proportion of the total amount of hydrogen consumed by all industries has declined sharply. Its proportion declined from 85 % in 2002 to 76 % in 2007. Among all industries, oil refining is still the second-biggest consumer of hydrogen, and the proportion of hydrogen used by that industry showed a small increase, rising from 10 to 11 % over a 5-year period. This can be mainly attributed to growing gasoline and diesel consumption caused by the increase in the number of vehicles on China's roads. Owing to a substantial increase in the output of methanol in China in recent years (from 2.1 million tons in 2002 to 10.72 million tons in 2007—an increase of about fourfold), hydrogen consumption has also increased rapidly. The proportion of hydrogen consumed in methanol production increased from 3 % in 2002 to 11 % in 27 %, which is almost as high as the consumption by oil refining.

Hydrogen in China mainly includes that produced from fossil fuels, such as coal, oil, and gas, that recovered from various industrial tail gases, by-product hydrogen produced by chlor-alkali plants, and hydrogen derived by water electrolysis. Owing to the difficulty in storage and transportation, most hydrogen is produced locally and consumed nearby. Figure 10.12 shows the circulation of hydrogen in production and consumption activities in China in 2007. Of all the hydrogen produced, 7,126,400 tons of hydrogen was produced from coal, 2,879,700 tons from natural gas, and 2,439,400 tons from heavy oil in petroleum. China has rich by-product hydrogen resources. A total of 5,299,400 tons of by-product hydrogen was produced in the coking and steel industries; 415,700 tons of by-product hydrogen was produced in the soda industry. If great quantities are not required, hydrogen can be produced by the electrolysis of water. In terms of consumption, hydrogen in China is mainly used

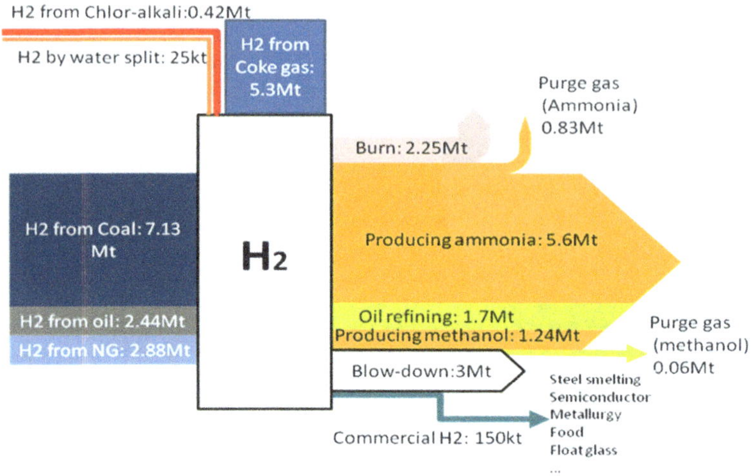

Fig. 10.12 Circulation of hydrogen in China in 2007

to produce ammonia, methanol, and oil products, and this usage consumed about 11,000,000 tons of hydrogen. Hydrogen is being increasingly used in such industrial areas as steel smelting, making of semiconductors, metallurgy, food processing, float glass, fine organic synthesis, and aerospace. In addition, over 5,000,000 tons of hydrogen is not properly used in such areas as coke oven gas and industrial purge gas: it is simply burned or automatically blown down. This represents a significant source of future commercial hydrogen.

10.4 China's Policy for Hydrogen as Transport Energy Source

10.4.1 National Guidelines on Medium- and Long-Term Program for Science and Technology Development (2006–2020): Basic Research Program for New-Energy Vehicles

In light of the degree of development of hydrogen energy and its technological applications, the National Guidelines on Medium- and Long-Term Program for Science and Technology Development (2006–2020), which was promulgated by the State Council of the People's Republic of China on February 9, 2006, outlined the following programs: 62 priority development themes, 22 cutting-edge technologies, and 10 basic research areas that had strategic significance for China. The guidelines included energy-saving and new-energy automobiles as one of the six priority development themes for the transport industry, hydrogen energy and fuel-cell technology as one of the four cutting-edge energy technologies, efficient energy material technology as one of the three cutting-edge new material technologies, and key scientific problems in energy sustainable development as one of the ten basic research areas with strategic significance for China.

10.4.1.1 Number 36 of 62 Priority Development Themes: Energy-Saving and New-Energy Automobiles

Efforts will be made in the following areas: research and development design; integration and manufacturing technology of hybrid electric vehicles, alternative-fuel vehicles, and fuel-cell vehicles; integration and control technology of power systems; automobile computing platform technology; technology for key parts of fuel-efficient, low-emission internal combustion engines, fuel-cell engines, power batteries, and drive motors; new-energy test vehicles; and infrastructure technology.

10.4.1.2 Number 15 of 22 Cutting-Edge Technologies: Hydrogen Energy and Fuel-Cell Technology

Efforts will be made in the following areas: studying efficient and low-cost technologies for hydrogen production from fossil fuels and renewable energy and for hydrogen storage and delivery; techniques for the preparation of key components for fuel cells and the assembly of stacks; techniques for electricity generation from fuel cells and assembly of automotive power systems; and technical specifications and criteria for hydrogen energy and fuel cells.

10.4.1.3 Number 11 of 22 Cutting-Edge Technologies: Efficient Energy Material Technology

Efforts will be made in the following areas: studying relevant materials for solar cells and their key technologies, fuel-cell key material technology, high-capacity hydrogen-storage material technology, efficient secondary battery materials and their key technologies, key materials for supercapacitors and preparation technology, and developing efficient energy-conversion and storage-material systems.

10.4.1.4 Number 6 of 10 Basic Research Areas with Strategic Significance for China: Key Scientific Problems in Energy Sustainable Development

Efforts will be made in the following areas: studying the physical and chemical basis for high efficiency and clean utilization and energy conversion of fossil energy; key scientific problems in high-performance conversion of heat into power and efficient energy conservation and storage; principles and new ways of large-scale utilization of renewable energy; theories on safe, stable, and economical operation of power grids; and the scientific basis for essential technology for large-scale nuclear energy and hydrogen technology.

10.4.2 China's Energy Conditions and Policies

China's Energy Conditions and Policies, which was released by the Information Office of the State Council of the People's Republic of China in December 2007, defined energy development strategies and objectives for China, including hydrogen and its application technology. In the section of this document entitled "Accelerating the Progress of Energy Technologies," the following areas were identified as frontier technology research: hydrogen production using renewable energy; efficient, low-cost technologies for hydrogen storage and delivery; techniques for the preparation of the key components of fuel cells and the assembly of stacks; and techniques for electricity generation from fuel cells and assembly of automotive power system. Also, basic theories on large-scale utilization of nuclear energy and hydrogen technology were designated an area of basic scientific research. The section of the document entitled "Strengthening International Cooperation in the Field of Energy" stated that it was necessary to "promote the cooperation of the international community in important energy technologies, such as renewable energy, hydrogen energy, and nuclear energy, and to establish a clean, economical, safe and reliable energy-supply system for the world in the future" and that this should be adopted as a goal of development and advanced technology.

10.4.2.1 Accelerating the Progress of Energy Technologies

Greater efforts will be made in studies related to cutting-edge technology. Cutting-edge technology can help tap the development potential of energy and assist the energy industry and energy technologies in achieving rapid development. China will focus on studies of technologies for hydrogen production from fossil fuels, biomass energy, and renewable energy; it will also examine efficient, low-cost hydrogen storage and delivery, techniques for preparing the key components of fuel cells and the assembly of stacks in addition to techniques for generating electricity from fuel cells and the assembly of automotive power systems.

Efforts will be made to conduct basic scientific research, which is the source of independent innovation. Such research determines the foundation and capacity of energy development. China will focus on studies of basic theory regarding high efficiency and clean utilization and conversion of fossil energy; it will also examine key principles of high-performance conversion of heat into power and efficient energy conservation and energy storage. China will examine basic theories regarding essential technologies related to the large-scale utilization of renewable energy and technologies of the large-scale use of nuclear and hydrogen energy.

10.4.2.2 Strengthening International Cooperation in the Field of Energy

Efforts will be made to form a development and extension system for advanced technology. Energy conservation and the promotion of energy diversification form part of a long-term plan to ensure global energy security. The international community needs to vigorously improve research, development, and promotion efforts regarding energy-saving technology; promote the comprehensive utilization of energy; and provide assistance to countries in improving their energy efficiency. Efforts should be made to actively promote cooperation in the efficient use of fossil fuels, such as with clean coal technology. It is also necessary to boost cooperation in the international community with important energy technologies, such as renewable energy, hydrogen energy, and nuclear energy, as well as establishing a clean, economical, safe, reliable energy-supply system for the future. From the viewpoint of sustainable human development, the international community should properly deal with such problems as capital input, intellectual property protection, and promotion of advanced technology to enable all countries to benefit from them and share the fruits of human progress.

10.4.3 Law of the People's Republic of China on Conserving Energy: Encourage Development and Utilization of Renewable Energy

The Law of the People's Republic of China on Conserving Energy was adopted at the 30th Session of the Standing Committee of the Tenth National People's Congress of the People's Republic of China on October 28, 2007. It came into force on April 1, 2008. The law notes, "The State encourages the development and utilization of new and renewable resources of energy" and "the State encourages the development and promotion of clean fuels and oil alternative fuels used by means of transport."

Article 7 declares that the state will implement an industrial policy that is conducive to energy conservation and environmental protection. It will restrict the development of industries characterized by high energy consumption and high pollution, and it will develop energy-saving and environmentally friendly industries.

The State Council and the governments of provinces, autonomous regions, and municipalities shall strengthen their energy conservation work and rationally adjust industrial structures, enterprise structures, product structures, and energy consumption structures. This will prompt enterprises to reduce their energy consumption for unit output and unit products; eliminate outdated production capacities; improve the development, processing, conversion, transmission, storage, and supply of energies; and improve their energy use efficiency. The state encourages and supports the development and utilization of new energies and renewable energies.

Article 45 declares that the state encourages the development, production, and use of energy-saving, environmentally friendly cars, motorcycles, railway locomotives, ships, and other transport vehicles. It will implement a system whereby old and outdated vehicles are abandoned and replaced. The state encourages the development and popularization of the use of clean fuel and oil substitute fuel for transportation.

10.4.4 Policy Outline of China Energy Conservation Technology (2006): Hydrogen and Its Application Technology

The Policy Outline of China Energy Conservation Technology, which was promulgated by the National Development and Reform Commission (NDRC) and the Ministry of Science and Technology in December 2006, discussed hydrogen and its application technology in terms of industrial energy saving and transport energy saving. These areas were also addressed in terms of energy conservation technology in petrochemical production, research and development of special high-performance metals and metal-matrix composites, popularization of alternative automobile fuel technology, and the development of new marine transport technology. The development of hydrogen optimization technology with hydrogen units was included as part of energy conservation technology in petrochemical production; promoting fuel cells and promoting rare-earth hydrogen-storage materials were included in research and development of special high-performance metals and metal-matrix composites; research and development of vehicles driven by new power, such as electric vehicles and hydrogen-powered vehicles, was included in popularization of alternative automobile fuel technology; and promoting the use of clean energy such as fuel cells in ships was included in development of new marine transport technology.

10.4.4.1 Energy Conservation Technology in Petrochemical Production

For atmospheric and vacuum distillation unit in refineries, technology should be adopted that optimizes energy-efficient processes, such as heat transfer and preflashing. For catalytic cracking units, it is necessary to promote technology that can reduce coke yield and the coking of units. For aromatic extraction processes, it is necessary to promote highly efficient solvent (such as tetraethylene glycol and sulfolane) technology. Hydrogen optimization technology has to be developed for hydrogen units. It is important to research and develop filter-absorption regeneration methods of low consumption. Extraction distillation techniques need to be promoted.

Efforts have to be made to research and develop optimization technology for the hot high-pressure separator process of hydrogenation units. In addition the following steps should be taken: adopt hydraulic turbines to recover pressure energy; develop and apply new hydrogenation catalysts, advanced inner structures of reactors, and circulating hydrogen desulphidation measures; promote large and equiflux heater technology for delayed coking units; and promote thermal combination technology among units.

10.4.4.2 Research and Development of Special High-Performance Metals and Metal-Matrix Composites

It is necessary to research and develop new highly efficient energy conversion and storage units and materials; the application and development of fuel cells, solar cells, metal-air cells, ultracapacitors, and relevant materials should be promoted. It is important to research, develop, and popularize Nd-Fe-B permanent magnetic materials, high-performance rare-earth luminescent materials, and rare-earth hydrogen-storage materials.

10.4.4.3 Popularization of Alternative Automobile Fuel Technology

According to local conditions, efforts should be made to popularize the technology for alternative fuels for vehicles, such as those using natural gas, alcohol fuels, synthetic fuels, and biodiesel. Research and development should be carried out for vehicles driven by new power sources, such as electric vehicles and hydrogen-powered vehicles.

10.4.4.4 Development of New Marine Transport Technology

It is necessary to conduct research into and promote marine transport technology using liquefied natural gas and compressed natural gas. It is also important to research, develop, and promote new alternative fuels for ships and moderately apply clean energy, such as fuel cells, to ships.

10.4.5 China National Plan for Coping with Climate Change: Fuel-Cell and Hydrogen Technology

According to the China National Plan for Coping with Climate Change, which was formulated by the NDRC in June 2007, fuel-cell and hydrogen technology is one of the technologies to reduce the discharge of greenhouse gas. Details of that plan are as follows. China is currently undergoing large-scale infrastructure construction, and it relies heavily on important technologies that will help reduce the emission of greenhouse gases. The main technologies it requires include advanced energy and manufacturing technology, environmental protection and resource comprehensive utilization technology, efficient transportation technology, new material technology, and new building material technology. The application and popularization of the following technologies in China will exert a great influence on reducing greenhouse gas emissions: coal-fired electricity-generation technology featuring high efficiency and low pollution, technology of large hydroelectric generating sets, new nuclear power technology, renewable energy technology, building energy-saving technology, technology for clean gas-fired vehicles and hybrid electric vehicles, urban mass transit technology, fuel-cell and hydrogen technology, oxygen-enrichment and coal injection smelting and long-life technology, technology for rebuilding small and medium-sized nitrogenous fertilizer production units, technology for new materials in road construction, and technology for new wall materials.

10.5 Development and Examples of Hydrogen-Powered Fuel-Cell Vehicle Technology

Since 2002, under the support of national projects promoted by the Ministry of Science and Technology, such as the Electric Vehicle Key Projects and Energy Conservation and New-Energy Vehicle Project, China began to vigorously conduct research and development of technology related to fuel-cell electric vehicles. This included fuel-cell city buses and other vehicles, and it involved conducting systematic studies of fuel-cell systems, hydrogen-storage systems, electric-drive systems, and vehicle integration and control technology.

10.5.1 Development and Products of Fuel-Cell City Buses

Since 2002, the R&D team of fuel-cell city buses at Tsinghua University has developed four generations of hydrogen fuel-cell city bus technology. As of December 2010, 15 fuel-cell buses of different types had been manufactured by various enterprises; among those, seven had been put into commercial operation. In addition, Shanghai Automotive Industry (Group) Corporation (SAIC) and Tongji University also developed fuel-cell buses in 2010. Figure 10.13 shows the development history of fuel-cell bus technology at Tsinghua University.

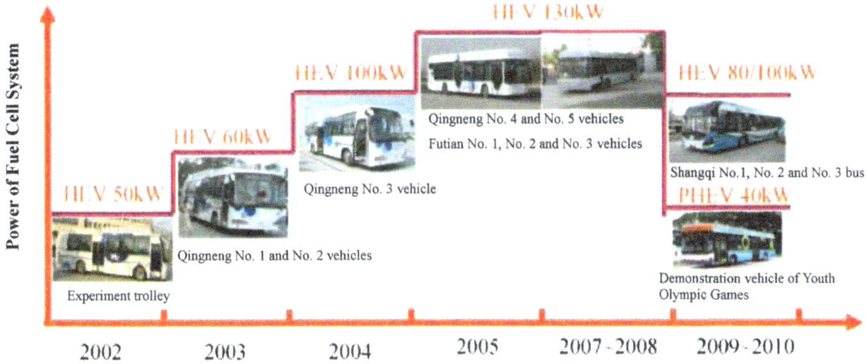

Fig. 10.13 Research and development process of fuel-cell electric bus powertrain

Development History of Fuel-Cell Bus at Tsinghua University (Fig. 10.13)

1. In 2002, Tsinghua University successfully developed a city experiment trolley powered by fuel cells and power batteries. The power of the fuel-cell system was 50 kW. NiMH batteries were adopted, and the storage pressure of the hydrogen was 20 MPa.
2. Based on this, the university successfully developed two fuel-cell buses—the Qingneng No. 1 and No. 2—in 2003. The power of the fuel cells was increased to 60 kW. NiMH power batteries were used with a capacity of 80 Ah.
3. Qingneng No. 3 fuel-cell buses were successfully developed in 2004; the power of the fuel-cell system was increased to 100 kW. The capacity of the power batteries was still 80 Ah.
4. In 2005, the university developed the second generation of fuel-cell bus, which was mainly characterized by the low-entrance design of the vehicle chassis. Two sample vehicles were developed—the Qingneng No. 4 and No. 5. The power of the fuel cells was further increased to 130 kW.
5. In 2007, the university successfully developed the third generation of fuel-cell bus, which was mainly distinguished by the low-floor technology adopted. The three fuel-cell buses produced—the Futian No. 1, No. 2, and No. 3—fulfilled their tasks during the Beijing Olympics and also carried passengers for 1 year.
6. In 2009, the university successfully developed the fourth generation of fuel-cell buses, which achieved single-stack independent operation of the fuel-cell systems. A double-stack structure (two 40-kW fuel-cell stacks) was adopted for the fuel-cell system. The total power of the fuel cells was reduced to some extent. Lithium-ion power batteries (capacity, 130 Ah) were adopted for the power batteries. The storage pressure of the hydrogen was 35 MPa. The three passenger cars developed—Shangqi No. 1, No. 2, and No. 3—provided passenger-carrying service during the 2010 Shanghai World Expo.

Table 10.1 Performance comparison among main fuel-cell city buses

Model	BJ6123C6N4D	SWB6129FC	KLQ6129GQH2
Manufacturer	Beiqi Foton	Shanghai Sunwin	Suzhou Jinlong
Year of approval	2008	2010	2009
Vehicle length (mm)	11,980	11,990	12,000
Unladen mass (kg)	14,000	14,500	14,500
Power (kW)	100	80 or 100	100
Battery and capacity	NiMH, 80 Ah	Li-ion, 80/110 Ah	Li-ion, 110 Ah
Max speed (km/h)	≥80	–	80
Range (km)	≥200	–	400
H_2 pressure (MPa)	20	35	35
Number of passengers	49	53	50

7. In 2010, the university successfully developed a chargeable fuel-cell hybrid electric vehicle—the Haige No. 1. This adopted a fuel-cell system with 40 kW power. Lithium-ion power batteries were adopted. The storage pressure of the hydrogen was 35 MPa. These fuel-cell buses were used in the first Youth Olympic Games held in Singapore in 2010.

In terms of technology, domestic fuel-cell buses were initially driven by hybrid power, combining fuel cells with power batteries. With improvements in fuel-cell system technology, the installed power of the fuel cells increased—from 60 kW in 2003 to 130 kW in 2008. In addition, a single-stack independent operational mode with a double fuel-cell stack was successfully demonstrated in 2010. Currently, the university is attempting to develop low-power fuel-cell stacks and chargeable hybrid electrical structures that feature low cost and high durability (like the demonstration vehicles used in the 2010 Youth Olympic Games).

Fuel-cell buses that have already been approved for demonstration operation thus far include the BJ6123C6N4D fuel-cell hybrid city bus manufactured by Beiqi Foton Motor Co., Ltd. (Beiqi Foton) and the SWB6129FC series fuel-cell city bus manufactured by Shanghai Sunwin Bus Corporation (Shanghai Sunwin). Their main technical parameters are given in Table 10.1, which also lists such details for fuel-cell vehicles manufactured by King Long United Automotive Industry (Suzhou) Co., Ltd. (Suzhou Jinlong).

10.5.2 Development and Products of Fuel-Cell Cars

The technological research and development work for fuel-cell cars in China is mainly carried out at Tongji University and Shanghai Fuel Cell Vehicle Powertrain Co., Ltd. As of December 2010, four generations of hydrogen fuel-cell cars had been successively developed, and the technology had been applied to almost 100 fuel-cell cars of different structures manufactured by various enterprises. In addition, a large number of cars using such technologies have been put into commercial demonstration operations at home and abroad. The main development of fuel-cell car technology at Shanghai Fuel Cell Vehicle Powertrain Co., Ltd. proceeded as follows:

1. In 2003, the company developed a sample fuel-cell car based on the Santana 2000—the Beyond No. 1. This sample car was mainly powered by a 20-kW fuel-cell engine and was equipped with a 50-Ah high-power lithium-ion battery. It had an electric-electric hybrid structure.
2. In 2004, the company successfully developed its second-generation sample fuel-cell car—the Beyond No. 2. This car joined the 2004 Challenge Bibendum in France for international clean-energy vehicles.
3. The Beyond No. 3 took part in the 2006 Challenge Bibendum for international new-energy vehicles. This adopted the Chery B11 as the prototype vehicle, which was equipped with the third-generation fuel-cell power system platform technology. The power of the fuel-cell engine was increased to 50 kW, and the capacity of the lithium-ion power battery was reduced to 15 Ah.
4. The new generation of fuel-cell car developed in 2007 used the Roewe automobile of the SAIC company as its basis. The capacity of the lithium-ion power battery was further reduced to 8 Ah and the power of the fuel cell increased to 55 kW; the maximum speed was 150 km/h and its continuous mileage rose to 300 km.
5. In 2010, 90 fuel-cell sedans went into demonstration operation at the Shanghai World Expo. The brands included the Roewe 750 of SAIC, the Lingyu of Shanghai Volkswagen, the Zhixiang of Chang'an, the Faw-Besturn, and the Eastar of Chery. Table 10.2 shows the main performance of domestic fuel-cell cars that took part in that operation.

10.5.3 Development of Fuel-Cell System

China's vehicle fuel-cell system is simultaneously dedicated to research and development with the technology and manufacture of products along two lines—constant pressure and pressurization system. The main research and development of pressurized fuel-cell system technology has been carried out by Dalian Chemical Physics Institute of the Chinese Academy and Sunrise Power Co., Ltd. In addition, Shanghai Shen-li High Tech Co., Ltd. and Wuhan University of Technology are involved in the research and development of constant-pressure systems.

Due to many years' development, China's vehicle fuel-cell systems have made great improvements in performance. This has particularly been the case after these technologies were identified as important with respect to energy-saving and new-energy automobiles in the Twelfth Five-Year Plan and fuel-cell vehicles went into demonstration operation at the Beijing Olympic Games and Shanghai World Expo. Such factors as the fuel-cell system's energy-conversion efficiency, unit mass and

Table 10.2 Main performance of fuel-cell cars in China (approved for demonstration in 2010)

Model		CA7904FC	SC7003EV	SQR7000 FEB11	CSA7000FCEV	CSA7001FCEV (PHEV)	CSA7002FCEV
Manufacturer		FAW Group	Changan	Chery	SAIC		
Unladen mass (kg)		1,800	1,780	1,875	1,975	1,846	1,881
Fuel cell	Power (kW)	55			45	25	55
	Supplier/model	Shanghai Shen-li SLHC-55K			Sunrise Power/STK-U-03A	Sunrise Power/STK-U-06D	GM FCS4.6
Battery	Capacity (Ah)	7.5Ah/li-ion				22Ah/ Li-ion	7Ah/ NiMH
	Supplier	Suzhou Phylion Battery				Johnson Controls	
H_2 pressure (MPa)		35	35	35	35	35	70

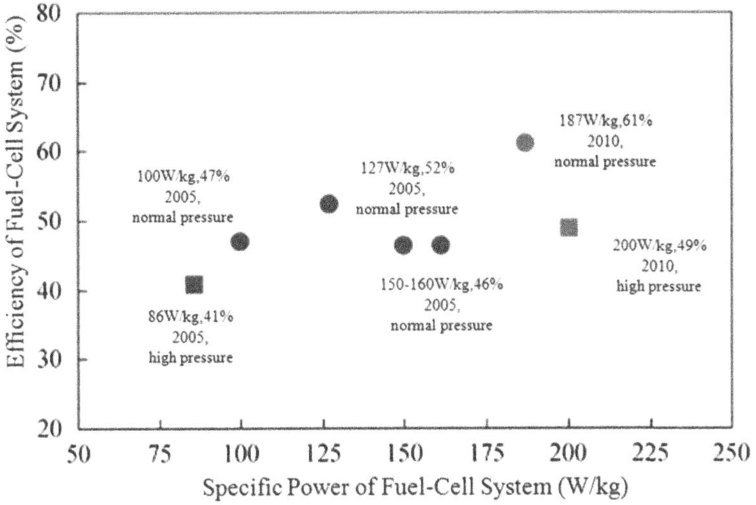

Fig. 10.14 Comparison between fuel-cell systems in terms of performance

power, durability, reliability, and environmental suitability are better able to handle harsh working conditions than they were in 2005. The vehicles can basically meet all operation demands in particular areas.

Figure 10.14 shows a comparison between two fuel-cell systems—the constant-pressure system and pressurized system—in energy efficiency and mass power density (Zeng Zijian 2011). It is evident from the figure that after 5 years of research and development, the vehicle fuel-cell system has shown remarkable improvement in terms of energy conversion efficiency. The efficiency of the pressurized system increased from 41 % in 2005 to 49 % in 2010, with a 20 % improvement in performance. Even the constant-pressure system is characterized by high system efficiency, showing an increase in efficiency from 46–52 % to 61 %—an improvement of over 20 %. In addition, the fuel-cell system is more compact and lighter, and it is clearly more efficient. For example, the mass specific power of the pressurized system more than doubled: it increased from 86 W/kg in 2005 to 200 W/kg in 2010. However, the mass specific power of the constant-pressure system also increased by more than 40 %.

Breakthroughs have also been made in the durability of vehicle fuel-cell systems, which are the main technical barriers to commercialization, and optimal solutions have been found (Yi Baolian and Hou Ming 2011). Fuel-cell buses have been put into operation on the Number 801 bus line in Beijing, where they have had over 1,500 h of service and traveled over 20,000 km. After analyzing the attenuation mechanism of fuel cells using a fast evaluation method that simulates the operating condition of dynamic load on the vehicle fuel cell, it has been found that potential scanning under high potential and dynamic operation conditions during the process of start-up, parking, and idling is the main cause of catalyst and carrier attenuation. Therefore, it was recommended that countermeasures be made in two areas—system control strategy and material improvement and innovation. Flexible operation can be adopted for system control strategy (Ouyang Minggao et al. 2009). In terms of material improvement and innovation, the following areas need to be studied: a highly stable electrocatalyst, corrosion-resistant catalytic carrier, antioxidant proton-exchange membrane, orderly membrane electrode assembly, and a conductive and corrosion-resistant metal bipolar plate.

Regarding environmental suitability, by studying the influence on materials and parts exerted by the icing of water in a proton-exchange membrane fuel cell at sub-zero temperatures, Chinese researchers have proposed the strategy of storage and start-up at such temperatures and achieved storage and start-up of fuel-cell system at –10 °C. At the same time, they have conducted studies into environmental conditions, e.g., studies of the effects of nitrogen oxides and sulfur oxides in the air on fuel cells and corresponding countermeasures.

10.5.4 Operation of Hydrogen Fuel-Cell Vehicles

The adaptability of hydrogen fuel-cell technology to actual road conditions and different geographic environments has been studied in vehicles worldwide with fuel-cell buses and sedans. Cases in point are projects related to fuel-cell electric vehicles in Europe (HyFLEET:CUTE project), the demonstration project of the California Fuel Cell Partnership, and the Japan Hydrogen & Fuel Cell Demonstration Project. In line with its own economic and technological development levels, China has since 2006 conducted commercial demonstration projects using domestic and foreign fuel-cell city buses in Beijing and provided special services at the Beijing Olympic Games. It has also operated such vehicles at the Shanghai World Expo and at the Singapore Youth Olympic Games in 2010. A total of 10 fuel-cell city buses and 110 fuel-cell sedans have taken part in demonstration projects throughout the country.

10.5.4.1 Commercial Operation of Fuel-Cell Buses in Beijing

A commercial demonstration project for fuel-cell buses in Beijing has been jointly supported by the United Nations Development Program and the Global Environment Fund and is a part of the HyFLEET:CUTE project of the European Union. The Beijing project was initiated in March 2003. After construction of the infrastructure and production of vehicles, three Mercedes-Benz Citaro low-floor fuel-cell buses, manufactured by Daimler Chrysler Automotive Co., Ltd., carried passengers from June 2006 to October 2007. Three BJ6123C6N4D low-floor fuel-cell hybrid electric buses, manufactured by Beiqi Foton Motor Co., Ltd., carried passengers from August 2008 to July 2009. Table 10.3 shows the basic performance parameters of the vehicles used in the project.

Table 10.3 Performance of fuel-cell buses used in Beijing Demonstration Project

Model	Citaro (Mercedes-Benz)	BJ6123C6N4D (Beiqi Foton)
Overall dimension ($L \times W \times H$) (m)	11.95 × 2.55 × 3.688	11.98 × 2.55 × 3.45
Unladen mass (kg)	14,300	14,200
Power of fuel-cell system (kW)	(Ballard HY205)/205	(Sunrise, Shen-li)/80 or 100
Efficiency of fuel-cell system (%)	43–38	–
H_2 pressure	35 MPa	25 MPa
Range (km)	150–250	>200

Table 10.4 Results of fuel-cell buses in Beijing Demonstration Projects

Index	2006.6–2007.10 (Citaro)	2008.8–2009.7 (BJ6123C6N4D)
Kilometers traveled	84,922	60,198
Running time of stack (h)	5,358	3,646
Number of passengers	56,851	39,995
Fuel consumption (kg)	16,621	5,753
Fuel consumption (kg/100 km)	19.57	9.56
Passengers per 100 km	66.9	66.4

During 16 months of operation, the three Citaro fuel-cell vehicles traveled a total of 95,116 km along the bus line (the distance traveled when carrying passengers was 84,922 km), carried 56,000 passengers, and consumed 16,621 kg of hydrogen (i.e., they consumed on average 19.57 kg of hydrogen per 100 km). The total running time of the fuel-cell system was 5,700 h (the effective time excluding that taken for debugging was 5,358 h). During 12 months of operation, the three Beiqi Foton fuel-cell vehicles traveled a total of 75,460 km along the bus line (the distance traveled when carrying passengers was 60,198 km), carried 40,000 passengers (i.e., they carried on average 66.4 passengers per 100 km), and consumed 5,753 kg of hydrogen (i.e., they consumed on average 9.56 kg of hydrogen per 100 km). The total running time of the fuel-cell system was 3,646 h. Details of the operation are given in Table 10.4 (Lun 2008).

With respect to vehicle length and unladen mass, the two types of vehicles were similar. Their running environments (both ran along the same line throughout the year; the average numbers of passengers they carried per 100 km was similar) were basically the same. Despite this, the fuel economy of the two was significantly different. The 9.56 kg of hydrogen consumed per 100 km by the Beiqi Foton was 48.9 % that consumed by the Citaro (19.57 kg). The economic efficiency of the fuel thus improved by over 50 %. The main reason for this was that the Citaro was powered entirely by fuel cells whereas the Beiqi Foton was driven using both fuel cells and power batteries (i.e., hybrid drive scheme). When vehicles are powered by both fuel cells and power batteries, the economic efficiency of the fuel is improved and the amount of hydrogen consumed per 100 km can be reduced. The hybrid drive scheme has several advantages. The requirements for fuel cells are not as high, and

Table 10.5 Main parameters of fuel-cell vehicles used during the 2008 Beijing Olympic Games

Model	PASSAT Lingyu fuel-cell vehicle (Shanghai Volkswagen)	BJ6123C6N4D (Beiqi Foton)
Overall dimensions ($L \times W \times H$) (m)	4.789 × 1.765 × 1.44	11.98×2.55×3.45
Unladen mass (kg)		14,200
Power of fuel-cell system (kW)	40	80 or 100 (Sunrise Power, Shanghai Shen-li)
Capacity of power battery (Ah)	8 (lithium-ion battery)	80 (NiMH battery)
Hydrogen-storage system (MPa)	35	20
Range (km)	>300	>200
Number of vehicles	20	3

Table 10.6 Operation of fuel-cell vehicles during the Beijing Olympic Games

Index	Fuel-cell sedan (July to September)	Fuel-cell bus (July to September)
Number of vehicles	20	3
Total running distance (km)	73,000	21,142
Daily running distance of each vehicle (km)	40	78
Hydrogen-refueling pressure (MPa)	35	20
Hydrogen refueling (number of times)	712	115
Average refueling duration each time (min)	3	10
Fuel consumption (kg/100 km)	1.2	10.8

so energy consumption of the auxiliary system required by the high-power fuel-cell system is reduced. The braking energy can be recycled effectively when the vehicle brakes. This advantage was adopted in the second generation of Citaro products.

10.5.4.2 Demonstration Operation in the 2008 Beijing Olympic Games

During the Beijing Olympic Games and Paralympic Games held from July to September 2008, 595 new-energy vehicles, including three fuel-cell city buses, 20 fuel-cell sedans, and other vehicles (e.g., hybrid electric vehicles and pure electric vehicles), were involved in a large-scale demonstration project. Among them, the fuel-cell electric vehicles were mainly used as guiding vehicles and as support cars in the marathon, in addition to serving Olympic officials. Table 10.5 shows the main parameters of the vehicles used in those operations. Table 10.6 presents data about the vehicles used during the demonstration period (the data derive from the Beijing hydrogen-refueling station).

Results of the operation showed that the traveling distance of the 20 Shanghai Volkswagen Lingyu fuel-cell sedans was 73,000 km and that of the three Beiqi Foton fuel-cell buses was 21,000 km. During the 90 days of operation, the sedans traveled an average of 40 km a day and the buses traveled an average of 78 km a day. During this period, the fuel-cell sedans consumed 1.2 kg of hydrogen per 100 km, whereas

Fig. 10.15 Two types of fuel-cell buses operated at the Shanghai World Expo that used different technologies (*left*, technology of Tsinghua University; *right*, technology of Tongji University)

the fuel-cell buses consumed 10.8 kg. Regarding the hydrogen-refueling time, the average duration for the fuel-cell sedan was 3 min (refueling pressure, 35 MPa), whereas that for the fuel-cell buses was 10 min.

10.5.4.3 Demonstrating Operation at the 2010 Shanghai World Expo

During the 2010 Shanghai World Expo, 196 fuel-cell vehicles, including six city buses, 90 sedans, and 100 sightseeing buses, were involved in a demonstration operation. The fuel-cell sedans included the Roewe of SAIC, the Linyu of Shanghai Volkswagen, the Chang'an Zhixiang, Faw-Besturn, and Eastar. The sedans used a new generation of fuel-cell power system platform technology. See Table 10.2 for their main performance parameters. The six fuel-cell city buses were manufactured by Shanghai Sunwin Bus Corporation, which uses the fourth generation of fuel-cell bus power technology developed at Tsinghua University and the *bybric* power system (dual fuel-cell system) technology developed at Tongji University. Figure 10.15 shows the fuel-cell buses, which rely on different technologies.

Wang Zhe (2010) summarized the running conditions of these 196 fuel-cell buses during the Shanghai World Expo. The basic conditions were as follows. The fuel-cell buses began running along the national exhibition line within the Expo site on May 18. The fuel-cell buses that used the technology of Tsinghua University traveled a total of 6,550 km from May to October 2010. They consumed an average of 12 kg of hydrogen per 100 km. The 90 fuel-cell sedans were used by the Administrative Center of the World Expo and Reception Office of the Municipal People's Government. They were used for guests who needed cars at the Jiyang Road base and Anting base in Shanghai. Vehicles serving at the Jiyang Road base included 50 Roewe sedans; those at the Anting base included Linyu, Besturn, Zhixiang, and Eastar sedans. The 100 fuel-cell sightseeing buses ran along the Elevated Pedestrian Walk and the North Ring Road within the expo site. The vehicles were dispatched 8,883 times, carried 1,540,130 people, and traveled a total of 487,629 km.

10.6 Analysis of Energy Flow

Starting with the commercial demonstration project of Beijing fuel-cell vehicles carried out in July 2008, three fuel-cell city buses manufactured by a domestic enterprise (Beiqi Foton) were put into operation for 1 year. The buses used in that operation adopted a hybrid power system. The routes used by the vehicles included regular routes for passengers and special routes for the Olympics (Fig. 10.16). Buses carrying passengers between the North Gate of the Summer Palace and Lianxiang Bridge along Line No. 801 traveled about 82 km every day. According to analysis (Table 10.7), the total running time of the project was 3,834 h and the running distance was 57,453 km; the average daily running time was 6.15 h, the average running distance was 92.07 km, and the average running speed was 14.97 km/h.

Fig. 10.16 Running routes of Beijing fuel-cell buses (*red*, routine routes for ordinary passengers; *green*, special routes for the Olympics) (Color figure online)

Table 10.7 Operation data of fuel-cell buses

Bus no.	Running days	Running hours	Kilometers traveled
No. 1	205	1,243	18,907
No. 2	218	1,357	19,810
No. 3	201	1,234	18,736
Total	624	3,834	57,453

Table 10.8 Hydrogen consumption (vehicle monitor vs. refueling station)

Bus no.	Average H_2 consumption (kg/100 km) (vehicle monitoring system)	Average H_2 consumption (kg/100 km) (hydrogen-refueling station)
No. 1	9.66	11.50
No. 2	10.54	12.02
No. 3	9.14	11.06

10.6.1 Analysis of Fuel Economical Efficiency

Based on the collected data from the vehicle monitoring system, the total fuel-cell energy efficiency of the three fuel-cell buses within each running period was 45.00 % for test vehicle No. 1, 47.49 % for No. 2, and 48.18 % for No. 3. Energy consumption by the three demonstration vehicles over 100 km per normal running day and their average hydrogen consumption during the whole operation are provided in Table 10.8.

To analyze the effect of changes in climate conditions and technical proposals, the average values of monthly fuel economy efficiencies of the vehicles were calculated. In addition, changes in fuel economy efficiency before and after the employment of regenerative braking were examined (Fig. 10.17). The following changes in average hydrogen consumption each month were observed:

1. The reason for the decrease in the average hydrogen consumption from August to October 2008 was that the use of the air conditioner decreased with the decrease in air temperature.
2. The increase in the average hydrogen consumption from October to December 2008 was due to the increased use of heating with the decrease in air temperature.
3. One of the reasons for the decrease in the average hydrogen consumption from February to March 2009 was that test vehicle No. 3 began to employ regenerative braking on March 16, 2009.

10.6.2 Analysis of Energy Flow in Vehicles

Deng Xue et al. (2010a) calculated the energy consumed by each part of the vehicle before and after technical improvements during the demonstration period. They derived an energy flow diagram of the vehicle under different running conditions

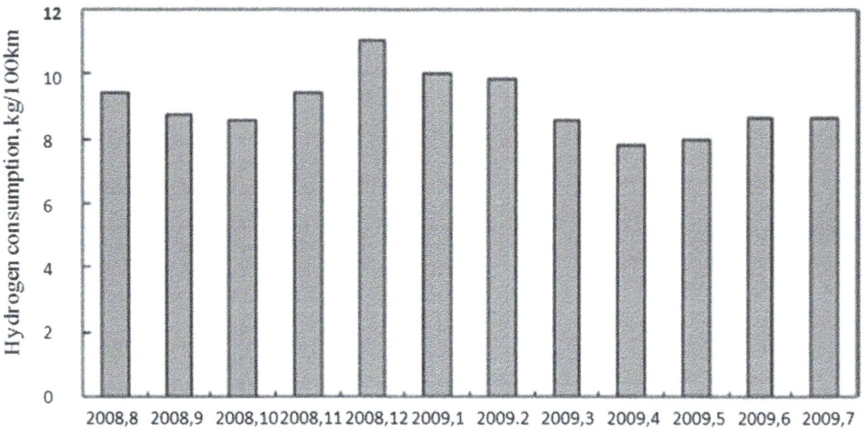

Fig. 10.17 Average hydrogen consumption per kilometer each month

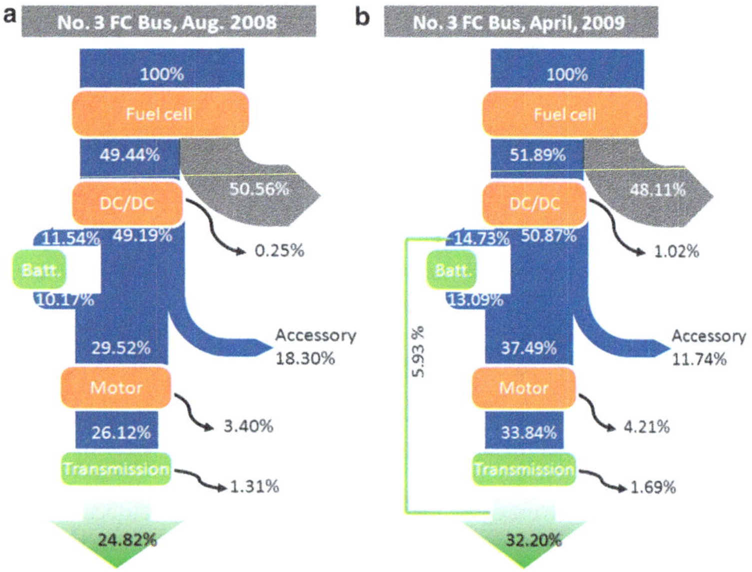

Fig. 10.18 Energy flow in fuel-cell bus. (**a**) Without braking energy recovery and with the air conditioner turned on. (**b**) With braking energy recovery and the air conditioner turned off

(whether the heating or air conditioner was used; whether braking energy was recovered), and this provides a useful reference toward further improving the fuel use efficiency of fuel-cell buses. Figure 10.18 presents a comparison of energy flow under two conditions. It is evident from the energy flow diagram that through the introduction of energy feedback technology, the efficiency of fuel-cell vehicles improved greatly: increased from 24.82 % in August 2008 to 32.20 % in April 2009.

10.7 Development Prospects for Hydrogen Energy Vehicles

Though a small number of vehicles using hydrogen internal combustion engines or hydrogen-natural gas internal combustion engines may be developed, hydrogen will mainly be employed in vehicles in the form of high-efficiency, low-emission fuel cells. Wang and Ouyang (2005, 2007) analyzed the status of hydrogen fuel-cell vehicles in China in 2020 and 2050. After maturation of the technology in 2020, hydrogen fuel-cell vehicles will become an important part of automobile technology.

In light of the differences in the development level of fuel-cell automobile technology, Deng Xue et al. (2010b) analyzed the potential for the hydrogen supply to meet the requirements for the development of fuel-cell vehicles. Their results showed the following. Taking the 6,000,000 tons of by-product hydrogen in 2007 as an example, in energy terms such hydrogen is equivalent to 18,120,000 tons of gasoline and diesel fuel; that accounts for 20 % of the amount of gasoline and diesel fuel consumed by vehicles in China. Calculated on the premise that 240 kg of hydrogen is consumed when a fuel-cell vehicle travels 20,000 km a year (consumption of 1.2 kg of hydrogen per 100 km), that would be enough to sustain the operation of *2.5 million* fuel-cell vehicles. That figure amounts to 45 % of the total vehicle population in 2009 in China and 6 % of the total predicted vehicle population in 2050.

If fuel-cell buses are put into commercial use in the future and the fuel economy efficiency of those vehicles increases from the present 9.56 kg/100 km to 6.32 kg/100 km, that fuel would be enough to sustain the operation of 6,860,000 fuel-cell buses (calculated according to the standard whereby 882.32 kg of hydrogen is consumed when a fuel-cell vehicle travels 14,400 km a year). The figure of 6,860,000 is much greater than the number of buses currently operating in China—1,700,000.

10.8 Conclusion and Recommendations

1. Hydrogen in China is still made mainly from fossil fuels. The annual output of hydrogen is 12,430,000 tons, of which 7,120,000 tons is produced from coal, 2,880,000 tons from natural gas, and 2,430,000 tons from oil. In addition, 6,000,000 tons of industrial by-product hydrogen is produced.
2. Most of China's hydrogen is produced and consumed in large synthetic ammonia, oil, refined methanol, and chlor-alkali chemical processes. In 2007, 880,000 tons of hydrogen was consumed in producing ammonia, 1,700,000 tons used in making oil, and 1,240,000 tons consumed in producing methanol. The amount of hydrogen directly used in transportation is tiny. Only fuel-cell vehicles employed in demonstration projects consume hydrogen.
3. China encourages the research and development of hydrogen fuel-cell automobile technology. Such moves as the National Guideline on Medium- and

Long-Term Program for Science and Technology Development, the China Energy Policy, the National Policy for Saving Energy Technology, and China's National Climate Change Program have all listed development projects for hydrogen and fuel-cell technology.

4. Considerable progress has been made in China's hydrogen fuel-cell vehicle and fuel-cell system technology. Core R&D teams investigating the technology related to fuel-cell vehicles have been established at Tsinghua University and Tongji University. Technology related to hybrid electric vehicles driven by both fuel cells and power batteries has also been developed. Fuel-cell vehicles have been put into demonstration operation in Beijing and Shanghai.
5. The efficiency and energy density of China's fuel-cell systems have improved greatly. The efficiency of pressurized fuel-cell systems has reached 49 %. The efficiency of the constant-pressure system has attained 62 %. The mass specific power of the pressurized system increased from 86 W/kg in 2005 to 200 W/kg in 2010. A fuel-cell system has been operational at storage and start-up temperatures of −10 °C.
6. Data analysis of demonstration operations of fuel-cell city buses in China has indicated that the buses consume 9.56 kg of hydrogen when they travel 100 km. Compared with the first generation of foreign fuel-cell vehicles put into demonstration operation over the same routes, these vehicles saved over 50 % hydrogen.
7. The 6,000,000 tons of industrial by-product hydrogen produced in China in 2007 is sufficient to power the operation of 25,000,000 fuel-cell cars or 6,860,000 fuel-cell buses.

References

Chen Linxin (2002) An analysis on the large-scale production of hydrogen in China [C]//Industry Gas Association. China Power Engineering Association 40th anniversary, 2002

Deng Xue, Wang HW, Huang HY et al (2010) Hydrogen flow chart in China. Int J Hydrog Energy 7:6475–6481

Deng Xue, Ye Yumin, Wang Hewu et al (2010a) Analysis of energy systems in fuel cell buses as new energy vehicles for Beijing Olympic games. J Tsinghua Univ (Sci Technol) 50(5):654–659

Deng Xue, Wang Hewu, Huang Haiyan (2010b) China's hydrogen supply potential for automotive transportation. Sci Technol Rev 28(9):96–101

Hou Zhen (2004) Investigation on modeling the hydrogen-consumption in the hydrocracking/hydrotreating process. Des Chem Eng 21(1):31–34

Lun JG (2008) Progress of the GEF-UNDP-CHINA cooperation project "Demonstration for Fuel Cell Bus Commercialization in China." In: 5th international fuel cell bus workshop, Reykjavik, Iceland

Mao Zongqiang (2006) Hydrogen energy: Green energy in 21st century. Chemical Industry Press, Beijing

Ouyang Minggao, Li Jianqiu, Yang Fuyuan et al (2009) Auto new power system: Configuration, modeling and control. Tsinghua University Press, Beijing

Ren Hongli (2008) Optimization of hydrogen resources for process flow of hydrogenated refinery. Des Chem Eng 18(3):15–21

Shangguan Fangqin (2009) Main energy-saving measures in steel production and the potential analysis of CO_2 emission reduction. Energy Metall Ind 28(1):3–8

Tang Lin, Chen Baosheng (2002) Recovery technology of hydrogen from discharge gas in ammonia synthesis. J Chem Ind Eng 23(4):16–18

Wang Zhe (2010) Shanghai World Expo Mode for FCV Demo and Post-Expo Strategic Deployment Chongming Island. http://iphe.net/docs/Events/China_9-10/2-1_Wang%20Zhe%200911.pdf. Accessed 17 Jan 2013

Wang HW, Ouyang MG (2005) Hydrogen and fuel cell development in China towards 2020. In: Proceedings of the 3rd international conference on fuel cell science, engineering, and technology, Ypsilanti, pp 667–671

Wang HW, Ouyang MG (2007) Transition strategy of the transportation energy and powertrain in China. Energy Policy 35(4):2313–2319

Xie Jidong, Li Wenhua, Chen Yafei (2007) Current development situation of hydrogen from coal. Clean Coal Technol 13(2):77–82

Yi Baolian, Hou Ming (2011) Solutions for the durability of fuel cells in vehicle applications. J Automot Saf Energy 2(2):91–100

Zeng Zijian (2011) Key technological innovation drives the formation of strategic emerging industries of new energy automobiles in China. Fortune world 2011 (07,08)

Zhang Xuelei, Wang Songling, Chen Haiping et al (2006) Economic evaluation of coke oven gas utilization projects. Mod Chem Indus 26(1):47–52

Zhao Xiang, Tong Junfang, Jiang Yanqing (2004) Methanol industry status quo of China. Fertil Ind 31(3):6–15

Zhu Qiming, Wei Junmei, Xu Boqing (2011) Production and application of hydrogen in China: present situation and prospects. http://atmsp.whut.edu.cn/resource/doc/1885.doc. Accessed 17 Jan 2013

Chapter 11
Life-Cycle Energy Consumption and Greenhouse Gas Emissions of Automotive Energy Pathways

Ou Xunmin and Zhang Xiliang

Abstract In this chapter, a life-cycle analysis (LCA) of vehicle-fuel pathways covering the stages of resource extraction, fuel production, and utilization is conducted to examine the macro impact of China's road transport energy supply and related greenhouse gas (GHG) emissions.

Original Chinese data on feedstock extraction and process efficiency, process fuel mix, transportation mode and average distance, fuel production efficiencies, process fuel mix with transport, storage and distribution (TSD) modes, and average distances for oil-, natural gas (NG)- and coal-based fuels, electricity, and H_2 pathways are all listed in the second part.

For different vehicle and fuel technology pathways, the fuel economy situation is presented by using gasoline spark ignition (SI) vehicles as the baseline.

By using a medium-size passenger car with an energy efficiency of 8 l of gasoline consumed per 100 km as the baseline model, the model calculates out the WTW fossil energy input and GHG emissions of the pathways for gas-based fuels, biofuels, coal-based fuels, electric vehicles, and fuel cell vehicles.

Keywords Automotive energy • Life-cycle analysis • China

11.1 Model and Methodology

11.1.1 Overall Introduction

In this chapter, a life-cycle analysis (LCA) of vehicle-fuel pathways covering the stages of resource extraction, fuel production, and utilization is conducted to examine the macro impact of China's road transport energy supply and related

Ou Xunmin (✉) • Zhang Xiliang
China Automotive Energy Research Center, Tsinghua University, Beijing 100084, P.R. China
e-mail: ouxm@mail.tsinghua.edu.cn

DOI 10.1007/978-3-642-36847-9_11,

greenhouse gas (GHG) emissions. Based on the GHG, Regulated Emission and Energy consumption of Transportation (GREET) model, the Tsinghua China automotive energy LCA Model (TLCAM) was constructed. TLCAM was designed as a computing platform for LCA of end-use energy and specific fuel and vehicle pathways in China using a computerized iterative method (Ou and Zhang 2011).

For the vehicle energy pathway in TLCAM, well to pump (WTP) and pump to wheels (PTW) are the two stages included in this well-to-wheels (WTW) energy consumption and GHG analysis. WTP can be used to study upstream production stages, including the exploitation of raw resources and feedstock plantation, feedstock transportation, fuel production, fuel transportation, storage, and distribution (TSD). PTW can be used to study the downstream fuel combustion process in the vehicle's engine (Wang 1999, 2001, 2004; Ou et al. 2009, 2010a, b, c, 2011, 2012).

11.1.2 System Boundary, Substages Divided, and Functional Units

WTP and PTW are the two stages included in this WTW energy consumption and GHG analysis. WTP can be used to study upstream production stages, including the exploitation of raw resources and feedstock plantation, feedstock transportation, fuel production, fuel transportation, storage, and distribution. PTW studies the downstream fuel combustion process in the vehicle's engine (Fig. 11.1).

11.1.3 WTW Calculation Methods

To calculate the WTW results, the two functional units are linked through the fuel economy of the vehicle:

$$E_{\mathrm{WTW}} = E_{\mathrm{WTP}} * \mathrm{FE} + E_{\mathrm{PTW}} \tag{11.1}$$

$$\mathrm{GHG}_{\mathrm{WTW}} = \mathrm{GHG}_{\mathrm{WTP}} * \mathrm{FE} + \mathrm{GHG}_{\mathrm{PTW}} \tag{11.2}$$

where E_{WTW} is the WTW primary fossil energy used per kilometer (MJ/km), E_{WTP} is the WTP primary fossil energy and is used to show the WTP overall conversion efficiency (MJ/MJ fuel), FE is the vehicle-fuel economy (MJ/km), E_{PTW} is the PTW direct primary fossil energy used (MJ/km), $\mathrm{GHG}_{\mathrm{WTW}}$ is the WTW GHG emission (g $CO_{2,e}$/km), $\mathrm{GHG}_{\mathrm{WTP}}$ is the WTP GHG emission (g $CO_{2,e}$/MJ), and $\mathrm{GHG}_{\mathrm{PTW}}$ is the PTW GHG emission (g $CO_{2,e}$/km).

The functional unit in the WTP stage is MJ of fuel supplied in the form of liquid fuel, gas, or electricity and in the PTW stage it is kilometers driven by a city bus.

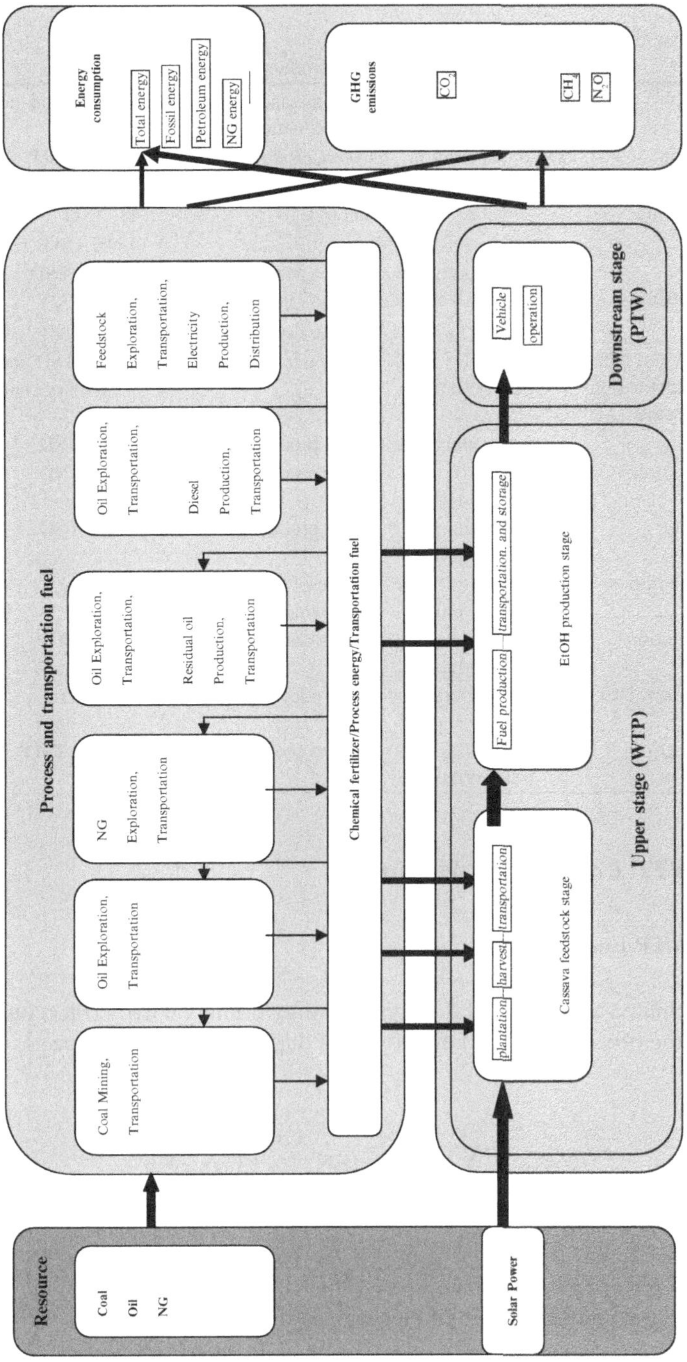

Fig. 11.1 System boundary for energy pathway LCA (cassava case)

Column 11.1 Substages for automotive energy pathways LCA (partial)

Exploitation of raw resources/feedstock plantation	Feedstock transportation	Fuel production	Fuel TSD
Oil exploitation	Oil transportation	Gasoline and oxygenates production, blend	Oxygenated gasoline TSD
Oil exploitation	Oil transportation	Diesel production	Diesel TSD
NG extraction and processing	NG transportation	CNG	CNG TSD
		NG to H_2	H_2 TSD
		LNG	LNG TSD
Coal extraction and processing	Coal transportation	Coal to MeOH	MeOH TSD
		Coal to DME	DME TSD
		Coal to liquid	CtL TSD
Oil, NG, coal, and other feedstock extraction and processing	Feedstock transportation	Power generation	Power TSD and battery charging
Corn plantation	Corn transportation	EtOH production	EtOH TSD
Cassava plantation	Cassava transportation	EtOH production	EtOH TSD
Sweet sorghum plantation	Sweet sorghum transportation	EtOH production	EtOH TSD
Soybean plantation	Soybean transportation	Biodiesel (BD) production	BD TSD
Jatropha plantation	*Jatropha* fruit transportation	BD production	BD TSD
Used cooking oil (UCO) collection	UCO transportation	BD production	BD TSD
Biomass collection	Biomass transportation	Biofuel production	Biofuel TSD

11.1.4 WTP Calculation Methods

11.1.4.1 WTP Energy Use

E_{WTP} is calculated as the sum of the corresponding primary energy (PE) consumption due to the process fuel (PF) directly used during each of the substages of the life cycle:

$$E_{\mathrm{WTP}} = \sum_{p=1}^{4}\sum_{j=1}^{9}\sum_{i=1}^{3}\left(\mathrm{EN}_{p,j} * \mathrm{EF}_{\mathrm{LC},j,i}\right) \tag{11.3}$$

where i is the PE type, j is the PF type, p is the substage number, $\mathrm{EN}_{p,j}$ is the used amount of PF type j during substage p for MJ fuel obtained (MJ/MJ fuel), and $\mathrm{EF}_{\mathrm{LC},j}$, is the LC energy used factor (EF) of PE type i for PF type j (MJ/MJ).

$EN_{p,j}$ can be derived through the direct amount of PF type j used during substage p ($EU_{p,j}$) combined with the product of each of the energy conversion efficiency factors from the following substages (η_p):

$$EN_{p,j} = \frac{EU_{p,j}}{\left(\eta_{p+1} * \ldots * \eta_4\right)} \quad \text{for} \quad (p = 1, 2, 3) \tag{11.4}$$

$$EN_{4,j} = EU_{4,j} \tag{11.5}$$

where the units of $EU_{p,j}$ are MJ/MJ feedstock/fuel obtained for $p = 1,3$ and MJ/MJ feedstock/fuel used for $p = 2,4$, respectively; η_p is the result of 1 divided by the sum of 1 plus all PF used for power and process feedstock or transportation fuel during substage p.

11.1.4.2 WTP GHG Emissions

The three key types of GHG emissions are converted to their CO_2 equivalents ($CO_{2,e}$) according to their global warming potential value (IPCC 2007) as shown by the following expression:

$$GHG_{WTP} = CO_{2,WTP} + 23 * CH_{4,WTP} + 296 * N_2O_{WTP} \tag{11.6}$$

where $CO_{2,WTP}$ is WTP CO_2 emissions (g/MJ fuel), $CH_{4,WTP}$ is WTP CH_4 emissions (g/MJ fuel), and N_2O_{WTP} is WTP N_2O emissions (g/MJ fuel).

During each of the substages, the life-cycle (LC) GHG emissions generally include the direct and indirect parts: the former refers to those emissions directly due to PF combustion and PF usage for process feedstock within the system boundary; the latter includes the emissions during the LC upstream stages of those PF utilized directly though they occur outside the entity boundary. For CH_4, an additional indirect part is considered for its LC GHG emissions—some emissions from noncombustion sources, including spills and other fugitive emission losses during the feedstock extraction stage (g/MJ fuel).

$CO_{2,WTP}$ is calculated as the sum of the direct and indirect parts:

$$CO_{2,WTP} = CO_{2,\text{direct}} + CO_{2,\text{indirect}} = \sum_{p=1}^{4} \sum_{j=1}^{9} EN_{p,j} * \left(CC_j * FOR_j * 44/12 + TCO_{2,j}\right) \tag{11.7}$$

where $CO_{2,\text{direct}}$ reflects the WTP direct CO_2 emissions (g/MJ fuel), $CO_{2,\text{indirect}}$ corresponds to WTP indirect CO_2 emissions (g/MJ fuel), CC_j is the carbon content factor of PF type j (g/MJ), FOR_j is the fuel oxidation rate of PF type j, $TCO_{2,j}$ is the LC indirect CO_2 emissions factor for PF type j (g/MJ), and 44/12 is the mass conversion rate from C to CO_2. The calculation for $CO_{2,indirect}$ is based on a carbon balance equation (Ou et al. 2008a; Wang and Weber 2001).

Calculations for $CH_{4,WTP}$ and N_2O_{WTP} are similar:

$$CH_{4,WTP} = CH_{4,direct} + CH_{4,indirect} = \sum_{p=1}^{4}\sum_{j=1}^{9} EN_{p,j} * \left(ERCH_{4,j} + TCH_{4,j}\right) + CH_{4,noncomb} \tag{11.8}$$

$$N_2O_{WTP} = N_2O_{direct} + N_2O_{indirect} = \sum_{p=1}^{4}\sum_{j=1}^{9} EN_{p,j} * \left(ERN_2O_j + TN_2O_j\right) \tag{11.9}$$

where $CH_{4,direct}$ is the WTP direct CH_4 emissions (g/MJ fuel), $CH_{4,indirect}$ is the WTP indirect CH_4 emissions (g/MJ fuel), $ERCH_{4,j}$ is the direct CH_4 emissions factor for PF type j (g/MJ), $TCH_{4,j}$ is the LC indirect CH_4 emissions factor for PF type j (g/MJ), $CH_{4,noncomb}$ corresponds to indirect CH_4 emissions from noncombustion sources, including spills and losses during the feedstock extraction stage (g/MJ fuel), N_2O_{WTP} corresponds to the WTP N_2O emissions (g/MJ fuel), N_2O_{direct} is the WTP direct N_2O emissions(g/MJ fuel), $N_2O_{indirect}$ is the WTP indirect N_2O emissions (g/MJ fuel), ERN_2O_j is the direct N_2O emissions factor for type PF j (g/MJ), and TN_2O_j is the LC indirect N_2O emissions factor for type PF j (g/MJ).

For the CH_4 noncombustion calculation:

$$CH_{4,noncomb} = \frac{CH_{4,feedstcok}}{(\eta_2 * \eta_3 * \eta_4)} \tag{11.10}$$

where $CH_{4,feedstock}$ corresponds to the indirect CH_4 emissions from noncombustion sources, including spills and losses during the feedstock extraction stage (g/MJ feedstock obtain).

11.2 Basic Data Collection and Processing

11.2.1 Data on Nonbiofuels

Table 11.1 lists original Chinese data on feedstock extraction and process efficiency, process fuel mix, transportation mode and average distance, fuel production efficiencies, process fuel mix with TSD modes, and average distances for oil-, natural gas (NG)- and coal-based fuels, electricity, and H_2 pathways.

Table 11.1 Basic parameters of oil-, NG-, and coal-based fuel pathways

Oil extraction
Extraction efficiency, 93.0 %
Process fuel mix: electricity (37 %), crude oil (20 %), NG (23 %), coal (10 %), diesel (8 %), residual oil (1 %), and gasoline (1 %)
Oil transportation mode
Sea tanker, 50 % (11,000 km); rail, 45 % (950 km); pipeline, 80 % (500 km); and waterways, 10 % (250 km)
Oil refinery
Process fuel mix: crude oil (50 %), coal (20 %), electricity (12 %), refinery still gas (10 %), residual oil (4 %), diesel (1 %), and gasoline (1 %)
Gasoline production efficiency, 89.1 %; diesel production efficiency, 89.7 %; and LPG production efficiency, 92.0 %
Gasoline and diesel TSD mode
Sea tanker, 25 % (7,000 km); railway, 50 % (900 km); waterways, 15 % (1,200 km); and road (short distance), 10 % (50 km)
LPG TSD mode
Sea tanker, 30 % (7,000 km); railway, 80 % (900 km); pipeline, 0 % (160 km); waterways, 15 % (1,200 km); and road (short distance), 10 % (50 km)
NG extraction and processing
Extraction efficiency, 96.00 %; process fuel mix for NG extraction: electricity (40 %), NG (23 %), residual oil (20 %), diesel (8 %), coal (7 %), and gasoline (2 %); NG processing efficiency, 94.00 %; and process fuel mix for NG processing: residual oil (40 %), NG (28 %), coal (20 %), electricity (10 %), diesel (1 %), and gasoline (1 %)
CNG, LNG, and GTL production
CNG production efficiency, 96.9 %; LNG production efficiency, 90.2 %; and GTL production efficiency, 54.2 %
NG and NG-based fuel transportation mode
For NG: pipeline, 100 % (1,500 km)
For CNG: pipeline, 100 % (300 km)
For LNG: sea tanker, 100 % (6,700 km) and road (short distance), 10 % (50 km)
For GTL: pipeline, 100 % (100 km to production site), similar to diesel TSD
Coal extraction and processing
Coal extraction efficiency, 97 %; coal processing efficiency, 97 %; and process fuel mix for crude coal extraction and processing: coal (80 %), electricity (16 %), diesel (2 %), gasoline (1 %), and NG (1 %)
Coal transportation
Railway, 50 % (1,000 km); waterways, 17 % (650 km); road (long distance), 8 % (310 km); and road (short distance), 100 % (50 km)
Coal-based fuel production
MeOH (50.22 %), DME (47.46 %), direct CTL (50.31 %), and indirect CTL (41.41 %)
Coal-based fuel transportation
Preproduction (50 km road) and postproduction, similar to gasoline and diesel TSD
Electricity supply mix
Coal (80.1 %), oil (1.8 %), NG (0.7 %), and many other sources (17.4 %)
Power-supply efficiency
Coal based (36 %), oil based (32 %), NG based (45 %), and others (assumed to be zero fossil energy input)
CCS will decrease the efficiency at the rate of 10 % with a capture rate of 90 % of CO_2 produced onsite

(continued)

Table 11.1 (continued)

Loss ratio during transmission and distribution
6.97 % in 2008 but 5.5 % in 2020
Hydrogen production efficiency
H_2 from NG, 71 %; H_2 from coal, 51.5 %; H_2 from water electrolysis, 80 %; H_2 from biomass, 42–48 %; and H_2 from high-temperature gas cool nuclear reactor, 50–60 % (2020)
Hydrogen processing and transportation
H_2 compression efficiency, 92.5 % and gaseous H_2 transportation, pipeline (1,000 km)

Notes:
1. The data are for 2007 unless otherwise specified
2. For hydropower, there is 5 g $CO_{2,e}$/MJ of life-cycle fossil energy used
3. For nuclear power, there is 6.506 g $CO_{2,e}$/MJ and 0.063 MJ/MJ of life-cycle fossil energy used and GHG emissions
4. For biomass power, there is 5.846 g $CO_{2,e}$/MJ and 0.064 MJ/MJ of life-cycle fossil energy used and GHG emissions

11.2.2 Data on Biofuels

The basic parameters of China's current biofuel pathways include corn to ethanol (CE), cassava to ethanol (KE), sweet sorghum to ethanol (SE), soybean to biodiesel (SB), Jatropha fruit to biodiesel (JB), and used cooking oil to biodiesel (UB) (Tables 11.2 and 11.3). Second-generation biofuel from agricultural and forestry residues is an important pathway of waste utilization. Thus, the energy used during plantation is not allocated to the energy input for biofuel life-cycle energy use and related GHG emissions. Though a great deal of electricity and heat (greater than with first-generation biofuels) is required for the pretreatment and hydrolysis of cellulosic ethanol, considerable lignin is available for use as fuel. There is thus no need to outsource large amounts of electricity or steam.

Since the biomass can be used for electricity feedstock, the biomass-to-liquid (BTL) pathway does not require external energy input. Data on the cellulosic ethanol and BTL pathways are presented in Table 11.4. The TSD situation is similar to that with EtOH (500 km) and biodiesel (200 km).

11.3 Assumptions for PTW

For different vehicle and fuel technology pathways, the fuel economy situation is presented in Table 11.5 using gasoline spark ignition (SI) vehicles as the baseline. It should be noted that the electric vehicle (EV) and plug-in hybrid electric vehicle (PHEV) pathways are gauged under hypothetical conditions with heating and air-conditioning in use. The fuel consumption in real operating conditions is about 15 % higher than in laboratory tests for inner combustion engine (ICE) vehicles and about 30 % for the electric drive mode.

Table 11.2 Basic parameters of EtOH biofuel pathways

Pathway	CE	Data source	KE	Data source	SE[h]
Production (tons/ha)	6.5	Data for Jilin in NSBC (2007a)	13.3	Dai et al. (2006)	64.5
Planting energy (MJ/ha)	4,047[a]	CATARC (2007)	1,572	Dai et al. (2006)[c]	2,800[i]
N fertilizer inputs (kg/ha)	162	Chai (2008)	100	Dai et al. (2006)	600
P fertilizer inputs (kg/ha)	13.3	Chai (2008)	100	Dai et al. (2006)	150
K fertilizer inputs (kg/ha)	131	Chai (2008)	200	Dai et al. (2006)	0
Pesticide inputs(kg/ha)	8	Chai (2008)	0	Chai (2008)	0
Collection radius (km)	125	Average of Zhang et al. (2008) and CATARC (2007)[b]	250	Dai et al. (2006)[d]	50
Conversion rate (tons of feedstock/tons of fuel)	3.2	CATARC (2007)	3.0	Dai et al. (2006)	18.8
Energy for extraction (GJ/ton)	25	Zhang et al. (2008)	13.9	Dai et al. (2006)[e]	20[j]
Distance transmission and distribution (km)	520	Zhang et al. (2008)	450	Dai et al. (2006)[f]	300
Sharing ratio of the by-product (%)	30.90	Wang (2007)	18.06	Dai et al. (2006)[g]	20

Notes:
[a]The energy mix is gasoline (7.16 %), diesel (86.62 %), and electricity (6.02 %)
[b]The values of Zhang et al. (2008) and CATARC (2007) are 100 and 150 km, respectively
[c]According to Dai et al. (2006), diesel fuel and electricity are 44 $l*ha^{-1}$ and 60,923 $kWh*year^{-1}$(200,000 ha), so the total planting energy can be determined based on LHV and the density of diesel—42.7 $MJ*kg^{-1}$ and 0.837 $kg*l^{-1}$
[d]Including 250 km in truck mode
[e]In Dai et al. (2006) the energy consumption per liter of EtOH is 11.898 MJ and almost 100 % of that is from coal
[f]Including 450 km in truck mode
[g]According to their average results
[h]The data of SE are based on a field visit of 2008 to Inner Mongolia, China
[i]The energy mix is gasoline (10 %), diesel (80 %), and electricity (10 %)
[j]The energy mix is coal (90 %) and electricity (10 %), and the energy efficiency from coal to steam is 80 %

11.4 Results for WTP

As noted above, the conversion efficiency of the WTP stage is defined as the ratio of the calorific value of the fuel to the total fossil energy input during the WTP stage (including the raw material input). This ratio is generally expressed as a percentage; however, it may exceed 100 % for those pathways that use nonfossil energy as feedstock and are therefore expressed as the ratio of output to input. In Table 11.6, all the WTP stage conversion rates are listed with error estimations.

Table 11.3 Basic parameters of biodiesel (BD) pathways

Pathway	SB	Data source	JB[d]	UB[g]
Production (tons/ha)	1.8	Data for Heilongjiang in NSBC (2007)[a]	5.0	–
Planting energy (MJ/ha)	4,494[a]	CATARC (2007)	800	–
N fertilizer input (kg/ha)	88	Chai (2008)	97	–
P fertilizer input (kg/ha)	33	Chai (2008)	27	–
K fertilizer input (kg/ha)	27	Chai (2008)	18	–
Pesticide input (kg/ha)	4	Chai (2008)	0	–
Collection radius (km)	200	Zhang et al. (2008)	250	35[h]
Conversion rate (tons of feedstock/tons of fuel)	5.9	CATARC (2007)	3.3[e]	20.0
Energy for extraction (GJ/ton)	12.9[b]	Chai (2008)	10[f]	7.5
Distance transmission and distribution (km)	200[c]	Field visit	300	100
Sharing ratio of the by-product (%)	27.50	Chai (2008)	40	0

Notes:
[a]The energy mix is gasoline (7.33 %), diesel (88.87 %), and electricity (3.80 %)
[b]The energy mix is coal (90 %) and electricity (10 %), and the energy efficiency from coal to steam is 80 %
[c]Currently in China, most BD is used in agricultural machines and fishing boats because its use as ordinary vehicle fuel is prohibited
[d]The data for JB were based on a field visit of 2009 to Hainan, China
[e]The data were confirmed by CATARC (2007)
[f]The data were confirmed by CATARC (2007)
[g]The data for UB are based on a field visit of 2009 to Beijing and CATARC (2007)
[h]The collection energy used for UCO was 30 MJ/ton (to collection points) and 135 MJ/ton (transported to processing plants)

11.5 Results for WTW

Using a medium-size passenger car with an energy efficiency of 8 l of gasoline consumed per 100 km as the baseline model, we can calculate the WTW fossil energy input and GHG emissions of such pathways as those for gas-based fuels, biofuels, coal-based fuels, electric vehicles, and fuel cell vehicles. The error bars that appear in Figs. 11.2 and 11.3 indicate the uncertainty of the efficiency of oil extraction, crude oil transport distance, and oil-refinery efficiency for gasoline and diesel pathways; they will be explained in detail for other pathways.

11.5.1 Gas-Based Fuel Pathways

The results for gas-based fuel pathways are shown in Figs. 11.2 and 11.3. The WTW fossil energy inputs for LPG, CNG, and LNG vehicles are about 2–7 % lower than

Table 11.4 Data for second-generation biofuels

Pathway	Cellulosic EtOH		BTL
Feedstock	Woody feedstock	Herbal feedstock	Mixed feedstock
Heat value (MJ/kg)	15.89	14.63	15.55
Energy used for collection (–)	Ignored	Ignored	Ignored
Feedstock transport distance (km)	100	100	100
Energy used during transportation (L diesel/100 ton km)	7.45	7.45	7.45
Conversion rate (ton feedstock/ton fuel)	4.1	3.7	6
External electricity used (GJ/ton fuel)	1.00	0.85	0
Process fuel	Coal (100 %)	Coal (100 %)	Biomass feedstock (100 %)
Electricity coproduct (kWh/ton fuel)	400	200	0
Distribution distance (km)	500	500	200

Table 11.5 Data for fuel economy of a combination of fuel and vehicle pathways

	Fuel economy (kilometers traveled per unit energy) (%)	Energy consumption per unit distance	Note
SI ICE-gasoline	100.0	1.00	
SI ICE-CNG	95.0	1.05	
SI ICE-LNG	95.0	1.05	
SI ICE-LPG	100.0	1.00	
SI ICE-methanol	107.0	0.93	
FFV-methanol	100.0	1.00	
SI ICE-ethanol	107.0	0.93	
FFV-ethanol	100.0	1.00	
SI ICE-hydrogen	120.0	0.83	
SI HEV-liquid fuel-gasoline	143.0	0.70	Energy-saving rate is 30 % of gasoline vehicles
SI HEV-hydrogen	160.0	0.62	Energy-saving rate is 25 % of H_2 SI vehicles
CI ICE-diesel	120.0	0.83	
CI ICE-DME	120	0.83	
CI ICE-biodiesel	120	0.83	
CI HEV-liquid fuel-diesel	170.0	0.59	Energy-saving rate is 30 % of diesel vehicles
EV-electricity	350.0	0.29	
SI PHEV-electricity	300.0	0.33	
SI PHEV-ICE mode-gasoline	143.0	0.70	
FCV-hydrogen	230.0	0.43	

Table 11.6 WTP efficiency

Pathway	Unit	WTP efficiency	Positive error	Negative error	Note
Part I: Fossil fuel as feedstock					
ICE-gasoline	%	73.60	3.68	3.68	Current situation; will not vary much in future
ICE-diesel	%	76.03	3.80	3.80	Current situation
ICE-LPG	%	78.84	3.94	3.94	Current situation
ICE-CNG	%	82.50	8.25	8.25	Current situation
ICE-LNG	%	78.99	3.95	3.95	Current situation
ICE-GTL	%	51.51	7.73	7.73	Current situation
ICE-MeOH	%	44.90	8.98	6.74	Current situation
ICE-DME (coal)	%	42.45	8.49	6.37	Current situation
ICE-CDL	%	44.99	6.75	4.50	Current situation
ICE-CTL	%	37.06	5.56	3.71	Current situation
ICE-MeOH (CCS)	%	35.30	8.83	7.06	When the technology is mature
ICE-DME (coal) (CCS)	%	35.00	8.75	7.00	When the technology is mature
ICE-CDL (CCS)	%	40.96	8.19	6.14	When the technology is mature
ICE-CTL (CCS)	%	33.24	6.65	4.99	When the technology is mature
EV-electricity (grid)	%	34.20	3.42	3.42	Current situation
EV-electricity (coal)	%	28.56	2.86	2.86	Current situation
EV-electricity (oil)	%	24.40	2.44	2.44	Current situation
EV-electricity (gas)	%	35.08	3.51	3.51	Current situation
EV-electricity (coal) (IGCC+CCS)	%	29.72	4.46	4.46	When the technology is mature
Part II: Nonfossil fuel as feedstock					
EV-electricity (nuclear)	Ratio of output to input	15.87	1.59	1.59	Current situation
EV-electricity (hydraulic)	Ratio of output to input	–	–	–	Current situation
EV-electricity (biomass)	Ratio of output to input	13.16	1.32	1.32	Current situation
ICE-EtOH (corn)	Ratio of output to input	0.65	0.13	0.13	Current situation
ICE-EtOH (cassava)	Ratio of output to input	1.15	0.23	0.23	Current situation
ICE-EtOH (sweet sorghum)	Ratio of output to input	1.38	0.14	0.69	Current situation
ICE-EtOH (herbs)	Ratio of output to input	7.29	0.73	1.46	When the technology is mature
ICE-EtOH (wood)	Ratio of output to input	4.55	0.45	0.91	When the technology is mature

(continued)

Table 11.6 (continued)

Pathway	Unit	WTP efficiency	Positive error	Negative error	Note
ICE-BD (waste oil)	Ratio of output to input	1.03	0.31	0.10	Current situation
ICE-BD (*Jatropha* oil)	Ratio of output to input	1.33	0.13	0.13	Current situation
ICE-BTL	Ratio of output to input	17.24	3.45	3.45	Current situation
FCV-H_2 (NG)	Ratio of output to input	0.53	0.05	0.05	Current situation
FCV-H_2 (coal)	Ratio of output to input	0.40	0.04	0.04	Current situation
FCV-H_2 (water electrolysis)	Ratio of output to input	0.26	0.03	0.03	Current situation
FCV-H_2 (biomass)	Ratio of output to input	3.37	0.34	0.34	When the technology is mature
FCV-H_2 (Nuclear)	Ratio of output to input	3.53	0.35	0.35	When the technology is mature
FCV-H_2 (coal) (CCS)	Ratio of output to input	0.37	0.06	0.06	When the technology is mature
FCV-H_2 (NG)	Ratio of output to input	0.33	0.03	0.03	Current situation
FCV-H_2 (coal)	Ratio of output to input	0.27	0.03	0.03	Current situation
FCV-H_2 (water electrolysis)	Ratio of output to input	0.20	0.02	0.02	Current situation
FCV-H_2 (biomass)	Ratio of output to input	0.67	0.07	0.07	When the technology is mature
FCV-H_2 (nuclear)	Ratio of output to input	0.68	0.07	0.07	When the technology is mature
FCV-H_2 (coal) (CCS)	Ratio of output to input	0.25	0.06	0.06	When the technology is mature

those of conventional vehicles; the rate of decrease is related to the gas-based fuel-transport distance. Since the carbon content of natural gas is lower than that of oil, the GHG emissions of gas-based fuels vehicles are 15 % lower than those of conventional vehicles. The analysis error of such gas-based fuel (LPG, CNG, and LNG) pathways arises from the transport distance.

It should be noted that since the production efficiency of gas to liquid (GTL) is relatively low, the WTW fossil energy input of the GTL pathway increases by 50 % compared with that of conventional diesel vehicles. The GHG emissions with the GTL pathway show a slight increase compared with those of conventional diesel vehicles owing to the low carbon content in natural gas. The future uncertainty related to production efficiency of GTL produces a significant analysis error interval.

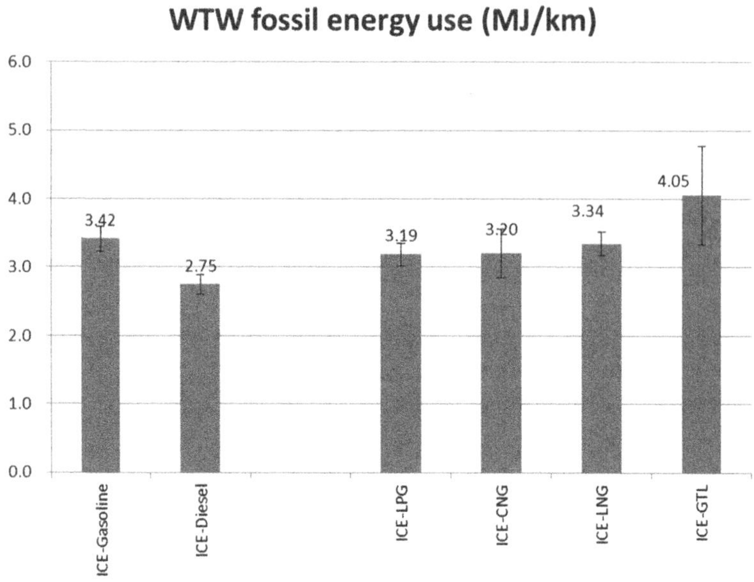

Fig. 11.2 WTW fossil energy input for gas-based fuel (MJ/km)

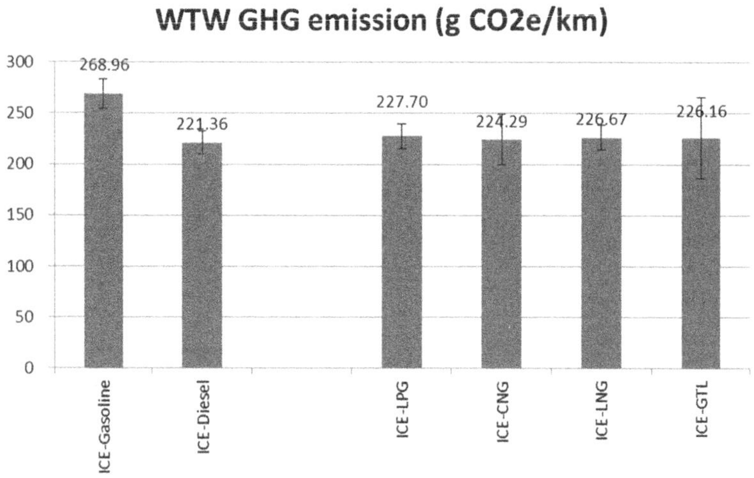

Fig. 11.3 WTW GHG emission for gas-based fuel (g $CO_{2,e}$/km)

11.5.2 Biofuel Pathways

The results for biofuel pathways appear in Figs. 11.4 and 11.5. As noted above, in terms of biofuel pathways (except for the waste cooking oil pathway), plants can absorb carbon dioxide during their growing period. As a consequence, the

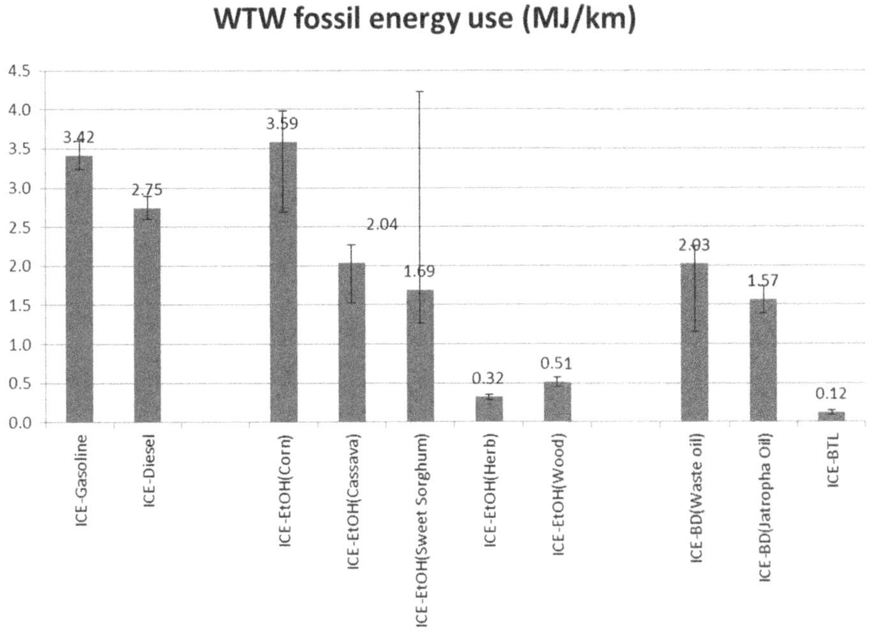

Fig. 11.4 WTW fossil energy input for biofuels (MJ/km)

carbon dioxide that derives from the fuel combustion stage can be regarded as carbon being recycled into the atmosphere from absorption during the growing period.

Although the raw material is not fossil fuels but biomass, the WTW fossil energy input and GHG emissions of the current first-generation biofuel-powered vehicles are high. This is a result of high gasoline and diesel consumption in the transport of raw materials and the large amount of coal consumed in fuel-processing plants for fermenting and distilling. Whether the energy consumption and GHG emissions can be reduced depends on the raw material pathways compared with conventional vehicle pathways. The corn-based ethanol pathway has 5 % more energy consumption and 82 % more GHG emissions; this is because a large amount of fertilizer is applied in plantations, and the high energy consumption in fertilizer production has to be considered in the LCA. The cassava-based ethanol pathway has 40 % less energy consumption and 17 % less GHG emissions. The sweet sorghum-based ethanol pathway has 50 % less fossil energy consumption and 40 % less GHG emissions compared with gasoline vehicles. This is because the sweet sorghum stalk residues in ethanol plants can produce the necessary amount of steam and electricity; they thus act as a substitute for external coal steam for electricity production.

It should be noted that there are great uncertainties in this analysis as a result of differences in factory processes, management levels, and whether or not by-products are synthetically used. Taking the sweet sorghum pathway as an example,

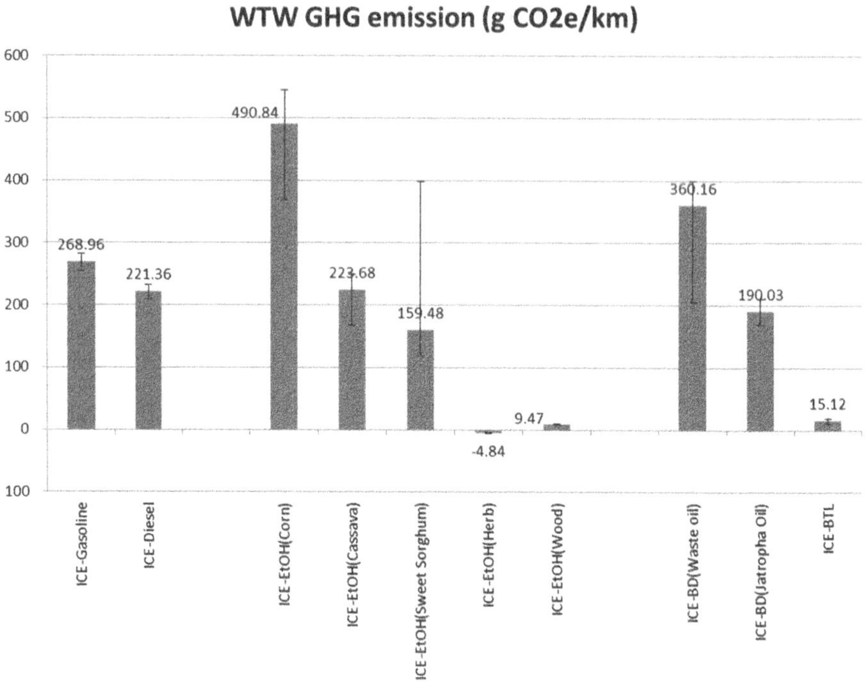

Fig. 11.5 WTW GHG emissions for biofuels (g $CO_{2,e}$/km). *Note*: It is assumed that the waste cooking oil pathway can store carbon and has no alternative usage

its energy consumption and GHG emissions are very high if this pathway has no comprehensive utilization of stalk. In such a case, this pathway operates poorly; the situation is exacerbated in the fuel ethanol pathways, as shown in Figs. 11.4 and 11.5.

The WTW fossil energy input of first-generation biodiesel fuel-powered vehicles is 26–43 % lower than that of conventional diesel vehicles. However, there is a high carbon content (g C/MJ) per calorific unit in waste cooking oil, and no planting process is involved, as it is with the other biofuel pathways. Thus, the WTW GHG emissions of biofuels based on waste cooking oil are 30 % lower than with the diesel pathway if oxidation of disposed waste cooking oil to carbon dioxide is assumed. The WTW GHG emissions of biofuels based on waste cooking oil are 63 % greater than with the diesel pathway if waste cooking oil is unused and simply put directly into the sewage system. The WTW GHG emissions of biofuels based on waste cooking oil are higher than with the diesel pathway since more GHG emissions result from the manufacturing of other products (such as glycerol and soap). Here, we take it that waste cooking oil is unused and simply discharged directly into the sewage system with no oxidation. In addition, the transport distance of the raw materials is the main factor in producing erroneous results regarding waste cooking oil. For example, the results may be reduced by 30 % if the transport distance decreases from the current average of 1,000 to 100–200 km.

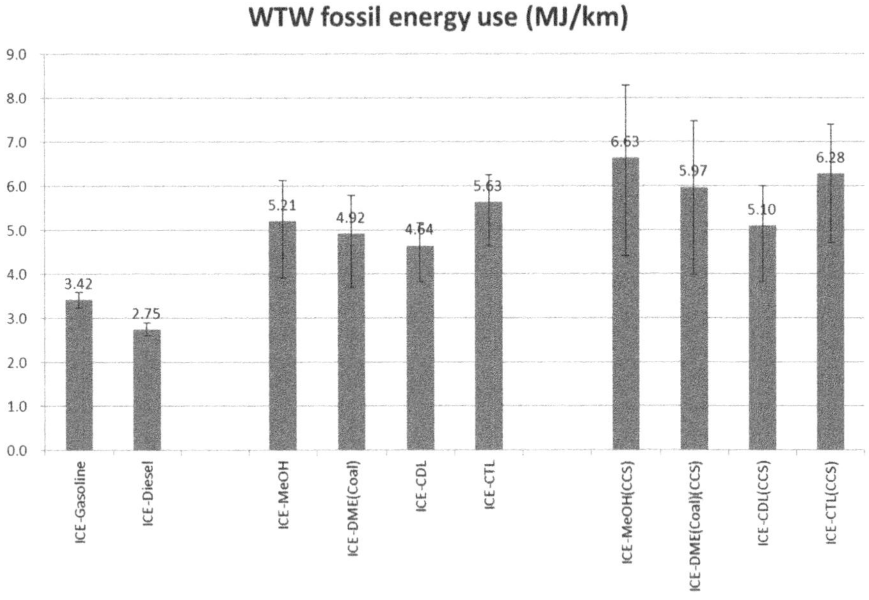

Fig. 11.6 WTW fossil energy input for coal-based fuel (MJ/km)

In terms of second-generation fuel ethanol and biodiesel pathways, there are both good energy-saving and good emission-reduction effects with less energy input in the production process: some lignin and residues can be used to produce electricity as fuel instead of large amounts of coal being consumed. Compared with the baseline vehicle, the fossil energy consumption input can be reduced by over 80 %, and the GHG reduction rate may be above 90 %—or even result in zero emissions.

11.5.3 Coal-Based Fuel Pathway

The results for coal-based fuel pathways appear in Figs. 11.6 and 11.7. Compared with the baseline vehicle, the WTW fossil energy input and GHG emissions in the coal-based fuel pathways without adopting CO_2 capture and storage (CCS) technology increase dramatically: a 50–105 % increase in fossil energy input and a 126–174 % increase in GHG emissions. The main reason is the relatively low conversion rate of coal-based fuel plants and the high carbon content of coal. Since the methanol and DME plants in China are quite dispersed and there is a great diversity in the technology level, errors in the analysis results are quite large.

The WTW fossil energy input in coal-based pathways increases by 90–129 % if CCS technology is adopted, whereas the GHG emission rate is reduced by 50–137 % compared with conventional gasoline and diesel vehicles. With the CCS technology

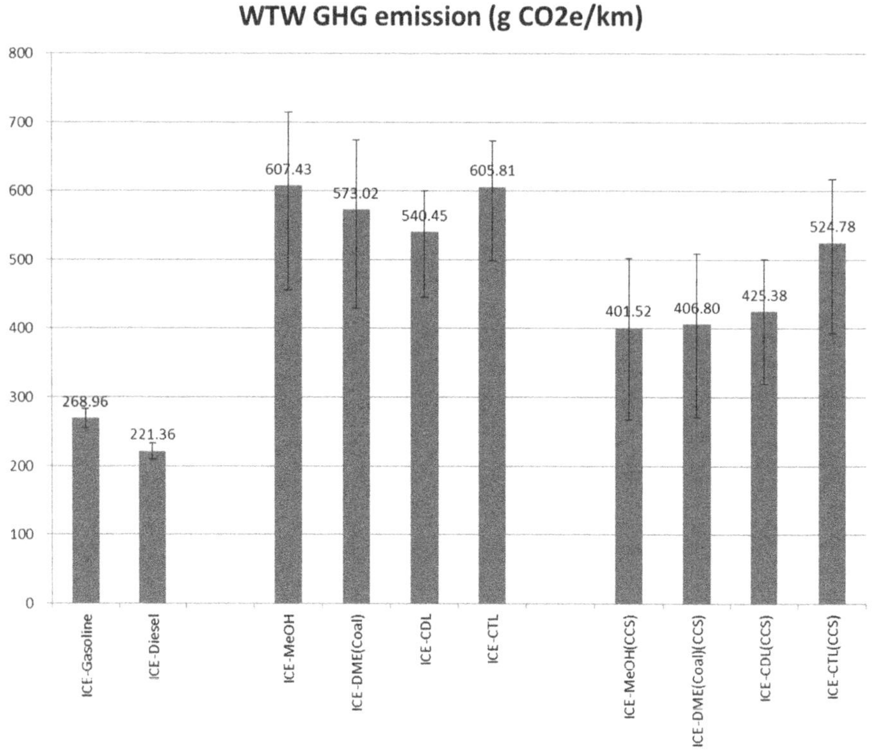

Fig. 11.7 WTW GHG emission for coal-based fuel (g $CO_{2,e}$/km)

pathway, the error bar is expanded owing to uncertainty in CCS technology energy consumption and the carbon capture rate. However, even with the most optimistic interpretation, the WTW fossil energy input and GHG emissions with coal-based fuel pathways are both greater than with the gasoline and diesel pathways.

There are two reasons for the application of CCS technology being unable to change the fact that coal-based fuels have greater GHG emissions than petroleum-based fuels: (1) GHG emissions in coal mining, disposing, and transport are higher than with the oil pathway, and (2) coal-based fuel plants have relatively low efficiency and high energy consumption, whereas the carbon content of the main fuel, coal, is higher than that of oil in oil-refining plants. As a result, the GHG emissions from plants are at a high level even though a carbon dioxide capture rate of 90 % is achieved.

11.5.4 EV Pathways

The results for EV pathways are shown in Figs. 11.8 and 11.9. The WTW fossil energy consumption with EVs adopting grid electricity as the power source is 62 %

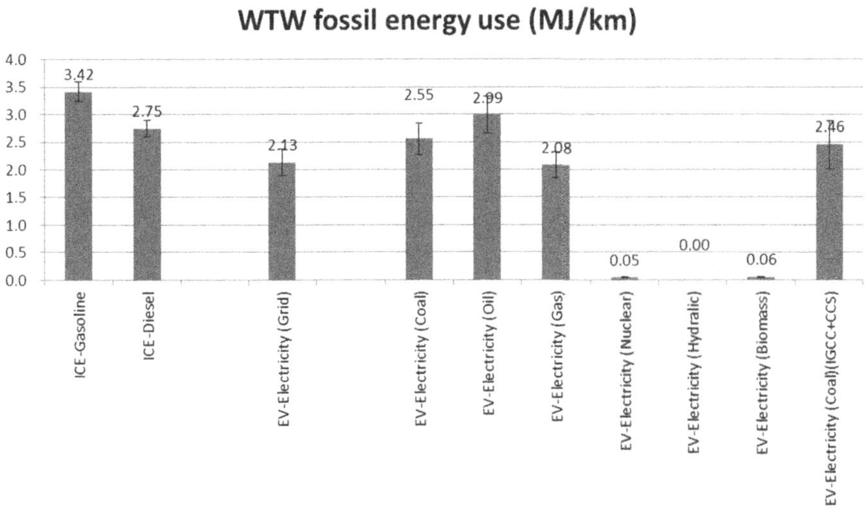

Fig. 11.8 WTW fossil energy input for EVs (MJ/km)

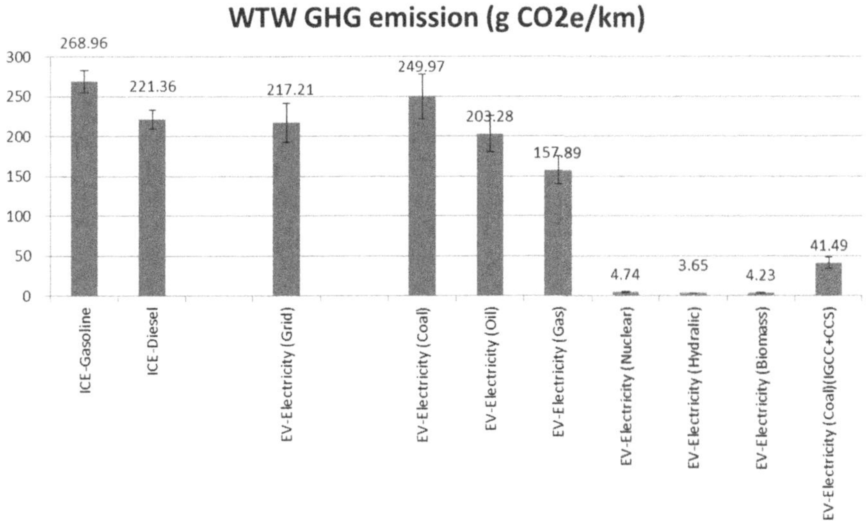

Fig. 11.9 WTW GHG emissions for EVs (g $CO_{2,e}$/km)

that of gasoline vehicles and 75 % that of diesel vehicles. This is because the energy efficiency of EVs is much higher than that of internal combustion engine vehicles. However, since the proportion of coal in China's power supply is over 80 %, the decline in GHG emissions with the EV pathway is only 20 % lower than with gasoline vehicles and under 10 % lower than that of diesel vehicles.

It is clearly the case that coal and oil power pathways deteriorate the energy-saving advantages of EVs if this kind of power source is used. This is particularly

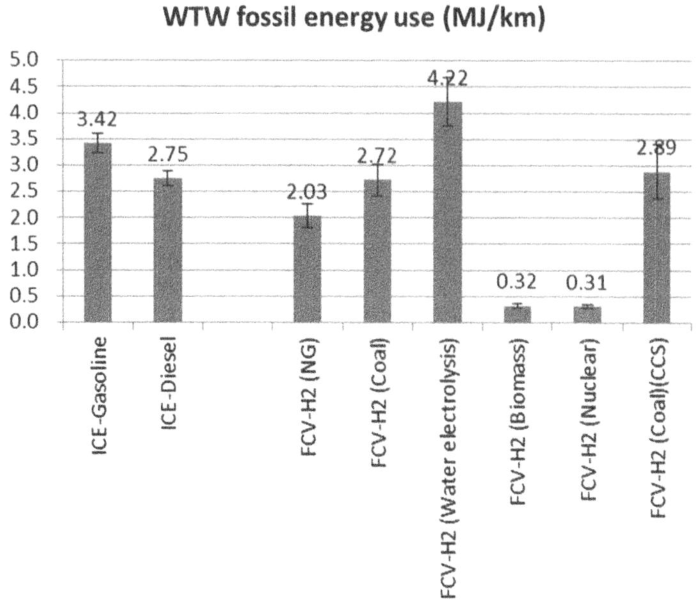

Fig. 11.10 WTW fossil energy input for FCVs (MJ/km)

true if the WTW fossil energy consumption of oil-based EV is 10 % more than that with the diesel pathway. Although the WTW GHG emissions with the coal power pathway are slightly lower than with the gasoline vehicle pathway (7 %), they are somewhat higher than with the diesel pathway (13 %).

If both integrated gasified cycling combustion (IGCC) and CCS technology are adopted in new coal-fired power plants, the WTW fossil energy input and GHG emissions with EV pathways are better than with the conventional gasoline and diesel vehicle pathways. There is the advantage of IGCC and CCS technology being efficient at capturing the carbon dioxide produced by power plants, with energy savings of 10–30 % and emission reductions of 80 %.

In terms of nuclear, hydro-, and biomass power, the WTW fossil energy input and GHG emissions are only 1–2 % lower than with the traditional gasoline and diesel pathways. The difference is thus almost negligible.

11.5.5 FCV Pathways

The results for the FCV pathways are shown in Figs. 11.10 and 11.11. Since the efficiency of FCVs is much higher than that of internal combustion engine vehicles, the WTW fossil energy consumption with the FCV pathway that uses hydrogen produced from natural gas is 40 % lower than with the gasoline pathway and 26 %

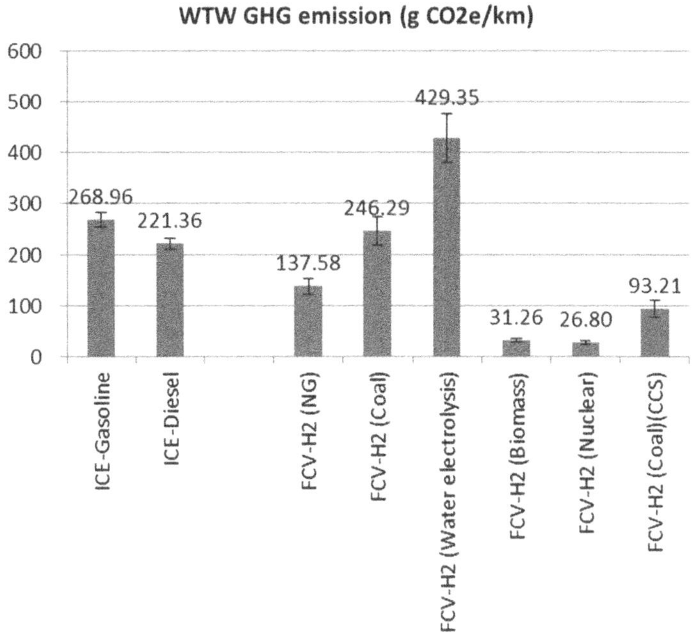

Fig. 11.11 WTW GHG emission for FCVs (g $CO_{2,e}$/km)

lower than with the diesel pathway; GHG emissions decrease by 49 and 38 %, respectively. In the FCV pathway that uses hydrogen produced from coal, the WTW fossil energy consumption is 20 % lower than with the gasoline pathway though the consumption is the same as with the diesel pathway; the GHG emissions decrease by 10 % and increase by 10 %, respectively. If the hydrogen is produced from biomass and future nuclear commercialization can be achieved, the WTW fossil energy consumption and GHG emissions with these two FCV pathways are only 1–2 % that of the traditional gasoline and diesel vehicles.

If coal-derived hydrogen plants adopt CCS technology in the future, the WTW fossil energy consumption in the hydrogen FCV pathway will decline by 15 % and increase by 5 %, respectively, compared with the gasoline and diesel pathways. The respective decline in GHG emissions will be 65 and 60 %.

11.6 Outlook for Vehicle-Fuel Combination LCA

Following the definition of the reference scenario and integrated policy scenario, this section will examine the LCA results for vehicle-fuel pathways in China. The WTP results for the energy pathways presented here will be used in Chap. 12; only the LC energy and future GHG emission levels are examined in this section. In combination

Table 11.7 LC fossil energy intensity for biofuels (MJ/MJ)

	2010	2015	2020	2030	2040	2050
Reference scenario						
First-generation EtOH	1.08	1.08	1.08	1.08	1.08	1.08
First-generation biodiesel	0.80	0.80	0.80	0.80	0.80	0.80
Second-generation EtOH			−0.01	−0.01	−0.01	−0.01
Second-generation biodiesel			0.07	0.07	0.07	0.07
Integrated policy scenario						
First-generation EtOH	1.08	1.05	1.03	0.97	0.92	0.86
First-generation biodiesel	0.80	0.78	0.76	0.72	0.68	0.64
Second-generation EtOH			−0.01	−0.01	−0.01	−0.01
Second-generation biodiesel			0.07	0.07	0.07	0.07

Note: Because of electricity coproduction as a substitute for coal-derived power, energy credit accrues for the second-generation EtOH pathway

Table 11.8 LC GHG emission intensity for biofuels (g $CO_{2,e}$/MJ)

	2010	2015	2020	2030	2040	2050
Reference scenario						
First-generation EtOH	115.0	115.0	115.0	115.0	115.0	115.0
First-generation biodiesel	79.0	79.0	79.0	79.0	79.0	79.0
Second-generation EtOH			−3.0	−3.0	−3.0	−3.0
Second-generation biodiesel			6.0	6.0	6.0	6.0
Integrated policy scenario						
First-generation EtOH	115.0	112.1	109.3	103.5	97.8	92.0
First-generation biodiesel	79.0	77.0	75.1	71.1	67.2	63.2
Second-generation EtOH			−3.0	−3.0	−3.0	−3.0
Second-generation biodiesel			6.0	6.0	6.0	6.0

Note: Because of electricity coproduction as a substitute for coal-derived power, GHG credits accrue for the second-generation EtOH pathway

with the prospects for future propulsion system technologies, the WTW results are presented to predict LC energy consumption and GHG emissions in future vehicle-fuel pathways.

11.6.1 WTP Results

For the period 2010–2050, the WTP results are assumed to be constant for the pathways for conventional gasoline, diesel, LPG, CNG, LNG, and GTL. However, the results differ for those pathways of bio- and coal-derived electricity and hydrogen is included. The results are presented in Tables 11.7, 11.8, 11.9, 11.10, 11.11, and 11.12.

Table 11.9 LC fossil energy intensity for coal-derived fuel (MJ/MJ)

	2010	2015	2020	2030	2040	2050
Reference scenario						
Coal to methanol	2.38	2.19	2.00	2.00	2.00	2.00
Coal to DME	2.56	2.34	2.11	2.11	2.11	2.11
Direct CTL	1.98	1.91	1.83	1.83	1.83	1.83
Indirect CTL	2.42	2.35	2.29	2.29	2.29	2.29
Integrated policy scenario						
Coal to methanol	2.38	2.19	2.00	1.90	1.81	1.76
Coal to DME	2.56	2.34	2.11	2.00	1.91	1.86
Direct CTL	1.98	1.91	1.83	1.74	1.65	1.61
Indirect CTL	2.42	2.35	2.29	2.17	2.07	2.02

Note: Under the integrated policy scenario, CCS is applied gradually from 2025 to 2050 (100 % of the application rate) for these coal-derived fuel pathways; the capture rates are all assumed to be 90 %

Table 11.10 LC GHG emission intensity for coal-derived fuel (g $CO_{2,e}$/MJ)

	2010	2015	2020	2030	2040	2050
Reference scenario						
Coal to methanol	238.0	232.0	225.9	225.9	225.9	225.9
Coal to DME	293.4	267.6	241.8	241.8	241.8	241.8
Direct CTL	236.3	227.2	218.1	218.1	218.1	218.1
Indirect CTL	285.5	278.0	270.4	270.4	270.4	270.4
Integrated policy scenario						
Coal to methanol	238.0	232.0	225.9	189.1	145.7	95.8
Coal to DME	293.4	267.6	241.8	202.4	156.0	102.5
Direct CTL	236.3	227.2	218.1	182.5	140.7	92.5
Indirect CTL	285.5	278.0	270.4	226.3	174.4	114.6

Note: Under the integrated policy scenario, CCS is applied gradually from 2025 to 2050 (100 % of the application rate) for these coal-derived fuel pathways; the capture rates are all assumed to be 90 %

Table 11.11 LC fossil energy intensity for electricity (MJ/MJ)

	2010	2015	2020	2030	2040	2050
Reference scenario	2.83	2.68	2.50	2.40	2.30	2.20
Integrated policy scenario			2.50	2.69	2.58	2.46

Notes:
1. Under the reference scenario, both the coal-to-power efficiency and low-carbon pathways improve in the future
2. under the integrated policy scenario, CCS is applied gradually from 2025 to 2035 (100 % of the application rate) for new-built coal power plants; up to 2050, CCS is applied for all running coal power plants; the capture rates are all assumed to be 90 %

Table 11.12 LC GHG emission intensity for electricity (g $CO_{2,e}$/MJ)

	2010	2015	2020	2030	2040	2050
Reference scenario	286.0	266.5	243.1	230.1	217.1	204.1
Integrated policy scenario			243.1	199.0	143.4	87.8

Notes:
1. Under the reference scenario, the coal-to-power efficiency and low-carbon pathways improve in the future
2. Under the integrated policy scenario, CCS is applied gradually from 2025 to 2035 (100 % of the application rate) for these newly built coal power plants; up to 2050, CCS is applied to all running coal power plants; the capture rates are all assumed to be 90 %

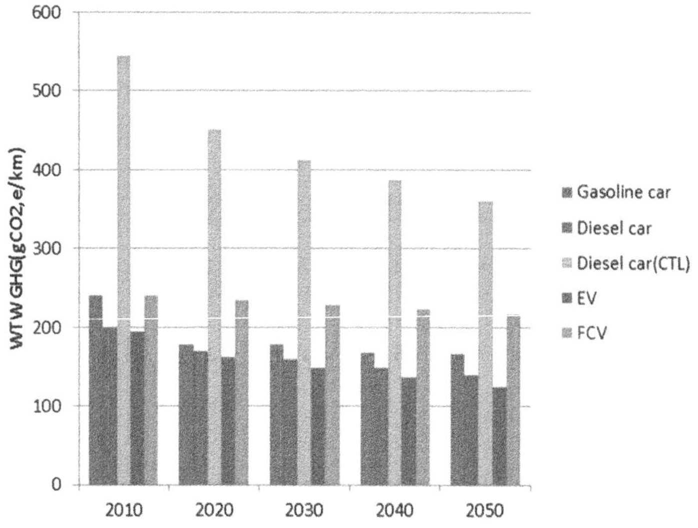

Fig. 11.12 WTW outlook under the reference scenario

11.6.2 WTW Results

With the combination of vehicle technology development (energy efficiency) in the future, the WTW results are examined under the reference scenario and integrated policy scenario. Here, the representative vehicle is a new passenger car with a fuel economy of 8 l of gasoline per 100 km.

11.6.2.1 Reference Scenario

As shown in Fig. 11.12, with technological improvement under the reference scenario, GHG emissions decrease for all pathways. With the gasoline passenger car

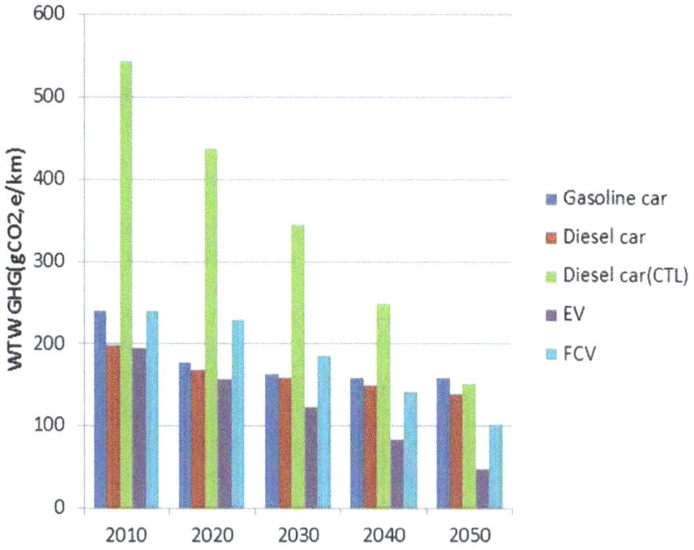

Fig. 11.13 WTW outlook in integrated policy scenario

pathway, GHG emissions decline from 240 g $CO_{2,e}$/km in 2010 to 170 g $CO_{2,e}$/km in 2050. With the EV pathway, since the technological improvement is smaller than with gasoline cars, the overall energy saving and GHG reduction advantage become progressively smaller. For FCV pathway, GHG emissions remain high throughout the whole period from 2010 to 2050. For the coal-derived pathways as coal to liquid (CTL), 100 % of GHG emissions will be more than that from gasoline cars, though the absolute level gradually decreases.

11.6.2.2 Integrated Policy Scenario

As shown in Fig. 11.13, with technological improvement under the integrated policy scenario, GHG emissions for all pathways decrease faster than under the reference scenario. With the gasoline passenger car pathway, GHG emissions decrease from 240 g $CO_{2,e}$/km in 2010 to 160 g $CO_{2,e}$/km in 2050; for EV pathway, through the application of CCS technology, the overall energy saving and reduction in GHG emissions increase, and there is a 70 % emission reduction compared with the gasoline car pathway by 2050; for FCV pathway, the disadvantage of GHG emissions are affected by the application of CCS technology in coal-derived hydrogen processes, and there is a 35 % reduction in emissions compared with the gasoline car pathway by 2050; for the coal-derived pathway as CTL, the absolute level of GHG emissions decreases quickly, and only about 10 % of GHG will be emitted than gasoline cars by 2050.

11.7 Overall Concluding Remarks

(a) China's current alternative fuel pathways are geographically unique. The pathways behave differently from LCA energy consumption and GHG emission analysis and conclusions reached for such places as the United States, the European Union, and Brazil.
(b) All current pathways are feasible in China in terms of energy security since they offer petroleum substitution, though they do not all have a clear energy-saving or GHG reduction effect.
(c) The present energy consumption and GHG situation can be improved by increasing productivity, reducing the use of resources, efficient energy conversion, and optimizing by-product use.
(d) A package of measures is required to achieve the potential of lower energy consumption and GHG emissions offered with alternative pathways. These measures include the following: speeding up improvements in battery technology for electric vehicles and establishing a charging infrastructure; finding solutions to the problems of high water consumption and high pollution emissions from coal chemical plants; and speeding up the R&D and commercial operation of low-carbon technologies, such as CCS.
(e) With the combination of vehicle technology development (energy efficiency) in future, the WTW results show that GHG emissions decrease with all pathways. With the gasoline passenger car pathway, GHG emissions are reduced from 240 g $CO_{2,e}$/km in 2010 to 160–170 g $CO_{2,e}$/km in 2050.
(f) Some policy measurements are suggested to promote efficient energy usage and reduce GHG emissions from a life-cycle perspective: (1) introduce measures to further enhance the energy efficiency of conventional gasoline and diesel vehicles to significantly reduce GHG emissions; (2) support the accelerated development of second-generation biofuels, the development of nonfossil energy sources of hydrogen production, and CCS technology; (3) improve the technology and promote demonstration projects for electric vehicles and fuel cell vehicles as well as supply options for effective energy-efficient vehicles with low GHG emissions; (4) make natural gas an option for use in vehicles; and (5) use coal-based fuel just as a short-term alternative fuel.

References

CATARC (China Automotive Technology and Research Center) and GM (General Motor) (2007) Well-to-wheels analysis of energy consumption and GHG emissions of multi vehicle fuel in future China. CATARC, Beijing

Chai QH (2008) Research on the Biomass-Derived Automotive Alternative Energy Industry. Ph.D. dissertation. Tsinghua University, Beijing [in Chinese]

Concawe, EUCAR, EC Joint Research Centre (2007) Well-to-wheels analysis of future automotive fuels and power trains in the European context. http://iet.jrc.ec.europa.eu/sites/about-jec/files/documents/WTW_App_2_010307.pdf. Accessed 17 Jan 2013

Dai D, Liu RH, Pu GQ, Wang CT (2006) Evaluation of energy production efficiency of biomass based fuel ethanol program. Trans Chin Soc Agric Eng 21:121–123 [in Chinese]
IPCC (Intergovernmental Panel on Climate Change) (2007) IPCC fourth assessment report: climate change 2007
NSBC (National Statistics Bureau of China) (2007) China energy statistical yearbook 2007. China Statistics Press, Beijing [in Chinese]
Ou X, Zhang X (2011) Life-cycle analysis of the automotive energy pathways in China. Tsinghua University Press, Beijing [in Chinese]
Ou XM, Zhang XL, Chang SY (2008a) Life-cycle analysis of energy consumption, GHG emissions and regulated pollutants emissions of automotive fuel pathways in China. Center of Automotive Energy Research Center, Tsinghua University, Beijing
Ou XM, Zhang XL, Chang SY et al (2009) Energy consumption and GHG emissions of six biofuel pathways by LCA in (the) people's Republic of China. Appl Energy 86:S197–S208
Ou X, Zhang X, Chang S (2010a) Alternative fuel buses currently in use in China: life-cycle fossil energy use, GHG emissions and policy recommendations. Energy Policy 38(1):406–418
Ou XM, Zhang XL, Chang SY (2010b) Scenario analysis on alternative fuel/vehicle for China's future road transport: life cycle energy demand and GHG emissions. Energy Policy 38:3943–3956
Ou X, Yan X, Zhang X (2010c) Using coal for transportation in China: life-cycle GHG of coal-based fuel and electric vehicle, and policy implications. Int J Greenhouse Gas Control 4(5):878–887
Ou X, Xiaoyu Y, Zhang X (2011) Life-cycle energy consumption and greenhouse gas emissions for electricity generation and supply in China. Appl Energy 88(1):289–297
Ou X, Yan X, Zhang X, Liu Z (2012) Life-cycle energy consumption and GHG emission intensities of alternative vehicle fuels in China. Appl Energy 90(1):218–224
Wang MQ (1999) GREET 1.5-transportation fuel-cycle model-volume 1: methodology, development, uses, and results. Center for Transportation Research, Argonne National Laboratory, Argonne
Wang MQ (2001) Development and uses of GREET 1.6. Center for Transportation Research, Argonne National Laboratory, Argonne
Wang MQ (2004) Allocation of energy use in petrol refineries to petrol products. Int J Life Cycle Assess 9(34):44
Wang MQ (2007) Operating manual for GREET version 1.7. http://greet.anl.gov/publications.html. Accessed 15 Jan 2013
Wang MQ, Weber T (2001) Well-to-wheels energy use and GHG emissions of advanced fuel/vehicles system – North American analysis. Center for Transportation Research, Argonne National Laboratory, Argonne
Zhang AL, Shen W, Han WJ, Chai QH (2008) Life cycle analysis of automotive alternative fuel. Tsinghua University Press, Beijing

Chapter 12
Scenario Analyses of Automotive Energy

Zhang Xiliang, Ou Xunmin, Zhang Jihong, Chai Qimin, Hao Han, Huo Hong, and He Jiankun

Abstract This chapter provides a further integrated analysis of these energy problems, introduces a comprehensive analytical framework and modeling tools, and focuses on an analysis and evaluation of several scenarios related to vehicle energy systems. The aim is to provide an understanding of the essential development of automotive energy in China as well as to offer a scientific, structured data basis for choice of technology and policy formulation regarding sustainable automotive energy.

This chapter presents five scenarios for the future development of China's automotive energy: a reference scenario, a scenario for developing electric vehicles, a scenario for developing fuel-cell vehicles, a scenario for developing biofuels, and an integrated policy scenario. A systematic and in-depth analysis is made about the role of innovations in key vehicle energy technology, such as with pure electric vehicles, fuel-cell vehicles, biofuels, and energy-saving technologies, in creating a sustainable vehicle energy system in China; this will include assessing the contribution to improving energy security, reducing GHG emissions, reducing transport costs, and upgrading the automotive energy industry. At the same time, conditions toward achieve the various scenarios and attendant problems will be examined. In addition, the possible role of traffic demand management in establishing sustainable vehicle energy is investigated in the integrated policy scenario.

Keywords Automotive energy • Scenario analysis • China

Zhang Xiliang (✉) • Ou Xunmin • Zhang Jihong • Chai Qimin • Hao Han • He Jiankun
China Automotive Energy Research Center, Tsinghua University, Beijing 100084, P.R. China
e-mail: zhang_xl@mail.tsinghua.edu.cn

Huo Hong
Institute of Energy, Environment and Economy, Tsinghua University, Beijing 100084, P.R. China

CAERC, Tsinghua University, *Sustainable Automotive Energy System in China*,
DOI 10.1007/978-3-642-36847-9_12,

In earlier chapters of this book, China's automotive energy problems were discussed in terms of vehicle traffic demand, power technology, vehicle fuel, the life-cycle (LC) energy efficiency of vehicle energy, greenhouse gas (GHG) emissions, and other aspects. This chapter provides a further integrated analysis of these energy problems, introduces a comprehensive analytical framework and modeling tools, and focuses on an analysis and evaluation of several scenarios related to vehicle energy systems. The aim is to provide an understanding of the essential development of automotive energy in China as well as to offer a scientific, structured data basis for choice of technology and policy formulation regarding sustainable automotive energy.

This chapter presents five scenarios for the future development of China's automotive energy: a reference scenario, a scenario for developing electric vehicles, a scenario for developing fuel-cell vehicles, a scenario for developing biofuels, and an integrated policy scenario. A systematic and in-depth analysis is made about the role of innovations in key vehicle energy technology, such as with pure electric vehicles, fuel-cell vehicles, biofuels, and energy-saving technologies in creating a sustainable vehicle energy system in China; this will include assessing the contribution to improving energy security, reducing GHG emissions, reducing transport costs, and upgrading the automotive energy industry. At the same time, conditions toward achieve the various scenarios and attendant problems will be examined. In addition, the possible role of traffic demand management in establishing sustainable vehicle energy is investigated in the integrated policy scenario.

12.1 Automotive Energy Model

12.1.1 General Framework of the Model

The Tsinghua China Automotive Energy Model (TCAEM) is an integrated assessment model for China's future automotive energy technology and policy developed by the China Automotive Energy Research Center, Tsinghua University. TCAEM consists of an automotive energy supply model, an automotive traffic demand model, and an integration and optimization model. The automotive energy supply model and automotive traffic demand model are made up of many submodules (Fig. 12.1). The model inputs are the driving and structural variables of automotive energy scenarios, such as economic development, population and geography, technological advances, and public policies that decide the future direction of automotive energy. The model outputs are characteristics of automotive energy scenarios, such as vehicle ownership, automotive traffic service capacity, automotive energy consumption and composition, vehicle power technology composition, and pollutant emissions of the automotive energy system. The main functions of the model include the following:

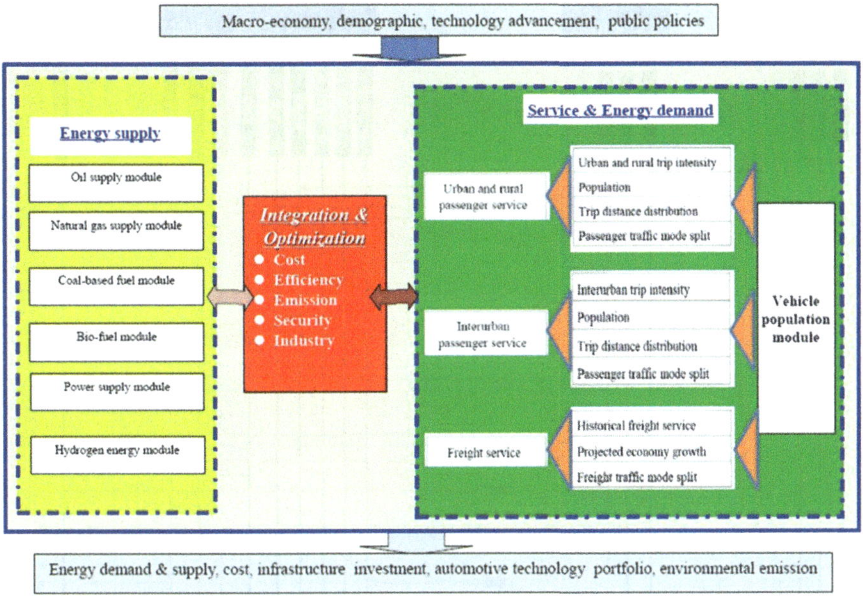

Fig. 12.1 Tsinghua China Automotive Energy Model (TCAEM)

1. Forecasts of vehicle ownership, automotive market structure, passenger turnover, and freight turnover
2. Comprehensive evaluation of technological progress and market penetration of vehicle power and alternative energy
3. Simulation of the proportions of different vehicle power technologies and alternative-energy sources to meet the automotive traffic service demand
4. Simulation of the changes in future automotive energy consumption and composition
5. Simulation of the total cost of future automotive traffic
6. Simulation of the effects of different automotive energy policies

12.1.2 Automotive Traffic Demand Forecasting Model

The automotive traffic demand forecasting model consists of four submodules: the vehicle ownership forecasting model, the urban passenger service forecasting model, the intercity passenger service forecasting model, and the freight service forecasting model. Macroeconomic conditions, population and geography, technological advances, and public policies are the driving variables in the automotive traffic demand forecasting model. Since the basic assumptions, model approach, and forecast results with the automotive traffic demand forecasting model were elaborated in Chap. 3, they will not be repeated here.

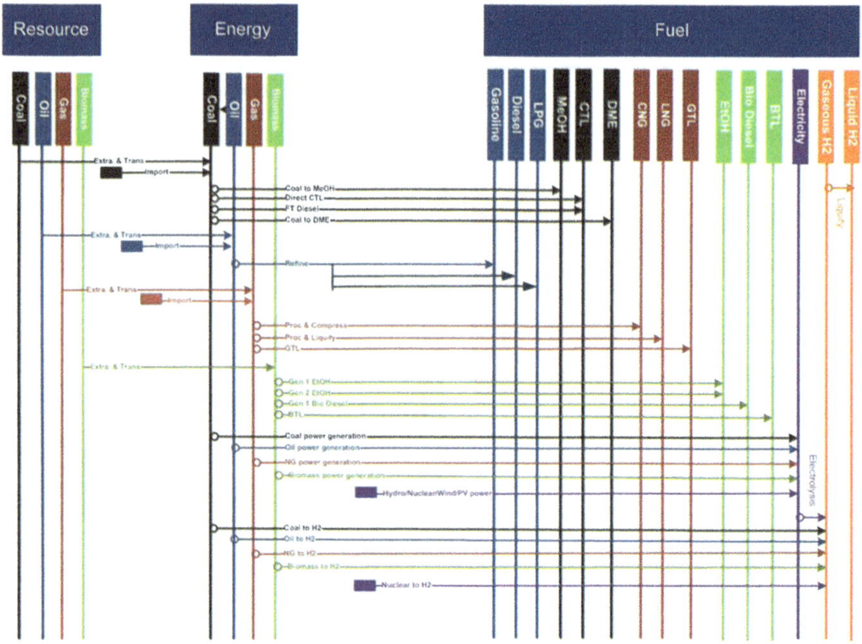

Fig. 12.2 Flowchart of the automotive energy system

12.1.3 Automotive Energy Supply Model

The automotive energy supply model consists of six submodules: petroleum-based fuels, natural gas-based alternative fuels, coal-based alternative fuels, bio-based alternative fuels, automotive electricity, and automotive hydrogen (Fig. 12.1). The six energy supply submodules produce several paths for automotive energy supply, and together they constitute the energy flow network of the automotive energy system (Fig. 12.2). The features and functions of the six submodules include the following:

1. Petroleum-based automotive fuels supply chain: estimate the available supply, cost, and CO_2 emission of automotive refined oil according to such factors as resource reserves, mining capacity, petroleum refining capacity, investment and cost of mining, investment and cost of petroleum refining, petroleum transport capacity, capacity of crude oil and refined oil input, and industrial policy.
2. Natural gas fuels supply chain: estimate the available supply, cost, and CO_2 emission of liquefied petroleum gas (LPG; petroleum-based LPG is also analyzed in this submodule since it is a fuel gas), compressed natural gas (CNG), and liquefied natural gas (LNG) for vehicles according to such factors as resource reserves, mining capacity, investment and cost of mining, transport capacity, capacity of natural gas input, and industrial policy.

3. Coal-based alternative fuels supply chain: estimate the available supply, cost, and CO_2 emission of automotive coal-based methanol, dimethyl ether (DME), and coal-to-diesel conversion according to such factors as resource reserves, mining capacity, investment and cost of mining, transport capacity of coal, capacity of coal input, investment and cost of coal-based alternative-fuel technologies, and industrial policy.
4. Bio-based alternative fuels supply chain: estimate the available supply, cost, and CO_2 emission of first-generation bioethanol, first-generation biodiesel, second-generation bioethanol, and second-generation biodiesel for vehicle according to such factors as the amount of biomass resources, transport capacity, investment and cost of bio-based alternative-fuel technologies, and industrial policy.
5. Automotive electricity supply module: estimate the available supply, cost, and CO_2 emission of automotive electricity according to such factors as electricity production capacity, power source structure, investment and cost of new-energy generation technologies, characteristics of the grid load, capacity of power transmission and distribution, charging infrastructure, and industrial policy.
6. Automotive hydrogen supply module: estimate the available supply, cost, and CO_2 emission of automotive hydrogen according to such factors as hydrogen production capacity, investment and cost of hydrogen production technologies, capacity of hydrogen transmission and distribution, hydrogenation infrastructure, and industrial policy.

12.1.4 Automotive Energy Integration and Optimization Model

The automotive energy integration and optimization model is built on the energy flow network of the automotive energy system (Fig. 12.2), the fuel network for vehicle technologies, and the vehicle network for transport services (Fig. 12.3). Through the automotive energy integration and optimization model, different automotive energy development scenarios are generated by optimizing, combining, and choosing different automotive energy supply paths and vehicle power technologies according to particular requirements regarding economy, energy efficiency, emissions, security, and fairness in the automotive energy system. Therefore, TCAEM can be used to generate both a reference scenario and an alternative policy scenario for automotive energy. The automotive energy integration and optimization model was developed using the MESSAGE platform, which was developed by the International Institute for Applied Systems Analysis. The economic goal was to minimize the cost of vehicle energy technology for the total transport service system from energy extraction to meeting traffic demand. The goals and requirements of energy and environmental emissions are reflected in the following constraint equations.

Function $c(q, t, f, x)$ indicates the integrated cost of traffic in China in 2050, mainly in terms of the energy cost and nonenergy cost of motor traffic. The nonenergy cost of motor traffic includes automobile acquisition, maintenance, insurance, taxes, and other expenses; the energy cost of motor traffic is the cost

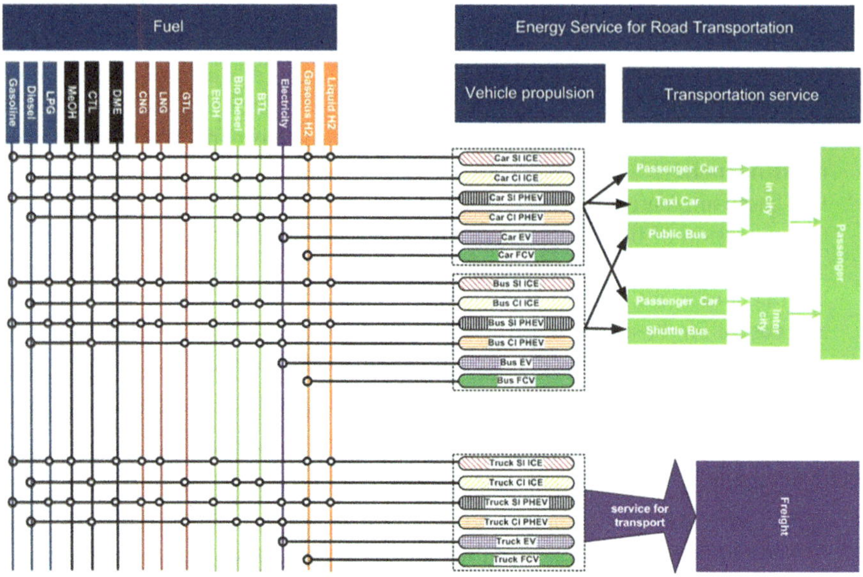

Fig. 12.3 Fuel and vehicle service chains

Table 12.1 Motor vehicle type number

	Passenger vehicle				Bus		Truck		
q	1	2	3	4	5	6	7	8	9
Type	Mini	Small	Medium	Large	Medium	Large	Light	Medium	Heavy

Table 12.2 Vehicle power technology number

t	1	2	3	4	5
Power technology	Spark ignition(SI)	Compressed ignition (CI)	Gas engine (GE)	Electric vehicle (EV)	Fuel-cell vehicle (FCV)

of fueling, which includes the costs of resources, transportation, investment in infrastructure, conversion, and fueling through the extraction-fueling process. $q \in \{1, 2, \ldots, 9\}$ indicates the type of vehicles used; the significance of the numbers is indicated in Table 12.1.

$t \in \{1, 2, 3, 4, 5\}$ indicates the type of vehicle power technology. Five kinds of power technology are considered in the model, and the significance of the numbers is indicated in Table 12.2.

Hybrid (including mini-hybrid and medium-hybrid) vehicles are not listed separately in this model but are included in two power technologies—SI and CI. Fuel-cell vehicles are mainly used as large buses and large passenger vehicles (such as SUVs). Pure electric vehicles and plug-in hybrid vehicles are mainly used in

Table 12.3 Relationship between motor vehicle type and vehicle power technology

t \ q	1	2	3	4	5	6	7	8	9
1	√	√	√	√	√		√		
2		√	√	√	√	√		√	√
3	√	√	√		√	√	√	√	
4	√	√	√	√	√	√			
5	√	√	√	√	√	√	√		

Table 12.4 Fuel number

f	1	2	3	4	5	6
Fuel	Gasoline	Diesel	LPG	CNG	LNG	Methanol
f	7	8	9	10	11	12
Fuel	Dimethyl ether	Bioethanol	Biodiesel	Coal-based diesel	Hydrogen	Electricity

Table 12.5 Relationship between vehicle power technology and fuel

Technology \ Fuel	Gasoline	Diesel	LPG	CNG	LNG	Methanol	Dimethyl ether	Bio methanol	Coal-based diesel	Hydrogen	Electricity
SI	√					√		√			
CI		√					√		√		
GE			√	√	√						
FCV										√	
EV											√

passenger transport, including passenger vehicles (taxis) and large buses (urban public transport). The correspondence between q and t is presented in Table 12.3.

$f \in \{1, 2, \ldots, 12\}$ indicates the type of fuel for motor vehicles; 12 kinds of fuels are considered in this model. The significance of the numbers is presented in Table 12.4.

Since the fuels used in different vehicle power technologies are not the same, the main combinations of vehicle power technology and fuels considered in the model appear in Table 12.5.

The operating mechanism of the model is to identify the combination of vehicle power technology and energy technology that amounts to the least integrated traffic cost while meeting the vehicle traffic demand; it aims to achieve this while meeting restrictions regarding various energy resources and GHG emission targets and taking into account the characteristics of the infrastructure. A brief mathematical description of the model is as follows:

$$\textbf{Min} \quad c(q,t,f,x) = \sum_i \sum_q \sum_t \sum_f \mathrm{Auto}_i(q,t,f) \cdot x_i(q,t,f)$$

$$\textbf{s.t.} \quad \sum_{t,f} x_i(q,t,f) \geq d_i(q), \quad x_i(q,t,f) \geq 0, \quad d_i(q) \geq 0;$$

$$\sum_{q,t} x_i(q,t,f) \cdot F_i(q,t,f) \leq E_i(f), \quad F_i(q,t,f) \geq 0, \quad E_i(f) \geq 0$$

$$\sum_{q,t,f} x_i(q,t,f) \cdot M_i(q,t,f) \leq \varphi_i, \quad M_i(q,t,f) \geq 0, \quad \varphi_i \geq 0;$$

$$\sum_f E_i(f) \cdot N_i(f) \leq \psi_i, \quad N_i(f) \geq 0, \quad \psi_i \geq 0 \tag{12.1}$$

$i \in \{2005, 2010, 2015, \ldots, 2050\}$ indicates the year; $\mathrm{Auto}_i\,(q,t,f)$ indicates the cost of combining vehicle type q, power technology t, and fuel f during time period i; $x_i\,(q,t,f)$ indicates the volume of traffic with the combination of vehicle type q, power technology t, and fuel f during time period i; $d_i(q)$ indicates the total volume of traffic of vehicle type q during time period i; $F_i\,(q,t,f)$ indicates the per-unit fuel economy of a vehicle (q,t,f); $E_i(f)$ indicates the upper limit of the supply of fuel f; $M_i\,(q,t,f)$ indicates the factor of per-unit GHG emissions of a vehicle (q,t,f) in its life cycle; φ_i indicates the life-cycle limit of GHG emissions by road traffic in year i; $N_i(f)$ indicates the factor of per-unit GHG emissions by fuel f during production and supply; ψ_i indicates the limit of the total GHG emissions during the production and supply of fuel f in year i.

The cost function $\mathrm{Auto}_i\,(q,t,f)$ is the integrated cost of the automobile transportation system, which signifies the total cost of vehicles (q,t,f) completing the per-unit volume of the traffic service. $\mathrm{Auto}_i\,(q,t,f)$ includes two items of cost—per-unit cost of ownership and the cost of utilization.

The cost estimation in the model, as shown in Eq. 12.2, is the dynamic levelized cost considering the discount rate factor:

$$\mathrm{Auto}_i(q,t,f) = \overline{\mathrm{Auto}_i}(q,t,f) + \mathrm{Energy}_i(f) \tag{12.2}$$

$\overline{\mathrm{Auto}_i}\,(q,t,f)$ indicates the nonenergy levelized cost of motor transport in year i; $\mathrm{Energy}_i(f)$ indicates the levelized cost of fuel f in year i.

The extended equation of $\overline{\mathrm{Auto}_i}\,(q,t,f)$ is shown in Eq. 12.3:

$$\overline{\mathrm{Auto}_i}(q,t,f) = \sum_{j=1}^{T1} w_{i-j}(q,t,f) * \overline{\mathrm{Auto}_{i-j}}(q,t,f) + w_{\mathrm{new}}(q,t,f) * \overline{\mathrm{Auto}_{\mathrm{new}}}(q,t,f) \tag{12.3}$$

T_1 indicates the useful life of motor vehicles; $w_{i-j}(q,t,f)$ indicates the proportion of newly purchased vehicles of type (q,t,f) in year $(i-j)$ among the total number of type (q,t,f) in year i. $w_i(q,t,f)$ indicates the proportion of newly purchased vehicles of type (q,t,f) in year i among the total number of type (q,t,f) in year i. $\overline{\text{Auto}_{\text{new}}}(q,t,f)$ indicates the levelized cost of newly purchased vehicles of type (q,t,f).

According to the levelized equation, the nonenergy cost of newly purchased motor vehicles is shown in Eq. 12.4:

$$\overline{\text{Auto}_{\text{new}}}(q,t,f) = \frac{\sum_{i=0}^{T1}(\text{Auto}I_i + \text{Auto}O_i)(1+r)^{-i} * \frac{r*(1+r)^{T1}}{(1+r)^{T1}-1}}{Q} \quad (12.4)$$

$\text{Auto}I_i$ indicates the investment cost of motor vehicles in year i, such as the cost of vehicle acquisition, obtaining a license, taxes, and future battery exchange. $\text{Auto}O_i$ indicates the operating cost of a motor vehicle in year i, such as compulsory insurance and maintenance costs. r indicates the interest rate. Q indicates the average total mileage of motor vehicles.

The extended equation of $\text{Energy}_i(f)$ appears as Eq. 12.5:

$$\text{Energy}_i(f) = \sum_{j=1}^{T2} m_{i-j}(f) * \text{Energy}_{i-j}(f) + m_{\text{new}}(f) * \text{Energy}_{\text{new}}(f) \quad (12.5)$$

T_2 indicates the useful life of the energy supply infrastructure. $m_{i-j}(f)$ signifies the proportion of fuel f produced by newly built fuel f supply facilities in year $(i-j)$ among the total amount of fuel f in year i. $m_{\text{new}}(f)$ indicates the proportion of fuel f produced by newly built fuel supply facilities in year i among the total amount of fuel f in year i. $\text{Energy}_{\text{new}}(f)$ signifies the levelized cost of fuel f supplied by newly built fuel f supply facilities.

From the levelized formula, the energy cost of newly purchased motor vehicles appears in Eq. 12.6:

$$\text{Energy}_{\text{new}}(f) = \frac{\sum_{i=0}^{T2}(\text{Energy}I_i + \text{Energy}O_i)(1+r)^{-i} * \frac{r*(1+r)^{T2}}{(1+r)^{T2}-1}}{P} \quad (12.6)$$

$\text{Energy}I_i$ indicates the investment cost of fuel f in year I, such as the initial investment in power plants and investment in charging stations. $\text{Energy}O_i$ signifies the operating cost of fuel production in year i, such as the cost of the resource itself and maintenance costs. P indicates the total amount of fuel supply.

The cost of fuel supply is the well-to-tank (WTT) cost, in which the conversion efficiency has to be considered in addition to the fixed investment. The extended equation of $\text{Energy}O_i$ appears in Eq. 12.7:

$$\text{Energy}O_i = \frac{\text{CERS}_I}{\beta_I} + \text{CCUR}_i + \text{CARC}_i - \text{CEXP}_i \quad (12.7)$$

β_i indicates the efficiency, which is the product of efficiency in every link in year i. CCUR indicates the costs of maintenance and operation except the cost of raw materials. CERS signifies the cost of raw materials. CEXP indicates the profit from fuel exports. CARC signifies all restraint costs, such as future carbon tax.

12.2 Reference Scenario

The reference scenario is the baseline scenario for the analysis of automotive energy development. Under the reference scenario, the government does not provide a clear target or requirement regarding the future development of automotive energy in China. In addition, no new policies affect the speed of innovation and development trends in vehicle and energy technologies. Thus, the current trends of vehicle power and energy technologies continue and so do existing market conditions. In the reference scenario, fundamental changes in automotive energy technologies do not occur in the short or medium term. The main purpose in building and analyzing the reference scenario is to investigate the future situation regarding automotive energy consumption and GHG emissions in China if positive technological innovation fails to appear and more stringent energy-saving policies are not implemented. This scenario attempts to determine the economic costs and assess whether the resulting situation is acceptable to both China and the world at large.

12.2.1 Key Scenario Assumptions

12.2.1.1 Demand for Auto Transport Services

In the reference scenario, the growth rates for different types of auto transport services are those presented in Chap. 3, and they appear in Table 12.6. No active policies are introduced on private auto transport demand or promotion of public transport service to achieve an effective transfer from private auto transport to public transport. There is likewise no rapid or effective transfer from truck-to-rail freight.

Table 12.6 Growth rates of auto transport services under the reference scenario

Type	2010–2020	2020–2030	2030–2050
Passenger vehicle (%)	15.33	3.84	0.02
Bus (%)	6.43	3.35	0.99
Truck (%)	9.96	5.55	1.59

Table 12.7 Fuel economy of newly purchased passenger vehicles

Type of passenger vehicle	2010	2020	2030	2050
Mini (displacement ≤1 L)	5.5	4.5	4.0	3.8
Small (displacement 1 to ≤1.6 L)	7.0	5.8	5.0	4.7
Medium (displacement 1.6 to ≤2.5 L)	9.0	8.0	7.0	6.3
Large (displacement ≥2.5 L)	12.0	10.5	9.5	8.8

Table 12.8 Rate of reduction in fuel consumption by the bus fleet (compared with 2010)

Year	2020	2030	2040	2050
Rate of reduction (%)	3	6	9	12

Table 12.9 Rate of reduction rate in fuel consumption by the truck fleet (compared with 2010)

Year	2020	2030	2040	2050
The rate of reduction (%)	5	10	15	20

12.2.1.2 Fuel Economy of Passenger Vehicles

The fuel economy of passenger vehicles will improve to some extent; however, this will apparently not be the case if there is a significant improvement. In the reference scenario, the assumptions regarding the fuel economy of newly purchased mini, small, medium, and large passenger vehicles are shown in Table 12.7.

12.2.1.3 Fuel Economy of Buses

The fuel economy of buses will improve to some extent if the existing development trend of vehicle energy-saving technologies continues. In the reference scenario, the assumptions about improvement in the fuel economy of the bus fleet are shown in Table 12.8.

12.2.1.4 Fuel Economy of Trucks

The fuel economy of trucks will improve to some extent if the existing development trend of vehicle energy-saving technologies continues. In the reference scenario, the assumptions about the improvement in the fuel economy of the truck fleet as L/(100 km t) are shown in Table 12.9.

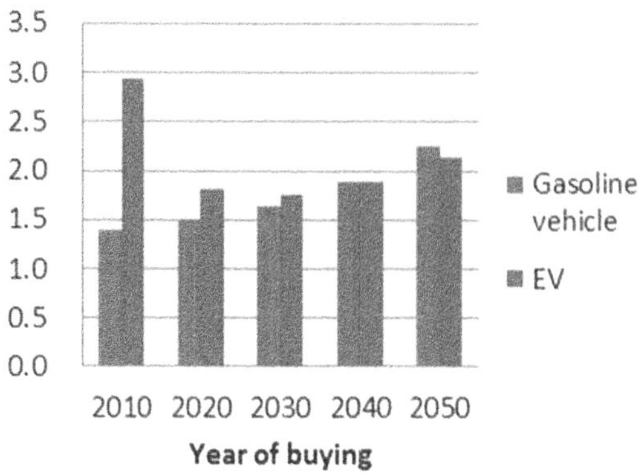

Fig. 12.4 Integrated costs of mini gasoline passenger vehicles and EV under the reference scenario

12.2.1.5 Electrification of Passenger Vehicles

Electric vehicles represent an important innovation in vehicle technologies. As with other technological innovations, the promotion of electric vehicles requires three stages, and their penetration rate displays an S-shaped curve. Under the reference scenario, the cost of the battery, motor, and electronic control of pure electric vehicle undergoes only a slow reduction. A comparison of the integrated costs of mini electric passenger vehicles and petroleum-based passenger vehicles is presented in Fig. 12.4. Basically, mini electric passenger vehicles cannot compete with petroleum-based passenger vehicles before 2040. Since pure electric mode vehicles will be unable to match the driving range of medium-sized and large passenger vehicles, plug-in hybrid electric vehicles (PHEV) and extended-range electric vehicles (EREV) will have to be employed. However, their integrated costs will still be higher than those of petroleum-based passenger vehicles.

12.2.1.6 Electrification of Buses

Under the reference scenario, the cost of the battery, motor, and electronic control will decrease slowly. The integrated cost of pure electric buses is unable to compete with that of petroleum-based buses before 2040 (Fig. 12.5).

12.2.1.7 Fuel-Cell Vehicles

Under the reference scenario, the cost of fuel-cell vehicle technologies falls slowly. The integrated cost is unable to compete with that of petroleum-based vehicles before 2050 (Fig. 12.6).

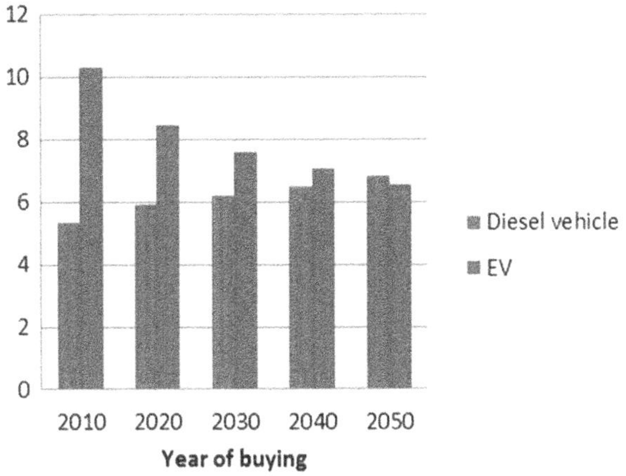

Fig. 12.5 Integrated costs of diesel city bus vehicles and EV under the reference scenario

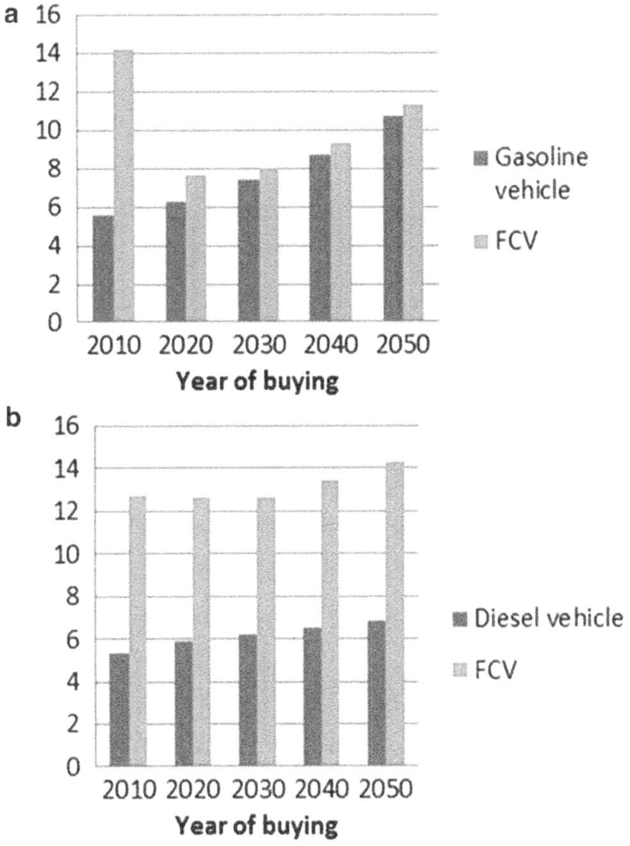

Fig. 12.6 Integrated costs of large gasoline passenger vehicles, diesel city buses, and fuel-cell vehicles under the reference scenario. (**a**) Large passenger cars. (**b**) City buses

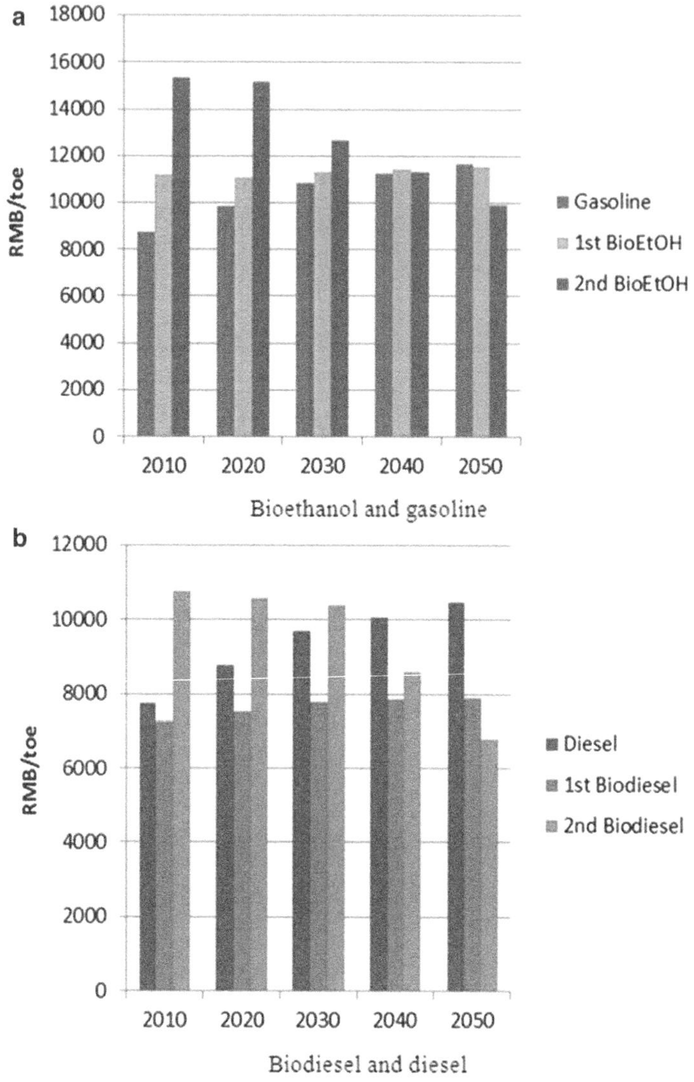

Fig. 12.7 Cost comparison between biofuels and petroleum-based fuels. (**a**) Bioethanol and gasoline. (**b**) Biodiesel and diesel

12.2.1.8 Biofuels

Before 2025, there is no fundamental breakthrough in second-generation biofuel technology. Limited quantities of first-generation biofuel vehicles are mainly used. After 2035, the cost of second-generation biofuels may show a greater decrease. A comparison of the cost between biofuels and petroleum-based fuels appears in Fig. 12.7.

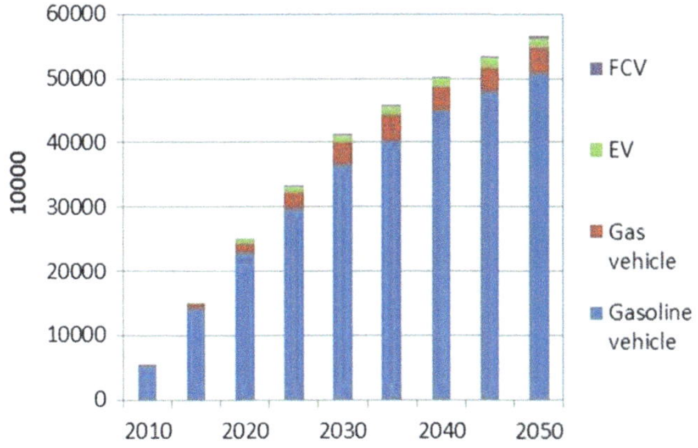

Fig. 12.8 Development of passenger vehicle power technologies under the reference scenario

12.2.2 Results of the Scenario

12.2.2.1 Power Technologies of Passenger Vehicles

The number of passenger vehicles with different power technologies under the reference scenario is shown in Fig. 12.8. It is clearly evident that in this scenario, gasoline vehicles absolutely dominate with respect to both passenger transport and freight. The proportion of gas-fuel vehicles is expected to increase significantly after 2020. Electric vehicles and fuel-cell vehicles are minor in terms of both number of vehicles and their share of passenger transport. Regarding the number of passenger vehicles, electric vehicles account for only 0.7, 2.2, and 2.8 % of vehicles, respectively, in 2020, 2030, and 2050; the respective proportion of gas-fuel vehicles is 5.6, 8.2, and 7.4 %. The proportion of fuel-cell vehicles is negligible. Under the reference scenario, since the proportion of new-energy vehicles is tiny, the passenger vehicle industry lacks motivation and the appropriate market conditions for upgrading industrial technology and improving competitiveness (Fig. 12.9).

12.2.2.2 Power Technologies of Buses

Under the reference scenario, the number of buses with different power technologies is shown in Fig. 12.10. The passenger transport share of buses with different power technologies appears in Fig. 12.11. Diesel and gasoline engines dominate both the passenger vehicle fleet and in their proportion of the passenger transport population under the reference scenario. For buses, electric and fuel-cell vehicles are minor both in terms of number of vehicles and their proportion of the bus population. Under the reference scenario, the proportions of gasoline vehicles and gas-fuel vehicles present an upward trend, while that of diesel vehicles presents a downward

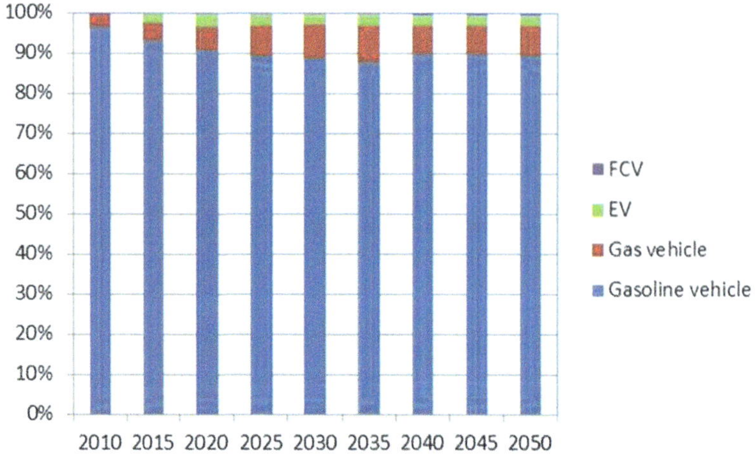

Fig. 12.9 Share of passenger vehicle power technologies under the reference scenario

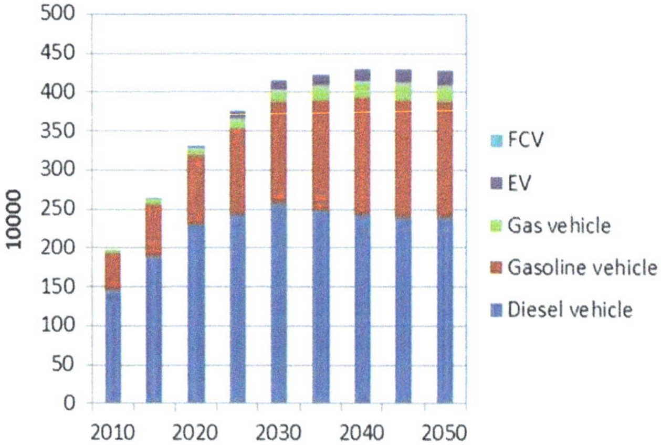

Fig. 12.10 Development of bus power technologies under the reference scenario

trend. The proportion of gasoline vehicles increases from 24 % (2010) to 31 % (2030); that of gas-fuel vehicles increases from 2.2 % (2010) to 3.9 % (2030) and thereafter remains stable. The proportion of diesel vehicles decreases from 74 % (2010) to 62 % (2030) and then remains stable. Among the number of passenger vehicles, the proportion of electric vehicles is 2.8 and 4.4 % in 2030 and 2050, respectively; the proportion of fuel-cell vehicles is negligible for those years. Under the reference scenario, since the proportion of new-energy vehicles is small, there is a lack of impetus in the bus industry and absence of appropriate market conditions for upgrading of industrial technology and improving competitiveness.

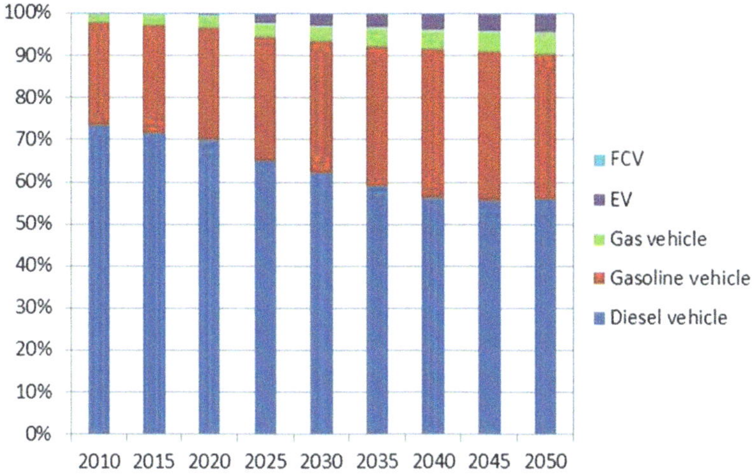

Fig. 12.11 Transport share of bus power technologies under the reference scenario

12.2.2.3 Power Technologies of Trucks

Under the reference scenario, the number of trucks using different power technologies is presented in Fig. 12.12. The proportion of trucks with different power technologies appears in Fig. 12.13. It is evident that under the reference scenario, diesel vehicles dominate both in terms of the freight fleet and road freight share. The road freight proportion of diesel trucks remains at about 80 % throughout the study period. The proportions of gasoline vehicles and gas-fuel vehicles show an increase. The proportion of gasoline vehicles rises from 18.35 % (2010) to 19.49 % (2030) and 18.36 % (2050). The proportion of gas-fuel vehicles grows from 0.96 % (2010) to 1.14 % (2030) and 1.6 % (2050). Since the proportion of new-energy vehicles is negligible under the reference scenario, the bus industry lacks impetus and the necessary market conditions for industrial upgrading and improving competitiveness.

12.2.2.4 Automotive Energy

The automotive energy consumption of China under the reference scenario until 2050 appears in Fig. 12.14. The automotive energy consumption of China continues to show an upward trend up to 2050. Automotive energy consumption increases most rapidly from 2020 to 2030, with an average annual growth rate of 11.3 %. The average annual increase from 2020 to 2030 is 3.5 and 0.6 % from 2030 to 2050. The total automotive energy consumption of China is 3.52, 4.79, and 5.39 hundred million tons of oil equivalent (toe) in 2020, 2030, and 2050, respectively; these figures amount to a 2.6-, 3.6-, and 4.0-fold increase compared with 2010 levels.

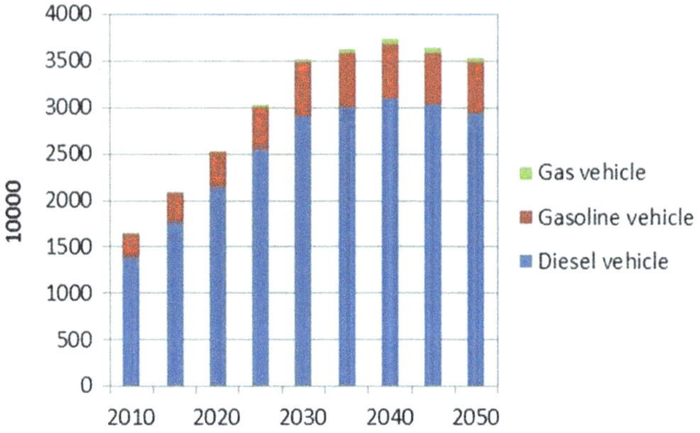

Fig. 12.12 Development of truck power technologies under the reference scenario

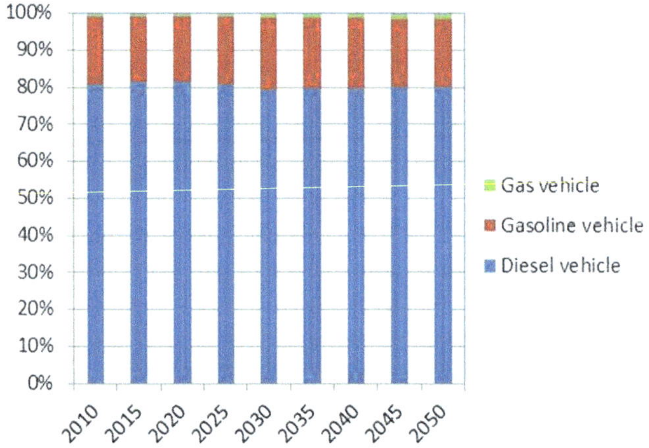

Fig. 12.13 Proportion of truck power technologies under the reference scenario

In terms of automotive energy, the total consumption of petroleum-based diesel and gasoline amounts to 3.28, 4.37, and 4.75 hundred million toe in 2020, 2030, and 2050, respectively, accounting for 93, 91, and 88 % of automotive energy consumption. Under the reference scenario, automotive energy is dominated by petroleum. The consumption of gasoline increases from 0.57 hundred million toe (2010) to 1.82 (2020), 2.31 (2030), and 2.40 hundred million toe (2050). The consumption of diesel increases from 0.71 hundred million toe (2010) to 1.46 (2020), 2.06 (2030) and 2.35 hundred million toe (2050).

Though the ratio of diesel-to-gasoline production in China has improved to 2:1, there is still a shortage of diesel. The ratio of diesel and gasoline consumption in motor transport was 1.26:1 in 2010. Under the reference scenario, the ratio of diesel and gasoline consumption in motor transport is expected to be 1:1 in 2035 and

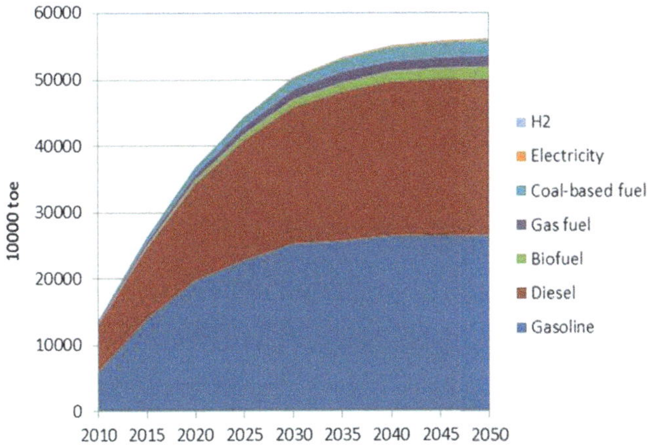

Fig. 12.14 Automotive energy in the reference scenario

remain stable thereafter. Thus, we can see that under the reference scenario, the development of automotive energy in the future will not alleviate the supply and demand imbalance between diesel and gasoline.

Under the reference scenario, the proportion of alternative fuels increases from 5 % (2010) to 7 % (2020), 9 % (2030), and 12 % (2050). Biofuels are a kind of strategic alternative fuel; their consumption is 5.34, 11.35, and 18.46 million toe in 2020, 2030, and 2050, respectively, accounting for 1.5, 2.4, and 3.4 % of total fuel consumption. Before 2025, biofuels are mainly first generation; additional biofuels after 2030 are mainly second generation, as depicted in Fig. 12.15. Since the market proportion of second-generation biofuels is less than 2 % until 2050 under the reference scenario, the biofuel industry lacks impetus and the necessary market conditions for industrial upgrading and improved competitiveness.

Figure 12.16 presents the well-to-wheel life-cycle consumption of fossil energy for vehicles until 2050 under the reference scenario. The consumption of petroleum for vehicles is 3.7, 5.0, and 5.4 hundred million toe in 2020, 2030, and 2050, respectively, accounting for 79, 78, and 76 % of the fossil energy consumption by vehicles. There will therefore be no change in the over-reliance on petroleum in automotive energy supply.

12.2.2.5 GHG Emissions

Figure 12.17 shows the well-to-wheel (WTW) CO_2 emissions of vehicles up to 2050 under the reference scenario. The CO_2 emissions related to energy maintain an upward trend up to 2050. CO_2 emissions by the road transport sector show the fastest increase between 2010 and 2020, with an average annual growth rate of 11.4 % for both direct and WTW emissions. Between 2030 and 2040, the growth slows and enters a stable phase. The CO_2 emissions by vehicles and WTW emissions amount

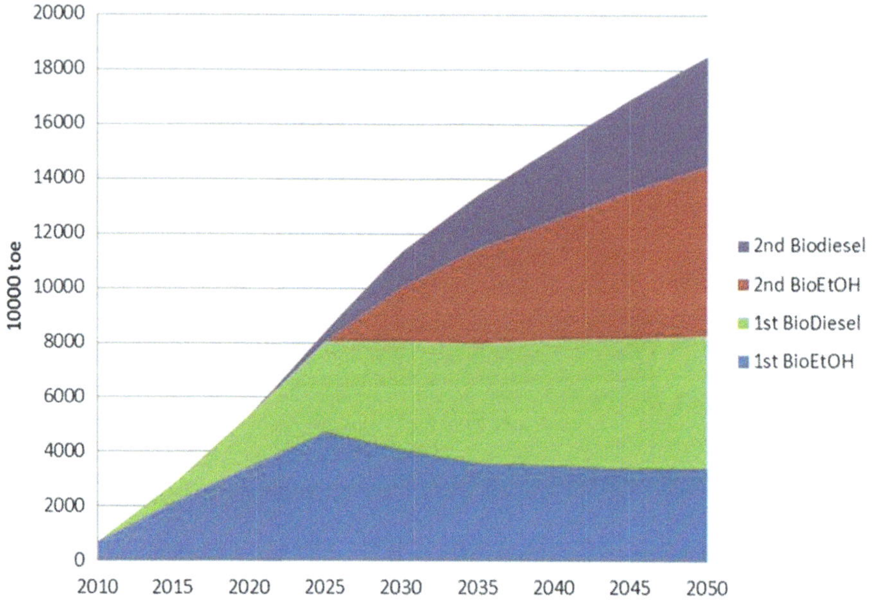

Fig. 12.15 Biofuel consumption by vehicles under the reference scenario

Fig. 12.16 Well-to-wheel fossil energy consumption for motor transport under the reference scenario

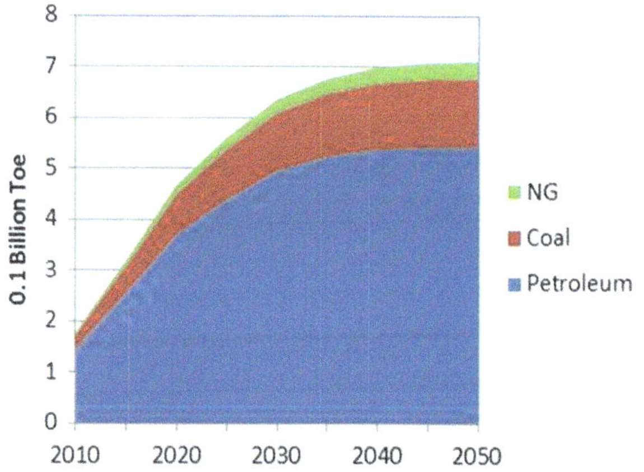

to 16.5 and 23.8 hundred million tons in 2050, respectively; these two figures represent a 300 % increase compared with 2010 levels. In terms of the controlled target of a global temperature rise of 2 °C, the global CO_2 emissions allowed are around 11 billion tons; in this context, the CO_2 emissions of 2.4 billion tons for the motor transport sector in China are unacceptable both nationally and internationally.

Figure 12.18 shows the composition of the WTW CO_2 emissions for motor transport up to 2050 under the reference scenario. The CO_2 emissions from passenger vehicles reach a peak in around 2030 and then remain stable. The CO_2 emissions

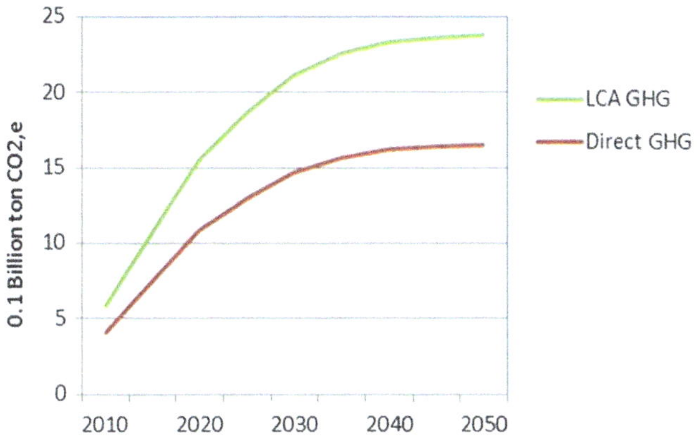

Fig. 12.17 CO_2 emissions of vehicles under the reference scenario

Fig. 12.18 CO_2 emissions of different types of vehicles under the reference scenario

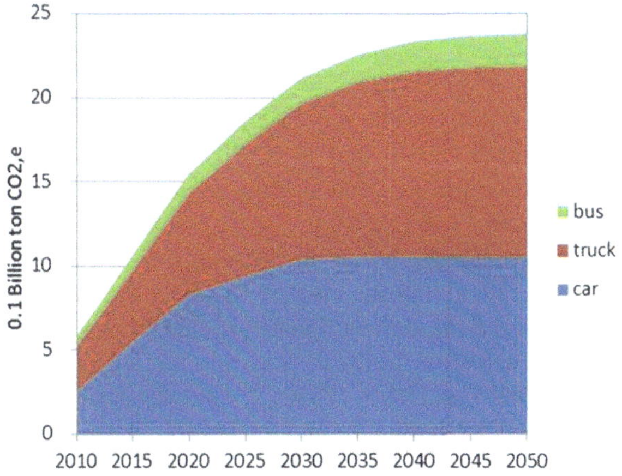

from motor freight continue to show an upward trend. The additional emissions from motor transport after 2030 derive mainly from freight. The CO_2 emissions from large buses present a slight increase. The number of passenger vehicles increases while the CO_2 emissions remain stable, mainly because the developments of hybrid and other energy-saving technologies for passenger vehicles constantly improve the fuel economy of the fleet and the annual mileage continues to decline.

12.2.2.6 Economy of Motor Transport

Figure 12.19 shows the integrated cost of motor transport under the reference scenario. The cost of motor transport shows a 2.5-, 4.5-, and 7.2-fold increase in 2020, 2030, and 2050, respectively, compared with 2010. The proportion of the costs of private passenger vehicles among all vehicles is 69 % (2020), 67 % (2030),

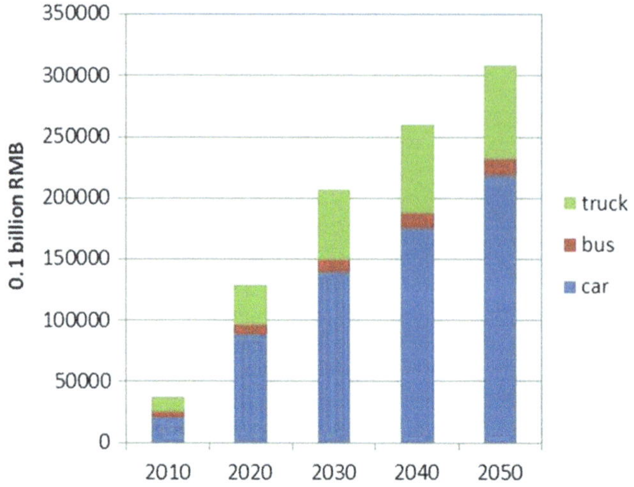

Fig. 12.19 Overall costs of automotive transportation under the reference scenario

and 71 % (2050). The proportion of the cost of fuel in the integrated cost is 29 % (2020), 31 % (2030), and 36 % (2050). The proportion of the expenditure of motor transport within the annual gross domestic product (GDP) is 16 % (2020), 16 % (2030), and 11 % (2050).

Figure 12.20 presents the cost of motor transport energy under the reference scenario, including the cost of gasoline, diesel, natural gas-based fuels, biofuels, coal-based fuels, and payments for electricity and hydrogen. The proportion of the cost of petroleum-based fuels within the total vehicle energy cost is 95 % (2020), 93 % (2030), and 91 % (2050). If 1 hundred million tons of the domestic petroleum production were used for motor transport, the petroleum imports for motor transport would be 2.7, 4.0, and 4.4 hundred million tons in, respectively, 2020, 2030, and 2050; this would amount to 1.7, 1.8, and 1.2 % of the GDP for those years.

12.2.3 Summary of Scenario Points

12.2.3.1 Energy

Automotive energy consumption continues increasing and does not reach a peak before 2050, though the growth rate decreases significantly after 2030. The concentration of automotive energy consumption is in diesel and gasoline. Total consumption of diesel and gasoline amounts to 4.75 hundred million tons in 2050, accounting for 88 % of automotive energy consumption, whereas the proportion of second-generation biofuels is still under 2 %. In 2050, the WTW petroleum consumption for vehicles exceeds 5.4 hundred million tons, which implies that the reliance on foreign

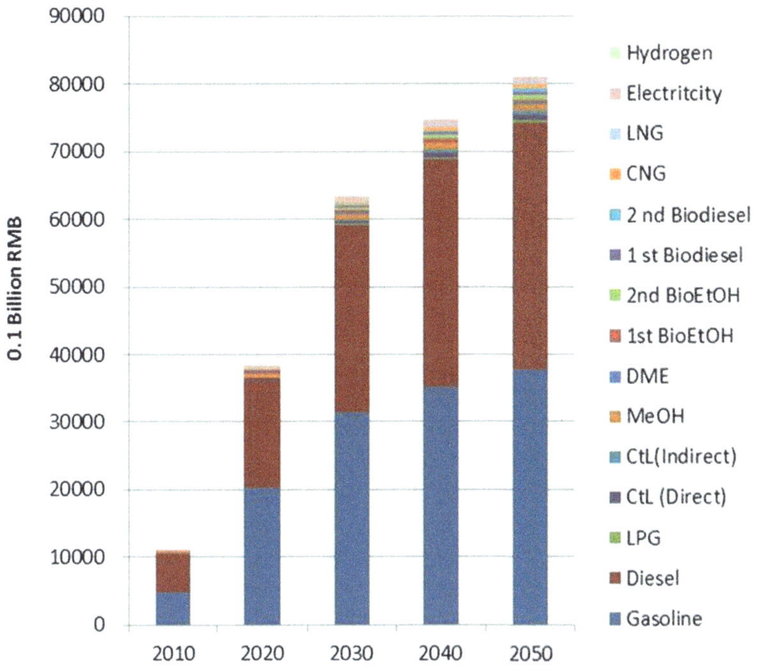

Fig. 12.20 Fuel cost of automotive transportation under the reference scenario

petroleum exceeds 80 %. The development of automotive energy in the future cannot address the imbalance between supply and demand for diesel and gasoline.

12.2.3.2 Automobiles

In 2050, the proportion of new-energy vehicles is under 4 % in the passenger vehicle fleet, less than 5 % in the bus fleet, and negligible in the truck fleet. Since the market scale of new-energy vehicles is small, the automotive industry lacks the market support necessary for industrial upgrading and transformation. Improvement in the proportion of gas-fuel engines is limited: 7 % in the passenger vehicle fleet in 2050, 5 % in the bus fleet, and 2 % in the truck fleet.

12.2.3.3 Environment

CO_2 emissions from the motor transport sector in China continue to rise, amounting to around 2.4 billion tons in 2050. According to the control target of a 2 °C global temperature rise, the global CO_2 emission allowed is around 11 billion tons, which means that the 2.4 billion tons emitted by China's motor transport sector is inacceptable both nationally and internationally.

12.2.3.4 Economy

Compared with 2010, the integrated cost of motor transport will show a 2.5- (2020), 4.5- (2030), and 7.2-fold (2050) increase. The proportion of the cost of petroleum imports for the motor transport sector in terms of annual GDP is 1.7 % (2020), 1.8 % (2030), and 1.2 % (2050).

12.3 Scenario for Developing Electric Vehicles

In terms of life-cycle energy efficiency and GHG emissions by various types of automotive fuel technology, electric vehicles have significant potential for reducing energy consumption and emissions compared with traditional petroleum-based vehicles. The purpose in building the scenario for developing electric vehicles is to examine the role of developing pure electric vehicles in attempting to solve the automotive energy problems in China. Specifically, the scenario analysis for developing electric vehicles addresses the following questions: (1) How can the development of electric vehicles ease automotive energy supply security? (2) What level of GHG emissions does the development of electric vehicles entail? (3) What is the role of developing electric vehicles in reducing automotive energy consumption? (4) What effect will developing electric vehicles exert on the supply and demand structure of the refined oil market in China? (5) What effect will developing electric vehicles have on upgrading the automotive industry and improving competitiveness? (6) What will be the significance of developing electric vehicles for the motor transport economy?

12.3.1 Key Assumptions in the Scenario

1. Demand for motor transport services: the demand for motor transport services is the same as in the reference scenario.
2. Fuel economy of passenger vehicles: improvement in the fuel economy of newly purchased passenger vehicles is the same as in the reference scenario.
3. Fuel economy of buses: improvement in the fuel economy of the bus fleet is the same as in the reference scenario.
4. Fuel economy of trucks: improvement in the fuel economy of the truck fleet is the same as in the reference scenario.
5. Electrification of passenger vehicles: under the scenario of developing electric vehicles, the R&D, demonstration, and promotion of battery, motor, and electronic control technologies of pure electric vehicles achieve major breakthroughs in the near and medium term, and it is supposed that the associated cost will quickly fall. A comparison of the integrated cost of mini pure electric passenger vehicles and petroleum-based passenger vehicles is shown in Fig. 12.21.

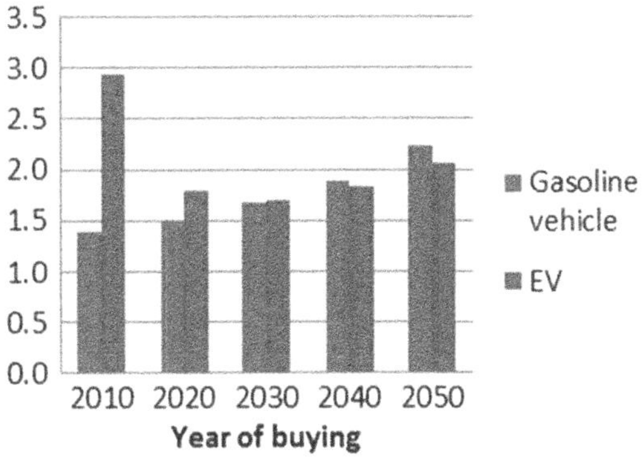

Fig. 12.21 Comparison of the integrated cost between mini gasoline passenger vehicles and electric vehicles under the scenario of developing electric vehicles

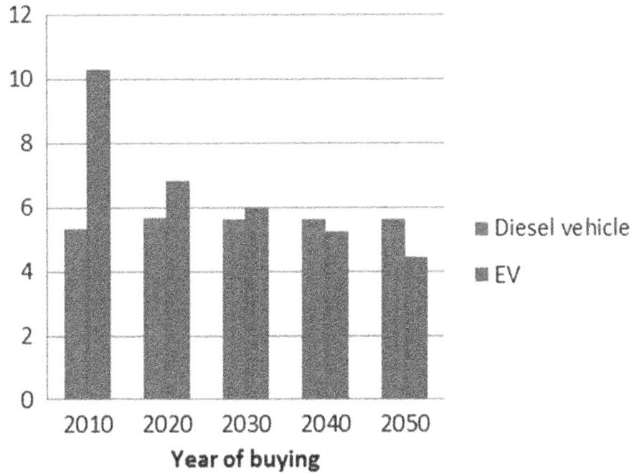

Fig. 12.22 Integrated cost of large electric buses and large petroleum-based buses under the scenario of developing electric vehicles

Basically, mini electric passenger vehicles are able to compete with petroleum-based passenger vehicles in around 2025; mini electric passenger vehicles then go into a phase of rapid development as small pure electric passenger vehicles. For medium-sized and large passenger vehicles, PHEVs and EREVs should be used, and their integrated cost can equal that of petroleum-based passenger vehicles by 2025; they will then go into a phase of rapid development.

6. Electrification of buses: pure electric vehicles are mainly used as large buses. The trend of the integrated cost of large pure electric buses and petroleum-based buses is shown in Fig. 12.22. The cost of pure electric buses can equal that of diesel buses by around 2025; pure electric buses then go into a phase of rapid development.
7. Fuel-cell vehicles: the assumption for fuel-cell vehicles is the same as in the reference scenario.
8. Biofuels: the assumption for biofuels is the same as in the reference scenario.

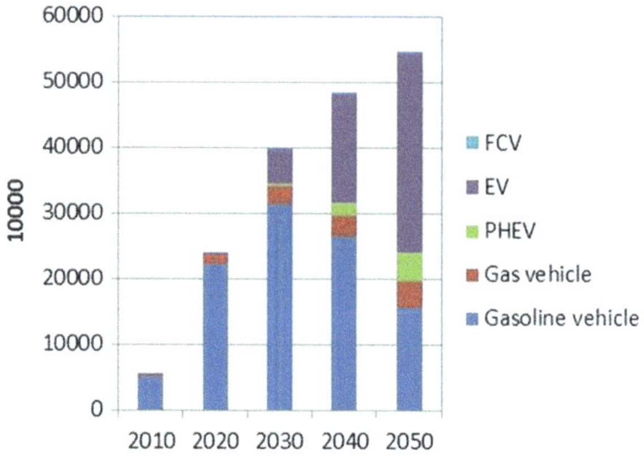

Fig. 12.23 Number of passenger vehicles using different power technologies under the scenario of developing electric vehicles

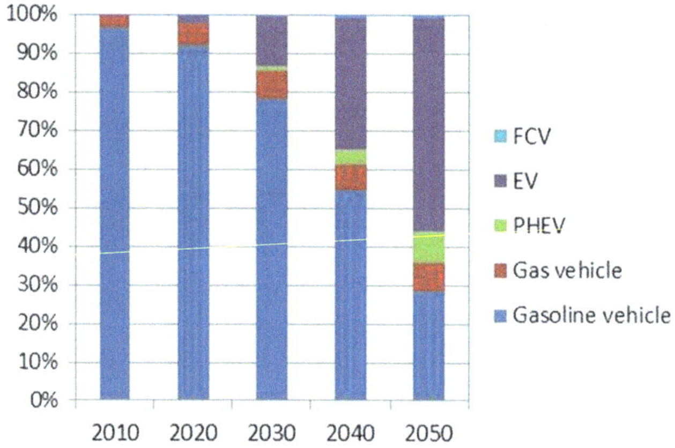

Fig. 12.24 Proportions of different passenger vehicle power technologies under the scenario of developing electric vehicles

12.3.2 Results of the Scenario

12.3.2.1 Power Technologies of Passenger Vehicles

The number of passenger vehicles using different power technologies in the future under the scenario of developing electric vehicles appears in Fig. 12.23. The proportion of passenger vehicles using different power technologies is shown in Fig. 12.24. The proportions of electric vehicles (EVs) among the number of passenger vehicles is 2, 14, and 60 % in 2020, 2030, and 2050, respectively; these figures represent increases of 1.3, 11.8, and 57.2 % compared with the reference scenario. The number of gasoline vehicles reaches a peak in 2030, and it then quickly decreases. By 2050, the proportion of mini, small, medium-size, and large EVs among newly purchased passenger vehicles is expected to reach 90, 70, 45, and

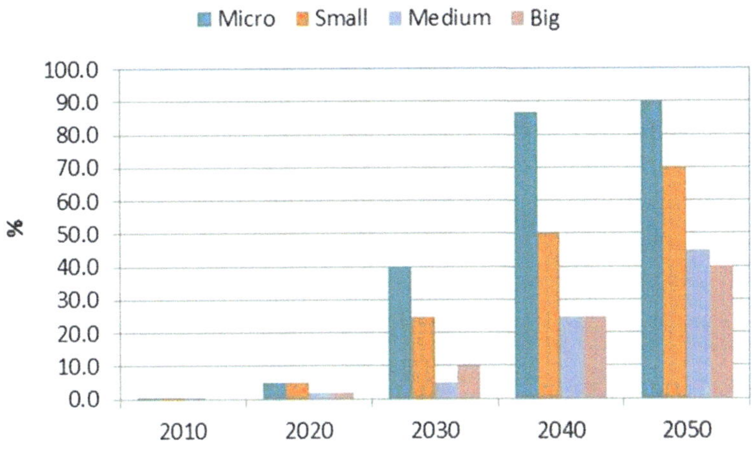

Fig. 12.25 Market penetration rate of newly purchased passenger vehicles of different types under the scenario of developing electric vehicles

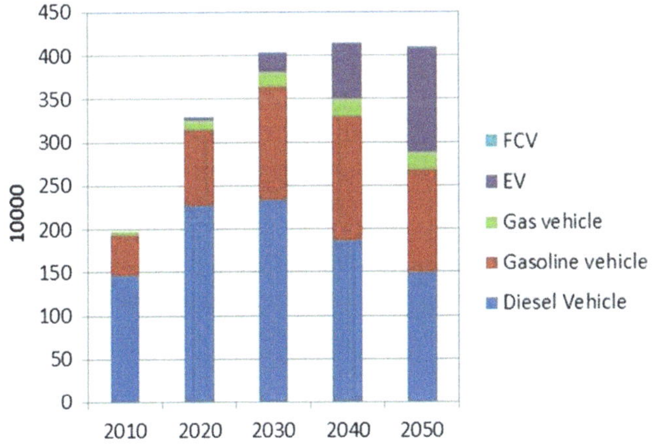

Fig. 12.26 Bus power technologies under the scenario of developing electric vehicles

40 %, respectively. Since the large-scale development of EVs provides necessary market support for upgrading and transforming the passenger vehicle industry, the development of the EV industry is very promising. The role of fuel-cell passenger vehicles and gas-fuel vehicles is the same as in the reference scenario (Fig. 12.25).

12.3.2.2 Power Technologies of Buses

The number of buses using different power technologies in the future appears in Fig. 12.26. The proportion of different power technologies for transport in the future is presented in Fig. 12.27. Under the scenario of developing EVs, the proportion of

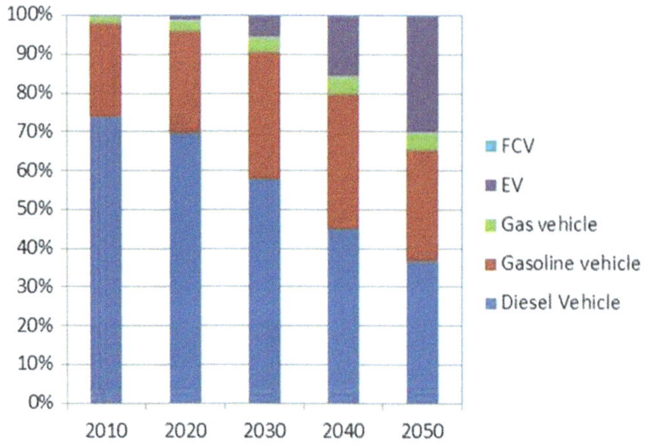

Fig. 12.27 Proportion of truck power technologies under the scenario of developing electric vehicles

EVs in the number of buses amounts to 1, 5, and 30 % in 2020, 2030, and 2050, respectively; these figures represent an increase of 0.5, 2.2, and 25.6 % compared with the reference scenario. The number of diesel buses reaches a peak in 2030 and thereafter shows a continual decline. In 2050, the proportion of diesel buses in the bus fleet falls below 40 %. The proportion of gasoline vehicles in the bus fleet increases from 24 % (2010) to 32 % (2030) and subsequently shows little change. Since the electrification of buses provides necessary market support for upgrading and transformation of the passenger vehicle industry, the development of the EV industry is a promising one. The role of fuel-cell and gas-fuel buses is the same as in the reference scenario.

12.3.2.3 Power Technologies of Trucks

Because of restrictions in technology and use, electrification is impractical for trucks. The composition of truck power technologies and the transport proportion are the same as in the reference scenario.

12.3.2.4 Automotive Energy

Figure 12.28 shows automotive energy consumption of China in 2050 under the scenario of developing EVs. Automotive energy consumption is 3.5, 4.6, and 4.5 hundred million toe in 2020, 2030, and 2050, respectively, and it reaches a peak in around 2040. Compared with the reference scenario, automotive energy

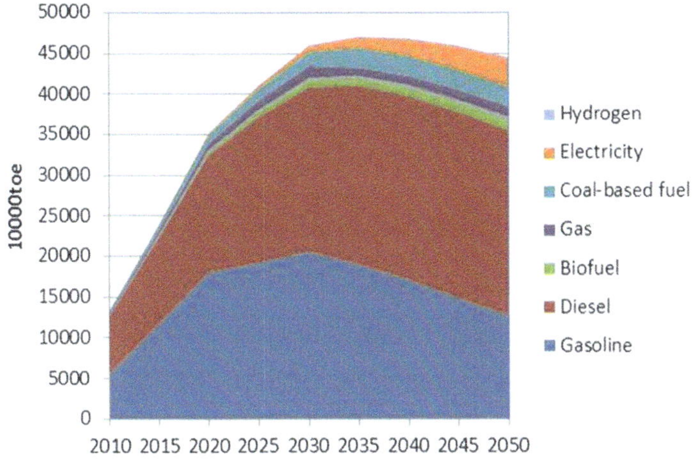

Fig. 12.28 Consumption of automotive fuels under the scenario of developing EVs

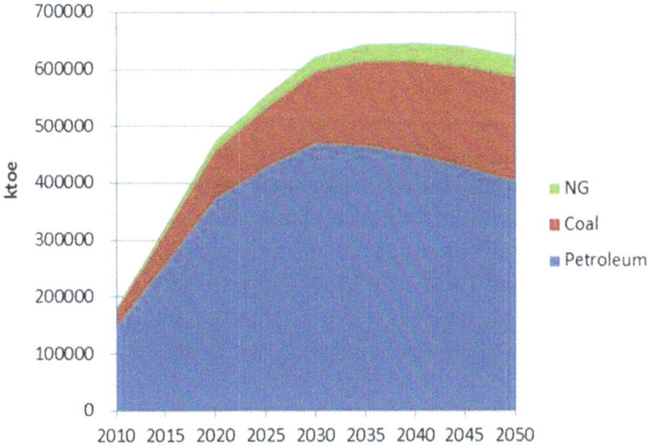

Fig. 12.29 Automotive WTW fossil energy consumption under the scenario of developing EVs

consumption is reduced by 4.5 and 17.1 %, respectively, in 2030 and 2050. Automotive gasoline consumption rapidly increases up to 2020, reaches a peak of 2.06 hundred million toe in 2030, and then falls to 1.27 hundred million toe by 2050. Compared with the reference scenario, automotive gasoline consumption is reduced by 11 and 47 % in 2030 and 2050, respectively, with a fall in automotive diesel consumption by 11 and 47 %. Thus, electrification mainly reduces gasoline consumption.

Figure 12.29 shows automotive WTW fossil energy consumption under the scenario of developing EVs. This consumption decreases by 2 and 12 % in 2030

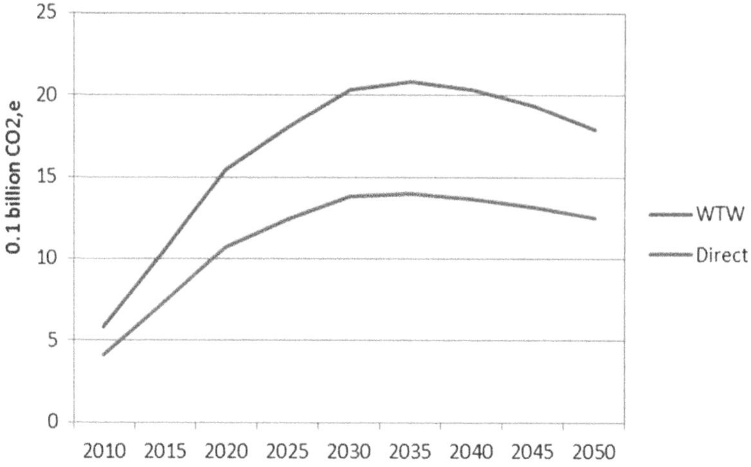

Fig. 12.30 GHG emissions under the scenario of developing EVs

and 2050, respectively, compared with the reference scenario; the consumption of petroleum-based fossil fuels falls by 5 and 26 %. The consumption of petroleum reaches a peak in 2030 and then declines. This indicates that the development of EVs can significantly improve the efficiency of the automotive energy system. In 2050, the reliance on foreign oil for the motor transport sector is reduced to 52 %; this compares with 82 % in the reference scenario.

Under the scenario of developing EVs, the ratio of automotive diesel and gasoline demand is 0.8:1, 1.0:1, and 1.8:1 in 2020, 2030, and 2050, respectively; these figures compare with 0.8:1, 0.9:1, and 1.0:1 in the reference scenario. This indicates that electrification may exacerbate the imbalance in the supply and demand of diesel and gasoline.

12.3.2.5 GHG Emissions

Figure 12.30 shows GHG emissions and motor transport WTW GHG emissions up to 2050 under the scenario of developing EVs. The total GHG emissions are 1.074, 1.384, and 1.255 billion tons in 2020, 2030, and 2050, respectively; these figures amount to reduction of 5.1 and 23.3 % in 2030 and 2050 compared with the reference scenario. WTW greenhouse gas emissions by motor transport are 1.550, 2.035, and 1.791 billion tons in 2020, 2030, and 2050, respectively; this amounts to a reduction of 2.9 and 24 % in 2030 and 2050 compared with the reference scenario.

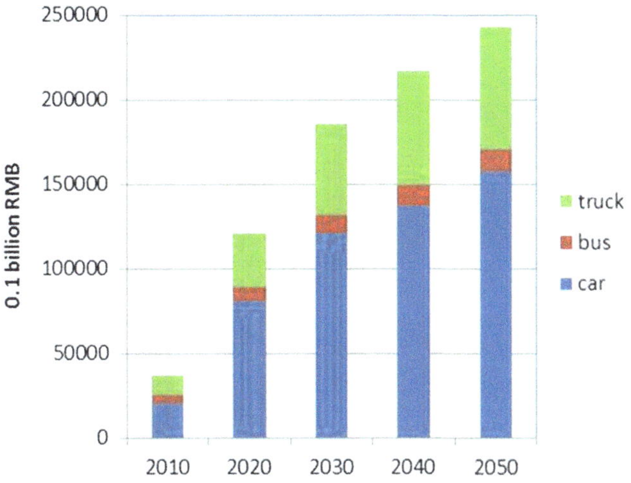

Fig. 12.31 Integrated cost of motor transport under the scenario of developing EVs

12.3.2.6 Economy of Motor Transport

Figure 12.31 shows the integrated cost of motor transport under the scenario of developing EVs, including the cost of purchase, energy, and maintenance. The integrated cost of motor transport is reduced by 6.6, 16.4, and 21.4 % in 2020, 2030, and 2050, respectively, compared with the reference scenario.

Figure 12.32 shows the cost of fuel under the scenario of developing EVs. The cost of fuel is reduced by 3.2 and 28.7 % in 2020 and 2050, respectively, compared with the reference scenario; this amounts to a fall in the cost of petroleum-based fuels by 6.5 and 35.4 %. The respective cost of petroleum import declines by 6.3 and 29.6 %.

12.3.3 Summary of Scenario Points

12.3.3.1 Energy

Under the scenario of developing EVs, automotive fuel and WTW petroleum consumption both reach a peak in 2030 and subsequently display a downward trend. The consumption of automotive fuel, automotive gasoline, and automotive WTW petroleum is reduced by 17, 47, and 26 %, respectively, in 2050 compared with the reference scenario; automotive diesel consumption decreases by only 3 %. This indicates that the introduction of EVs significantly improves the efficiency of the automotive energy system and reduces the reliance on foreign oil. It should be noted

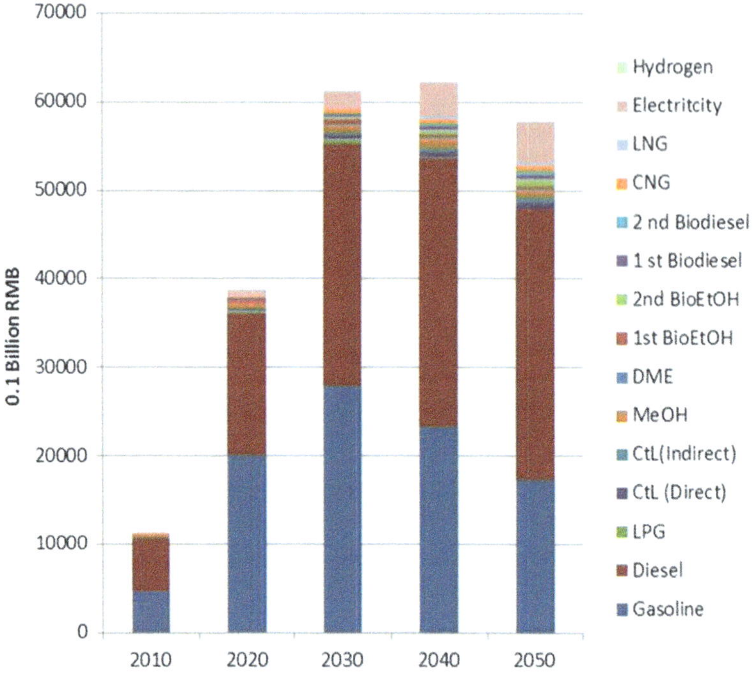

Fig. 12.32 Cost of fuel under the scenario of developing EVs

that the reduction in automotive fuel is mainly observed with gasoline, whereas the effect on diesel is limited; this may exacerbate the imbalance in the supply and demand of diesel and gasoline.

12.3.3.2 Automobiles

The number of gasoline passenger vehicles reaches a peak in 2030 and thereafter rapidly decreases. In 2050, the proportion of gasoline vehicles in the passenger vehicle fleet falls below 30 %. The proportion of mini, small, medium-sized, and large EVs among newly purchased passenger vehicles is expected to reach 90, 70, 45, and 40 %, respectively, in 2050. The number of diesel buses attains a peak in 2030; it then shows a decline. In 2050, the proportion of diesel buses in the bus fleet falls below 40 %, while the proportion of electric buses exceeds 30 %. Both the passenger vehicle and bus industry receive the market support to develop EVs.

12.3.3.3 Environment

WTW greenhouse gas emissions by motor transport reach a peak in around 2035 and afterward show a continued decline. WTW greenhouse gas emissions by motor transport are reduced by 24 % compared with the reference scenario.

12.3.3.4 Economy

Compared with the reference scenario, the integrated cost of motor transport decreases by 21.4 %. The cost of petroleum-based fuel is reduced by 35.4 %, and the cost of petroleum imports falls by 29.6 % in 2050.

12.4 Development Scenarios for Fuel-Cell Vehicles

From the life-cycle analysis of energy efficiency and GHG emissions using different automotive fuel technologies presented in Chap. 11, compared with traditional oil-based technologies, fuel-cell vehicles have the potential to reduce energy consumption and GHG emissions. The objective in building rapid development scenarios for fuel-cell vehicles was to investigate the importance in developing such vehicles to solve China's automotive energy issues. Specifically, an analysis of the development scenarios for fuel-cell vehicles is intended to answer the following questions: (1) How important is the development of fuel-cell vehicles in easing the crisis facing China's automotive energy supply security? (2) What reductions in GHG emissions can the development of fuel-cell vehicles in the automotive transportation sector achieve? (3) How important is the development of fuel-cell vehicles to reducing fossil energy consumption? (4) What effect will the development of fuel-cell vehicles have on the supply and demand of different energy sources in China's automotive industry? (5) What influence will the development of fuel-cell vehicles have on the restructuring and upgrading of China's automotive industry? (6) What significance will the development of fuel-cell vehicles have on the economy of automotive transportation?

12.4.1 Scenario Assumptions

12.4.1.1 Demand for Vehicle Transport Services

The demand for vehicle transport services in the scenario of developing fuel-cell vehicles is the same as in the reference scenario.

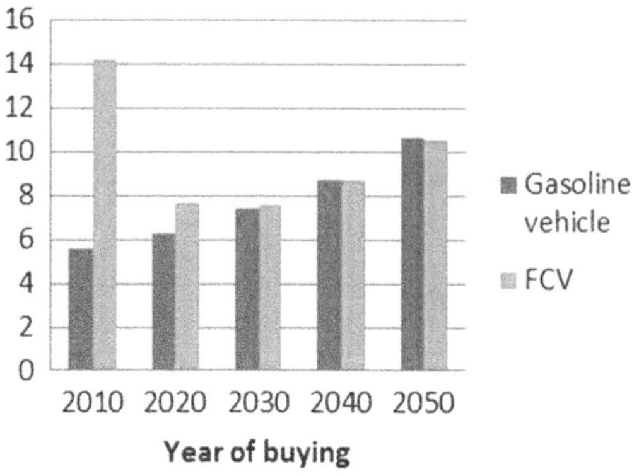

Fig. 12.33 Transport costs of large fuel-cell and petroleum-based passenger vehicles in the scenario of developing fuel-cell vehicles

12.4.1.2 Passenger Vehicle Fuel Economy

In the scenario of developing fuel-cell vehicles, the improvement in passenger vehicle fuel economy is the same as in the reference scenario.

12.4.1.3 Bus Fuel Economy

The improvement in the bus fuel economy in the scenario of developing fuel-cell vehicles is the same as in the reference scenario.

12.4.1.4 Truck Fuel Economy

In the scenario of developing fuel-cell vehicles, the improvement in the truck fuel economy is the same as in the reference scenario.

12.4.1.5 Fuel-Cell Passenger Vehicles

Large and medium-sized passenger vehicles are suitable for fuel-cell power. Fuel-cell passenger vehicles use hydrogen supply technology, which derives hydrogen from coal with the carbon capture and storage (CCS) technology. In the scenario of developing fuel-cell vehicles, medium-sized and large fuel-cell passenger vehicles enter a stage of rapid development in around 2035. The trends of transport costs for large fuel-cell and petroleum-based passenger vehicles appear in Fig. 12.33.

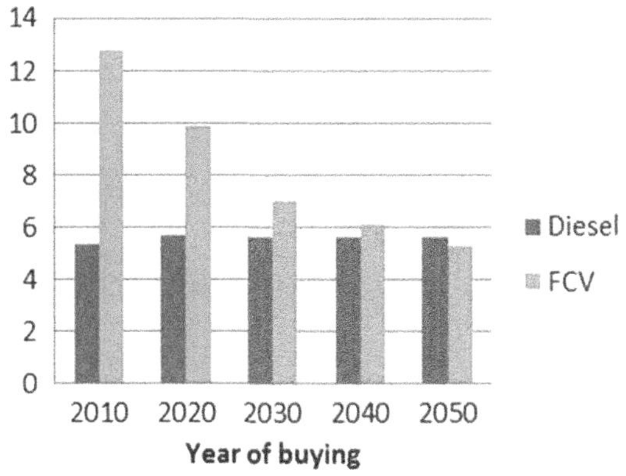

Fig. 12.34 Transport costs of large fuel-cell and petroleum-based buses in the scenario of developing fuel-cell vehicles

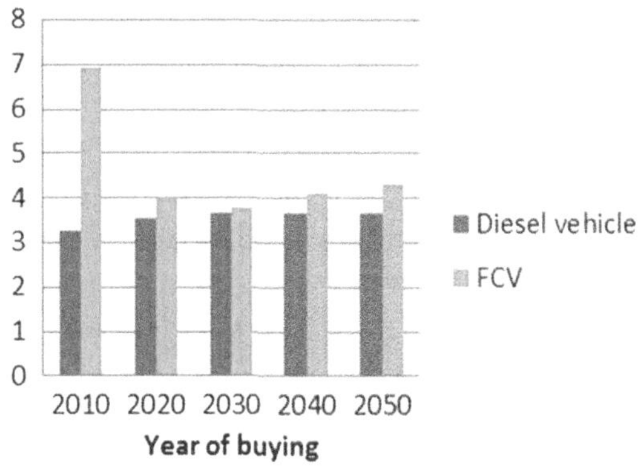

Fig. 12.35 Transport costs of medium-sized fuel-cell and petroleum-based trucks in the scenario of developing fuel-cell vehicles

12.4.1.6 Fuel-Cell Buses

Large buses are suitable for fuel-cell power. Fuel-cell buses use a hydrogen supply, which derives hydrogen from coal with CCS. In the scenario of developing fuel-cell vehicles, large fuel-cell buses enter a stage of rapid development in around 2035. The trends of transport costs with large fuel-cell and petroleum-based buses are presented in Fig. 12.34.

12.4.1.7 Fuel-Cell Trucks

Medium-sized trucks are suitable for fuel-cell power. Fuel-cell trucks use a hydrogen supply technology, which derives hydrogen from coal using CCS. In the scenario of developing fuel-cell vehicles, medium-sized fuel-cell trucks enter a stage of rapid development in around 2030. The trends of transport costs for medium-sized fuel-cell and petroleum-based trucks appear in Fig. 12.35.

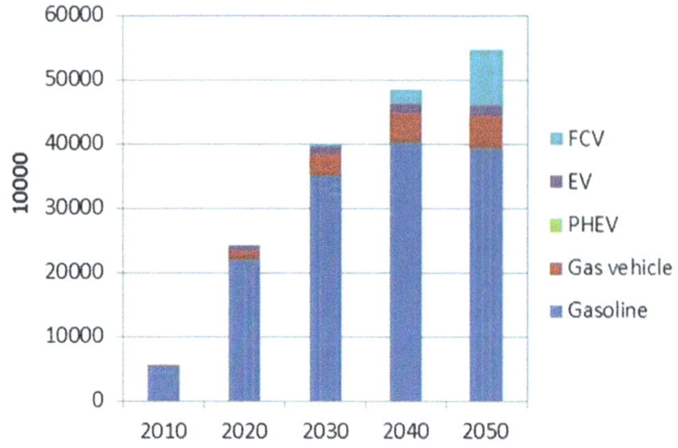

Fig. 12.36 Development of driving technologies for passenger vehicles in the scenario of developing fuel-cell vehicles

12.4.1.8 Electric Vehicles

The assumption for EVs in the scenario of developing fuel-cell vehicles is the same in the reference scenario.

12.4.1.9 Biofuels

The assumption for biofuels in the scenario of developing fuel-cell vehicles is the same as in the reference scenario.

12.4.2 Scenario Results

12.4.2.1 Driving Technology for Passenger Vehicles

In the scenario of developing fuel-cell vehicles, vehicle ownership according to different driving technologies is as shown in Fig. 12.36.

In the scenario of developing fuel-cell vehicles, the proportion of fuel-cell vehicles among all passenger vehicles is 16 % in 2050, which is 15 % higher than in the reference scenario. The proportion of passenger transport with different driving technologies is presented in Fig. 12.37. The proportion of fuel-cell vehicles in private transport is over 16 %, which is 15 % higher than in the reference scenario. In the scenario of developing fuel-cell vehicles, the market penetration of different-sized fuel-cell vehicles appears in Fig. 12.38. In 2050, the proportion of mini, small, medium-sized, and large fuel-cell vehicles is expected to be 10, 30, 50, and 50 %, respectively. In this scenario, the development of fuel-cell passenger vehicles takes

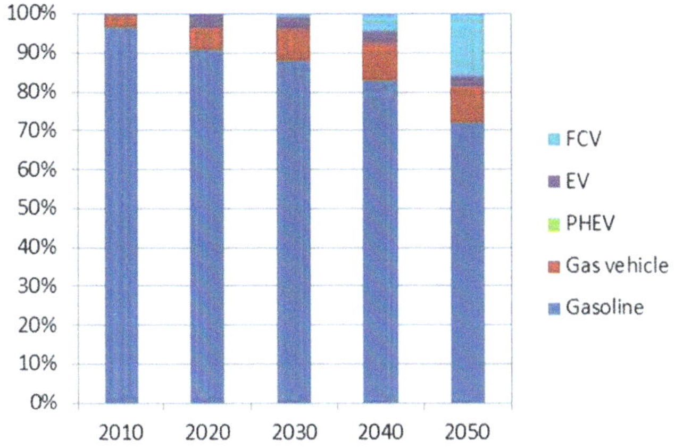

Fig. 12.37 Proportion of passenger transport with different driving technologies in the scenario of developing fuel-cell vehicles

Fig. 12.38 Market penetration of different-sized fuel-cell vehicles in the scenario of developing fuel-cell vehicles

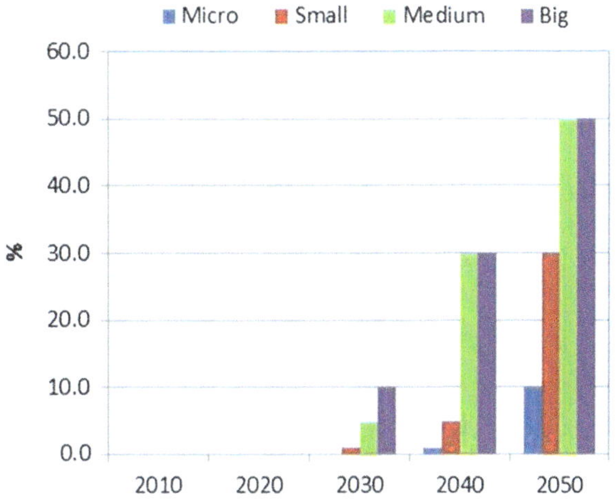

place on a considerable scale, which is promising for the development of China's fuel-cell vehicle industry. Passenger vehicles powered by electricity and fuel gas are the same as in the reference scenario.

12.4.2.2 Driving Technology for Buses

In the scenario of developing fuel-cell vehicles, the proportion of buses using different driving technologies appears in Fig. 12.39. In 2030 and 2050, the proportion of fuel-cell vehicles among all buses is, respectively, 4 and 36 %; this is 4 and 35.9 % higher than in the reference scenario. The future proportion of buses using

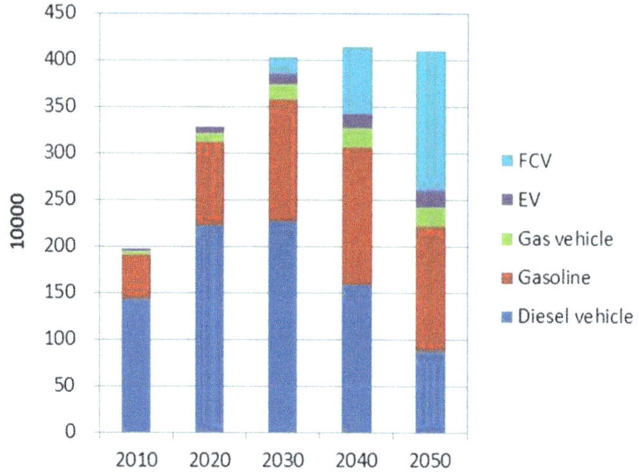

Fig. 12.39 Development of driving technologies for buses in the scenario of developing fuel-cell vehicles

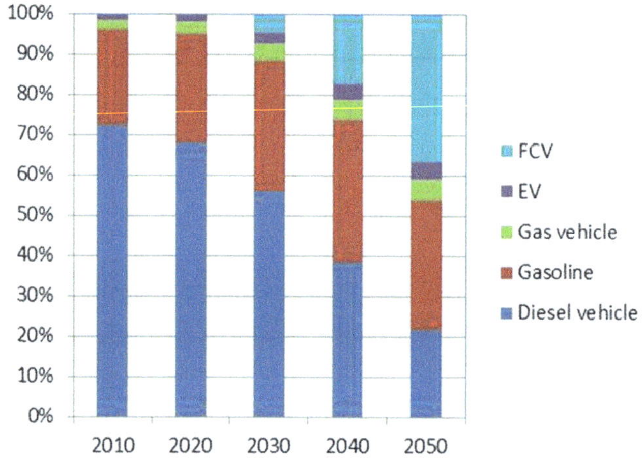

Fig. 12.40 Proportion of passenger vehicles using different driving technologies in the scenario of developing fuel-cell vehicles

different driving technologies is presented in Fig. 12.40. The proportion of fuel-cell vehicles among the passenger vehicle population in 2030 and 2050 is, respectively, 4 and 36 %, which is 4 and 36 % higher than in the reference scenario. In the scenario of developing fuel-cell vehicles, the market penetration of fuel-cell vehicles among new bus sales in 2030 and 2050 is, respectively, 4 and 36 %, as evident in Fig. 12.41. In this scenario, there is considerable development of fuel-cell buses, which is advantageous for the development of China's fuel-cell vehicle industry. Buses using electricity and fuel gas are the same as in the reference scenario.

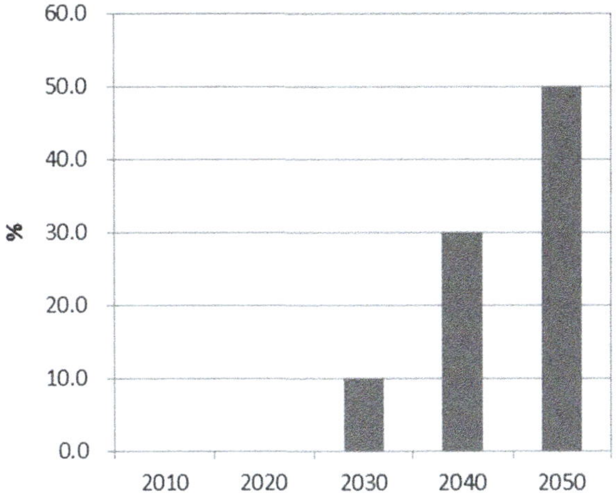

Fig. 12.41 Market penetration of fuel-cell vehicles among new bus sales in the scenario of developing fuel-cell vehicles

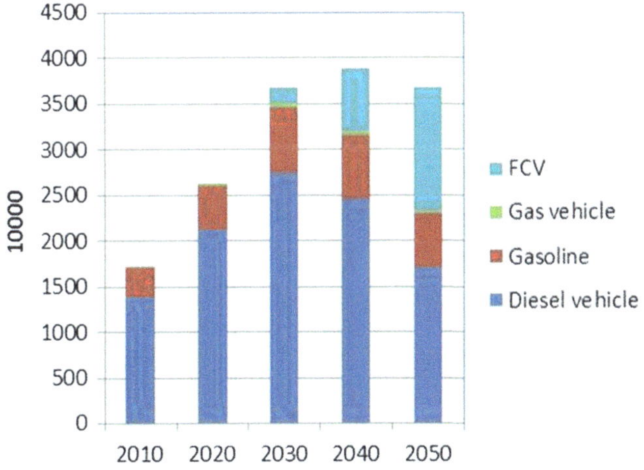

Fig. 12.42 Development of driving technologies for trucks in the scenario of developing fuel-cell vehicles

12.4.2.3 Driving Technology for Trucks

In the scenario of developing fuel-cell vehicles, the number of trucks using different driving technologies appears in Fig. 12.42. In 2030 and 2050, the proportion of fuel-cell vehicles among all trucks is, respectively, 1 and 16 %; this is 0.8 and 15.5 % higher than in the reference scenario. The freight-sharing proportion of trucks using different driving technologies is presented in Fig. 12.43. Among freight services, the proportion of fuel-cell vehicles in 2030 and 2050 is, respectively, 1 and 16 %; this is 0.8 and 15.5 % higher than in the reference scenario.

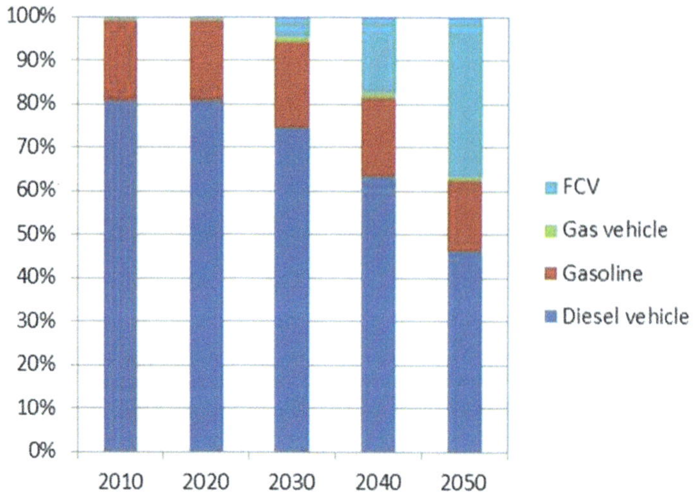

Fig. 12.43 Freight-sharing proportion of trucks using different driving technologies in the scenario of developing fuel-cell vehicles

Fig. 12.44 Market penetration of fuel-cell vehicles among new truck sales in the scenario of developing fuel-cell vehicles

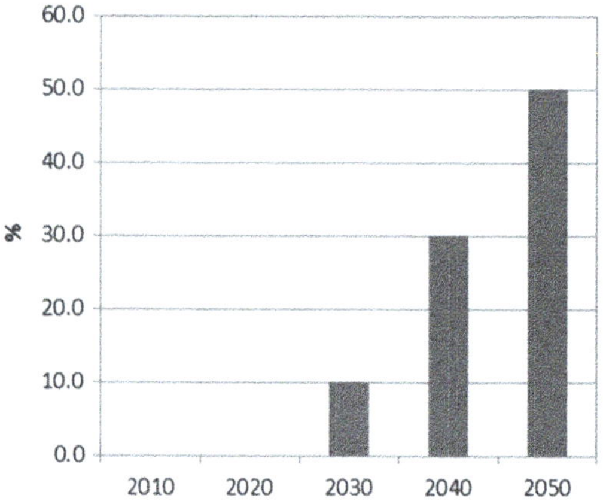

In the scenario of developing fuel-cell vehicles, the market penetration of fuel-cell vehicles among new truck sales in 2030 and 2050 is, respectively, 10 and 50 %, as indicated in Fig. 12.44. In this scenario, there is considerable development of fuel-cell trucks, which is beneficial for the development of China's fuel-cell vehicle industry. Trucks powered by electricity and fuel gas are the same as in the reference scenario.

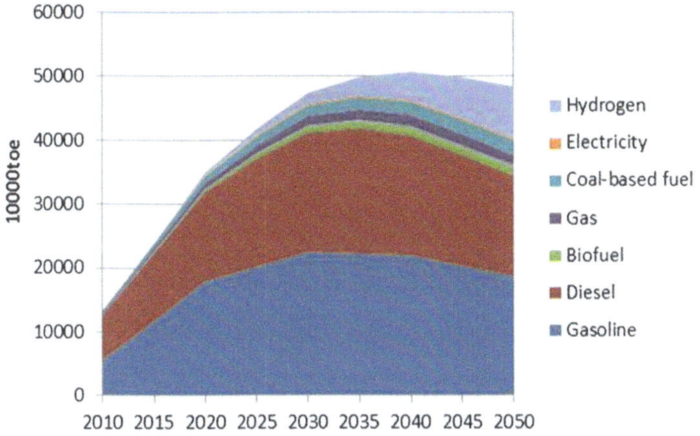

Fig. 12.45 China's vehicle energy consumption in the scenario of developing fuel-cell vehicles

12.4.2.4 Vehicle Energy

China's vehicle energy consumption from 2010 to 2050 in the scenario of developing fuel-cell vehicles shows a rapid increase, as shown in Fig. 12.45. In this scenario, China's vehicle energy consumption in 2020, 2030, and 2050 is, respectively, 0.35 billion, 0.478 billion, and 0.485 billion toe; the peak of 0.51 billion toe occurs around 2035. In the scenario involving the development of fuel-cell vehicles, China's vehicle energy consumption in 2030 and 2050 declines by, respectively, 1.4 and 9.7 % compared with the reference scenario. Vehicle gasoline consumption rapidly rises until 2020; there is a peak of 0.225 billion toe in around 2030 and subsequently there is a slow decline to 0.187 billion toe by 2050. In this scenario, China's vehicle gasoline consumption in 2030 and 2050 decreases, respectively, by 2.6 and 22.1 % compared with the reference scenario. Vehicle diesel consumption peaks at 0.196 billion toe in around 2035 before falling to 0.157 billion toe by 2050. In this scenario, China's vehicle diesel consumption in 2030 and 2050 falls by, respectively, 9.2 and 33.2 % compared with the reference scenario.

WTW fossil energy consumption for China's vehicles from 2010 to 2050 in the scenario of developing fuel-cell vehicles appears in Fig. 12.46. As a result of using hydrogen from coal by means of CCS technology in this scenario, WTW fossil energy consumption by China's vehicles in 2030 and 2050 increases, respectively, by 2.9 and 5.3 % compared with the reference scenario. However, oil consumption for those years falls, respectively, by 3.7 and 27.4 %. By 2050, the dependence on oil imports by the automotive transport sector is reduced from 82 % in the reference scenario to 75 % in the scenario of developing fuel-cell vehicles. This indicates that fuel-cell vehicle technology will play a significant role in improving the security of oil supply in the future.

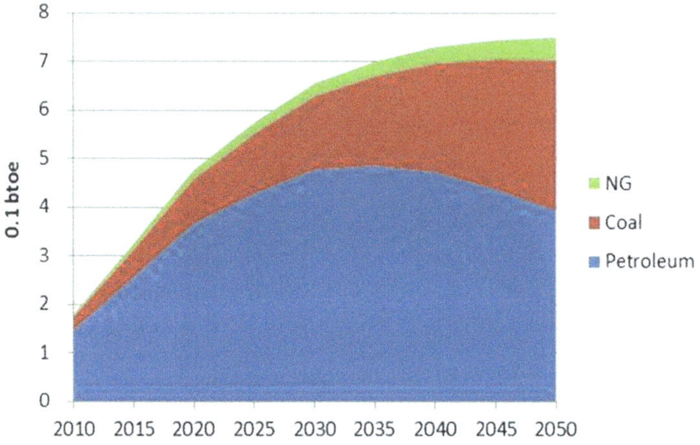

Fig. 12.46 WTW fossil energy consumption by China's vehicles in the scenario of developing fuel-cell vehicles

In the scenario of developing fuel-cell vehicles, the demand ratio of vehicle diesel and gasoline in 2020, 2030, and 2050 is 0.8:1, 0.8:1, and 0.9:1; this compares with 0.8:1, 0.9:1, and 1.0:1 in the reference scenario. This indicates that the introduction of fuel-cell vehicle technology will to some extent help address the imbalance in the supply and demand for diesel and gasoline.

12.4.2.5 GHG Emissions

Vehicle GHG emissions in China from 2010 to 2050 in the scenario of developing fuel-cell vehicles are presented in Fig. 12.47. The total amount of GHG emissions from China's vehicles in 2020, 2030, and 2050 is, respectively, 1.054 billion, 1.394 billion, and 1.222 billion tons in this scenario; this amounts to a reduction by, respectively, 1.7, 4.4, and 25.3 % compared with the reference scenario.

In the scenario of developing fuel-cell vehicles, WTW greenhouse gas emissions from China's vehicles for 2010–2050 also appear in Fig. 12.47. The total amount of such emissions for vehicles in China in 2020, 2030, and 2050 is, respectively 1.54 billion, 2.08 billion, and 1.75 billion tons; this amounts to a decrease by 0.6 and 25.9 % for 2030 and 2050, respectively, compared with the reference scenario.

12.4.2.6 Vehicle Transport Economy

The comprehensive cost of vehicles, including purchase price, energy use, and vehicle maintenance, in the scenario of developing fuel-cell vehicles is indicated in Fig. 12.48. The comprehensive cost of vehicles in 2020, 2030, and 2050 is reduced by 6.4, 9.9, and 17.6 %, respectively, compared with the reference scenario.

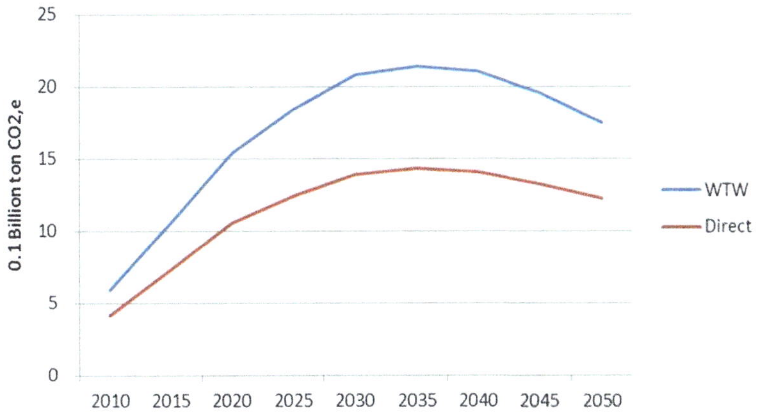

Fig. 12.47 GHG emissions by China's vehicles in the scenario of developing fuel-cell vehicles

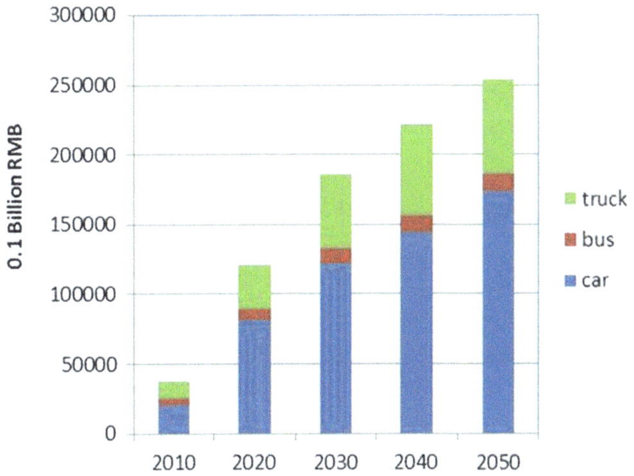

Fig. 12.48 Comprehensive vehicle costs in the scenario of developing fuel-cell vehicles

The energy cost in the scenario of developing fuel-cell vehicles appears in Fig. 12.49. Compared with the reference scenario, vehicle fuel costs increase by 0.7 % in 2020 with the rapid development of fuel-cell vehicles; these costs begin to decline up to 2030 and decrease by 1.1 and 45.4 % in 2030 and 2050, respectively. Petroleum-based fuel costs decrease by 5.7 and 57.2 % in 2030 and 2050, respectively, compared with the reference scenario; this leads to a reduction in the costs of imported oil by 8.0 and 56.2 % in 2030 and 2050.

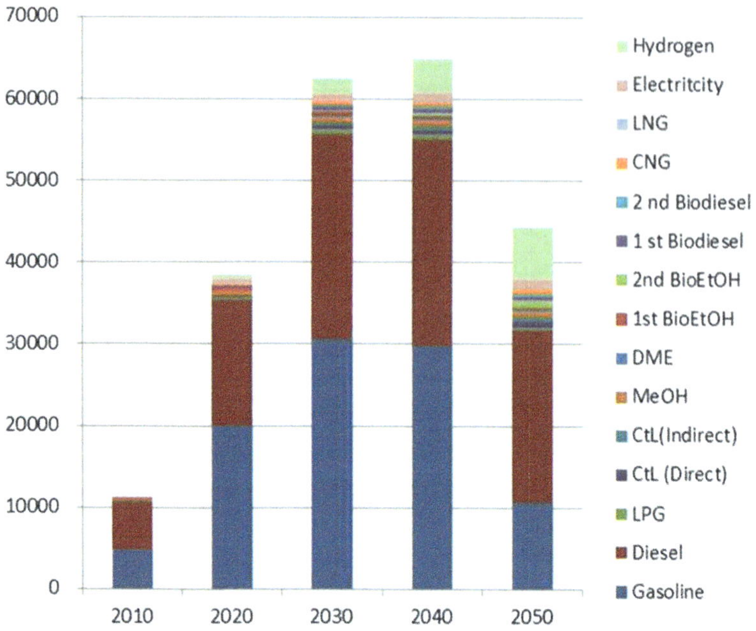

Fig. 12.49 Vehicle energy costs in the scenario of developing fuel-cell vehicles

12.4.3 Scenario Summary

12.4.3.1 Energy

In the scenario of developing fuel-cell vehicles, vehicle fuel and vehicle WTW oil peak between 2035 and 2040 and then begin to decline. In 2050, vehicle fuel, vehicle gasoline, vehicle diesel, and vehicle WTW oil consumption are reduced by, respectively, 9.7, 22.1, 33.2, and 27.4 % compared with the reference scenario; by contrast, vehicle WTW fossil energy increases by 5.3 %. This indicates that the introduction of fuel-cell technology for vehicles will improve the efficiency of the automotive industry while at the same time producing a notable reduction in the dependence on oil imports. However, as a result of using hydrogen from coal and CCS technology, the efficiency of the overall vehicle energy system will decrease. Moreover, the introduction of fuel-cell vehicle technology will reduce the demand for vehicle gasoline and diesel, which to some extent will address the imbalance in the supply and demand of these two fuels.

12.4.3.2 Vehicles

The gasoline passenger vehicle population peaks in 2040 and thereafter declines. The proportion of gasoline vehicles among the entire passenger vehicle population falls below 72 % in 2050, and the proportion of gasoline vehicles is under 70 % by 2050. Among new passenger vehicle sales in 2050, the proportion of mini, small, medium-sized, and large fuel-cell vehicles (FCVs) is, respectively, 10, 30, 50, and 50 %. The market penetration of FCVs among new bus and truck sales in 2050 is expected to reach 50 %. The development of FCVs will receive market support in terms of passenger vehicles, buses, and trucks in China.

12.4.3.3 Environment

The WTW GHG emissions by vehicles show a peak in 2035 and subsequently decline. Compared with the reference scenario, such emissions are reduced by 26 % in the scenario of developing FCVs.

12.4.3.4 Economy

The comprehensive cost of vehicles decreases by 21.4 %. Compared with the reference scenario, reductions in petroleum-based fuel are 35.4 % and those in the cost of oil imports are 29.6 %.

12.5 Development Scenarios for Biofuels

According to the life-cycle analysis of energy efficiency and GHG emissions from automotive fuel technologies presented in Chap. 11, compared with conventional oil-based technologies, biofuel technology—especially second-generation biofuel technology—has the potential to reduce energy consumption and GHG emissions. Biofuels thus represent a strategic form of vehicle energy technology. The objective in building rapid development scenarios for biofuels is to examine the importance of developing second-generation biofuels to solve the automotive energy issues that China faces. Specifically, analysis of the development scenarios for biofuel is an attempt to address the following questions: (1) How important is the development of biofuels in easing the crisis facing China's automotive energy supply security? (2) To what extent can GHG emissions be reduced in the automotive transportation sector by the development of biofuels? (3) To what extent can the development of biofuels reduce the consumption of fossil energy? (4) In what way would the development of biofuels affect imbalances in the supply and demand of different forms of automotive energy in China? (5) How would the development of biofuels affect the restructuring and upgrading of China's automotive industry? (6) What effect would the development of biofuels have on automotive transportation economy?

12.5.1 Scenario Assumptions

1. Demand for vehicle transport services: in the scenario of developing biofuels, the demand for vehicle transport services is the same as in the reference scenario.
2. Vehicle fuel economy: improvement of vehicle fuel economy in the scenario of developing biofuels is the same as in the reference scenario.
3. Vehicle driving technology: in the scenario of developing biofuels, improvement in vehicle driving technology is the same as in the reference scenario.
4. Biofuels: in the scenario of developing biofuels, second-generation bioethanol and second-generation biodiesel will have entered a stage of rapid development by 2025. The comprehensive vehicle costs for biofuel and petroleum-based fuel are presented in Fig. 12.50.

12.5.2 Scenario Results

12.5.2.1 Driving Technology for Passenger Vehicles

According to the scenario assumptions for developing biofuels, the development of various kinds of vehicle driving technology is almost the same as in the reference scenario. Therefore, these details will not be repeated here.

12.5.2.2 Vehicle Energy

China's vehicle energy consumption for 2010–2050 in the scenario of developing biofuels appears in Fig. 12.51. In that scenario, vehicle biofuel energy consumption in China in 2020, 2030, and 2050 is, respectively, 16.7 million, 48.8 million, and 73.5 million toe; this accounts for 4.7, 10.1, and 13.5 % of the total vehicle energy consumption. The proportion of biofuels in 2020, 2030, and 2050 increases, respectively, by 3.2, 7.8, and 10.2 %. Under the same scenario, the market penetration of various kinds of biofuel in vehicle energy consumption is shown in Fig. 12.52. After 2020, the main biofuel supply is from second-generation biofuel, and the biofuel industry will have a relatively good opportunity to upgrade. It should be noted that the development of biofuels does not fundamentally change vehicle energy consumption.

WTW fossil energy consumption by China's vehicles from 2010 to 2050 in the scenario of developing biofuels appears in Fig. 12.53. Under this scenario, such energy consumption is reduced by 0.6, 4.3, and 8.6 % in 2020, 2030, and 2050, respectively; oil consumption declines by 2.1, 6.6, and 9.8 %. In 2050, the dependence on oil imports by the automotive transport sector is reduced from 82 % in the reference scenario to 79 % in the scenario of developing biofuels. This

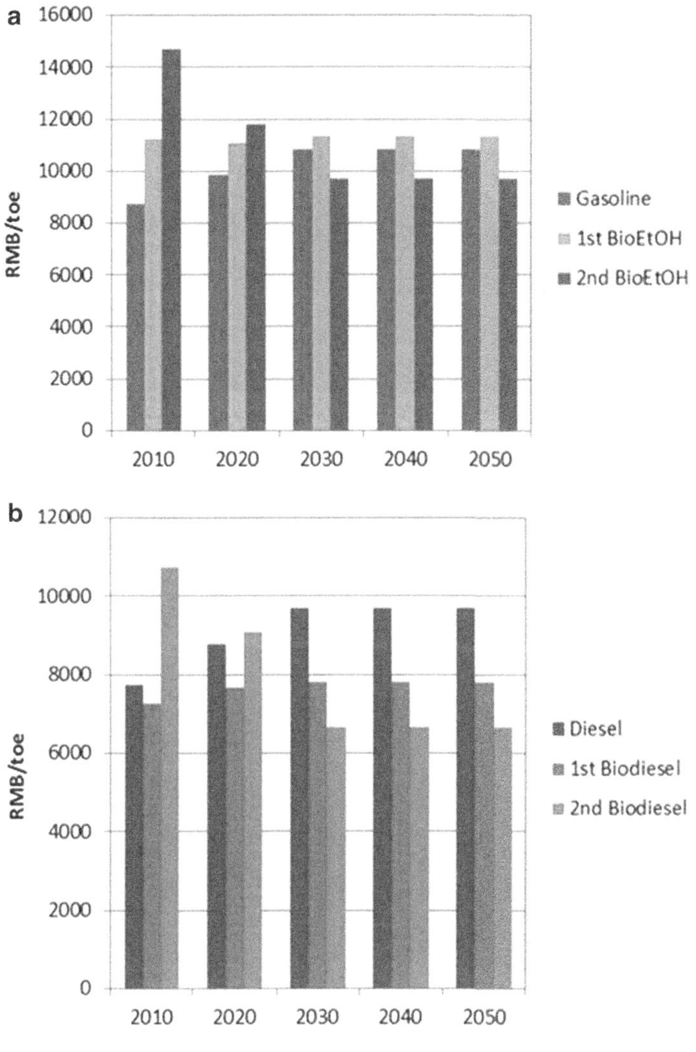

Fig. 12.50 Comprehensive costs of biofuel- and petroleum-based vehicles. (**a**) Bioethanol and gasoline. (**b**) Biodiesel and diesel

indicates that biofuels will play a significant role in improving the security of the oil supply in the future.

In the scenario of developing biofuels, the vehicle diesel-to-gasoline demand ratio for 2020, 2030, and 2050 is 0.9:1, 0.8:1, and 0.8:1, respectively; this compares with 0.8:1, 0.9:1, and 1.0:1 in the reference scenario. This indicates that the problem with the diesel-gasoline ratio will not be solved under this scenario. However, the imbalance in the supply and demand for diesel and gasoline will be alleviated to some extent in the biofuel scenario.

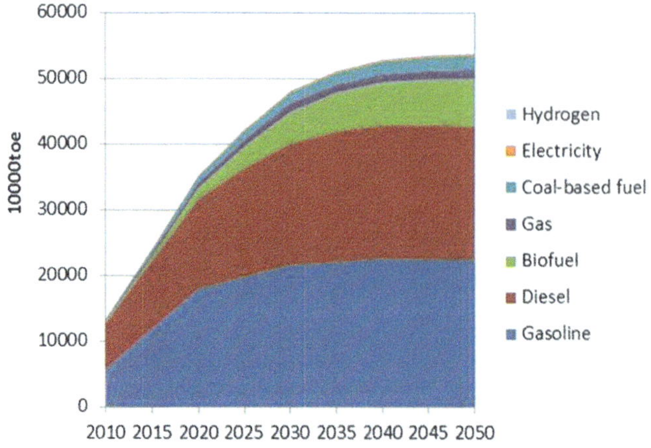

Fig. 12.51 Vehicle energy consumption in China from 2010 to 2050 in the scenario of developing biofuels

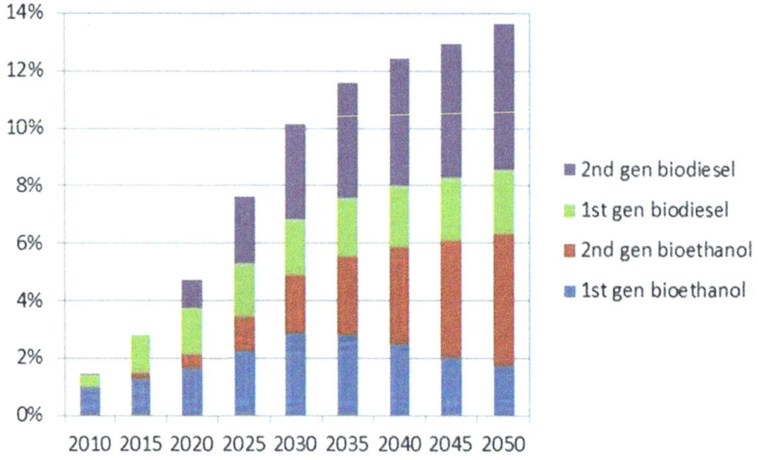

Fig. 12.52 Market penetration of biofuel in the scenario of developing biofuels

12.5.2.3 GHG Emissions

WTW GHG emissions by China's vehicles for 2010–2050 in the scenario of developing biofuels appear in Fig. 12.54. Under this scenario, the total amounts of such emissions in 2020, 2030, and 2050 are, respectively, 1.53 billion, 2.00 billion, and 2.03 billion tons; these represent respective reductions by 1.1, 4.9, and 14.2 %, compared with the reference scenario. This indicates that biofuels will play a significant role in reducing GHG emissions in the future.

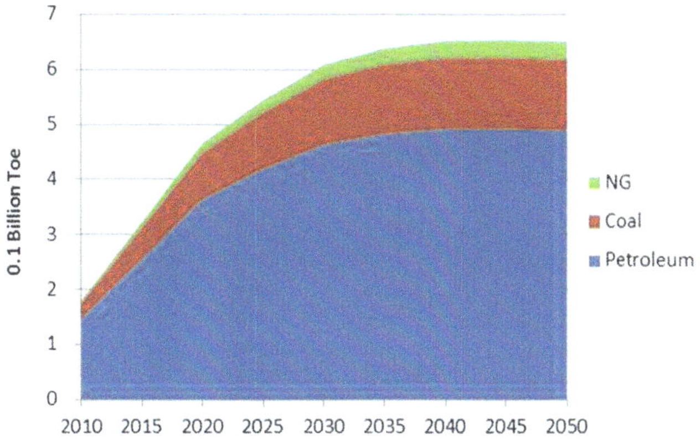

Fig. 12.53 WTW fossil energy consumption by vehicles in the scenario of developing biofuels

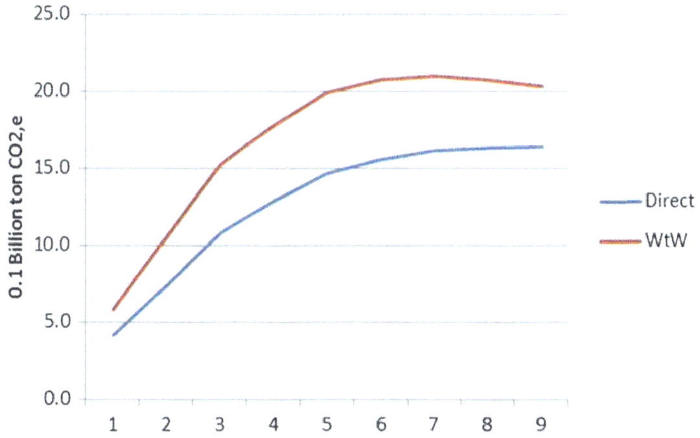

Fig. 12.54 GHG emissions of different kinds of vehicles in the scenario of developing biofuels

12.5.2.4 Vehicle Transport Economy

The comprehensive cost of vehicles including purchase price, energy use, and maintenance, in the scenario of developing biofuels is shown in Fig. 12.55. The comprehensive cost of vehicles in 2020, 2030, and 2050 is respectively reduced by 6.7, 10.4, and 17.3 % compared with the reference scenario.

The energy costs in the scenario of developing biofuels are presented in Fig. 12.56. Vehicle fuel costs increase respectively by 1.7 and 0.1 % in 2020 and 2030. FCVs develop rapidly compared with the reference scenario, and their fuel costs fall by 13.5 % in 2050. Petroleum-based fuel costs are reduced by 3.4, 7.4,

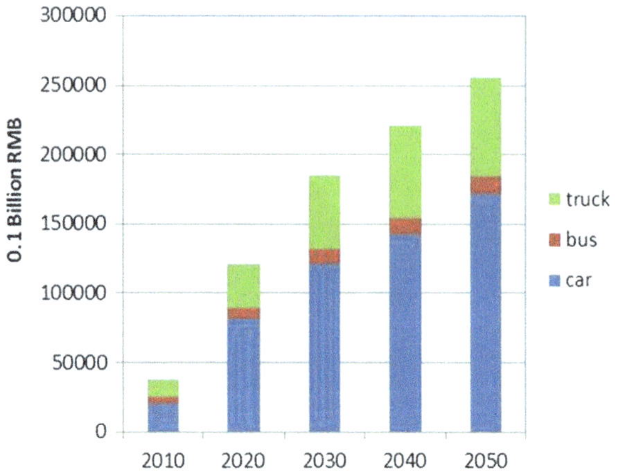

Fig. 12.55 Comprehensive cost of vehicles in the scenario of developing biofuels

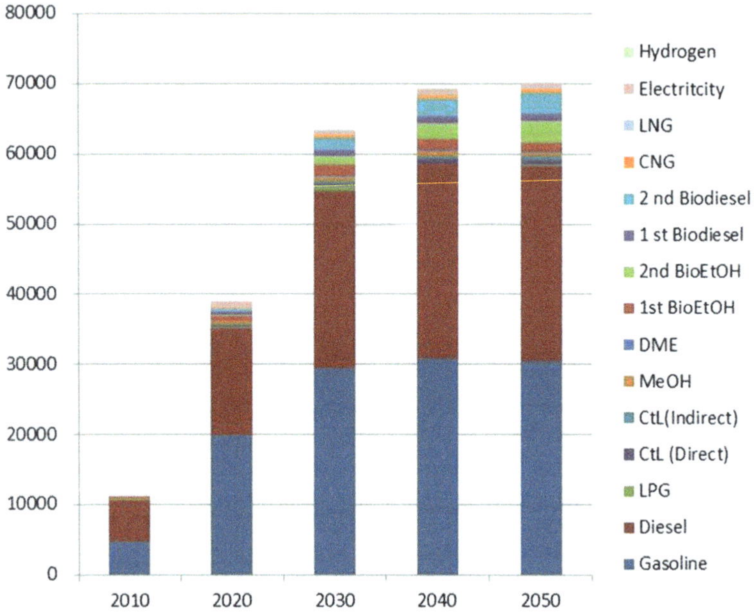

Fig. 12.56 Vehicle fuel costs in the scenario of developing biofuels

and 21.4 %, respectively, in 2020, 2030, and 2050; this results in a decline in oil import costs by 5.2, 14.3, and 20.1 % in 2020, 2030, and 2050 respectively. This indicates that the development of biofuels will produce medium- and long-term macroeconomic benefits.

12.5.3 Scenario Summary

12.5.3.1 Energy

The consumption of biofuels amounts to 16.7 million, 48.8 million, and 73.5 million toe in, respectively, 2020, 2030, and 2050; this accounts for 4.7, 10.1, and 13.6 % of the total vehicle fuel consumption. Compared with the reference scenario, WTW fossil energy consumption by China's vehicles falls by 0.6, 4.5 and 8.6 % in, respectively, 2020, 2030, and 2050 in the scenario of developing biofuels; petroleum-based energy consumption is reduced by 2.1, 6.6, and 9.8 %. Biofuels can improve vehicle energy supply security to some extent over the long term. Moreover, the introduction of biofuel technology can reduce the proportion of vehicle gasoline and diesel, which will to some extent also alleviate the imbalance in the supply and demand for diesel and gasoline.

12.5.3.2 Vehicles

The development of biofuels to replace gasoline and diesel will produce no obvious change in the composition of vehicle power technology.

12.5.3.3 Environment

The WTW GHG emissions by vehicles show a peak in 2040 and then decline. When compared with the reference scenario, such emissions fall by 14 % in the scenario of developing biofuels.

12.5.3.4 Economy

The comprehensive cost of vehicles falls by 6.7, 10.4, and 17.3 % in 2020, 2030, and 2050, respectively. Moreover, the cost of oil imports decreases by 5.2, 14.3, and 20.1 % in 2020, 2030, and 2050, respectively, compared with the reference scenario.

12.6 Integrated Policy Scenario

The integrated policy scenario is a target scenario of vehicle energy development. This scenario sets clear target requirements for China's vehicle energy systems in 2050 regarding aspects of energy efficiency, GHG emissions, energy supply security, balance in fuel supply and demand, vehicle transport economy, and industrial development. The integrated policy scenario thus incorporates innovations

Table 12.10 Main policy objectives for China's vehicle energy in 2050

Policy target	Index	Unit	Value of index in 2050
Energy efficiency	WTW fossil energy consumption by passenger transport	MJ/100 km·person (equivalent of gasoline, L/100 km·person)	≤75(2.4)
	WTW fossil energy consumption by freight transport	MJ/100 km·ton (equivalent of diesel, L/100 km·ton)	≤55(1.6)
Energy security	Oil import dependency of motor transport	Ratio of vehicle oil imports and vehicle oil consumption (%)	≤50
GHG emission	Per capita vehicle GHG emission	Tons CO_2/person	≤0.7
	Per capita WTW vehicle GHG emission	Tons CO_2/person	≤1
Balance in fuel supply and demand	Vehicle diesel and gasoline consumption	Ratio of vehicle diesel and gasoline	≤1.2:1
Transport economy	Integrated cost of motor transport	Proportion of GDP (%)	≤10
	Oil imports cost of motor transport	Proportion of GDP (%)	≤1

in vehicle power and energy technology in addition to demand management of transport services. The main objective in performing an analysis using this scenario was to identify the effect of technological development and various vehicle energy policy objectives in 2050 and also to provide a scientific basis for policies that promote sustainable vehicle energy systems.

12.6.1 Scenario Objective and Assumption

12.6.1.1 Objective of Vehicle Energy Policy

The policy objective of the integrated policy scenario is to provide a scientific basis for promoting sustainable vehicle energy systems by the year 2050. In accordance with that objective, some policy objective indicators are suggested, the details of which appear in Table 12.10.

Table 12.11 Average growth rate of vehicle transport services in the integrated policy scenario

Type	2010–2020	2020–2030	2030–2050
Passenger vehicles (%)	12.51	3.64	0.36
Buses (%)	6.43	3.35	0.99
Trucks (%)	9.51	4.97	1.07

Table 12.12 Fuel economy of passenger vehicles

Type of passenger vehicle	2010	2020	2030	2050
Mini (displacement ≤1 L)	5.5	4.5	3.5	3.5
Small (displacement 1 to ≤1.6 L)	7.0	5.75	4.5	4.5
Medium-sized (displacement 1.6–2.5 L)	9.0	8.0	7.0	6.0
Large (displacement ≥2.5 L)	12	10.5	9.5	8.5

Table 12.13 Decline in bus fuel consumption compared with 2010

Year	2020	2030	2040	2050
Rate of reduction (%)	5	10	15	20

12.6.1.2 Vehicle Transport Demand

The growth rate in vehicle transport demand is analyzed in Chap. 3, and those results appear in Table 12.11. A comparison of vehicle transport demand under the integrated policy scenario and reference scenario is given in Chap. 3.

12.6.1.3 Fuel Economy of Passenger Vehicles

Compared with the reference scenario, the fuel economy of passenger vehicles is promoted in the integrated policy scenario, as indicated in Table 12.12.

12.6.1.4 Fuel Economy of Buses

Compared with the reference scenario, the fuel economy of buses is significantly promoted in the integrated policy scenario. The promotion of fuel economy is presented in Table 12.13.

Table 12.14 Decline in truck fuel economy compared with 2010

Year	2020	2030	2040	2050
Rate of reduction (%)	20	30	35	40

12.6.1.5 Fuel Economy of Trucks

The fuel economy of trucks in the integrated policy scenario is significantly promoted in comparison with the reference scenario. The promotion of fuel economy appears in Table 12.14.

12.6.1.6 Vehicle Electrification

In the integrated policy scenario, the electrification of various types of vehicles is the same as in the scenario of developing EVs.

12.6.1.7 Fuel-Cell Vehicles

In the integrated policy scenario, the development of different FCVs is the same as in the scenario of developing FCVs.

12.6.1.8 Biofuels

In the integrated policy scenario, the development of biofuels is the same as in the scenario of developing biofuels.

12.6.2 Scenario Results

12.6.2.1 Driving Technology for Passenger Vehicles

In the integrated policy scenario, the population of vehicles with different driving technologies appears in Fig. 12.57. It is evident that there is a peak in 2030 followed by an immediate decline. In 2050, the proportion of gasoline vehicles among passenger vehicle population is under 30 %. Gas vehicles show a relatively fast development up to 2030: their proportion amounts to a peak of 9 % in 2030—up from 3 % in 2010. That proportion is reduced to 4 % in 2050.

In the integrated policy scenario, the proportion of EVs among the passenger vehicle population amounts to 14 and 57 %, respectively, in 2030 and 2050; those figures are 11.8 and 54 % higher than in the reference scenario. The market penetration of mini, small, medium-sized, and large electric passenger vehicles is

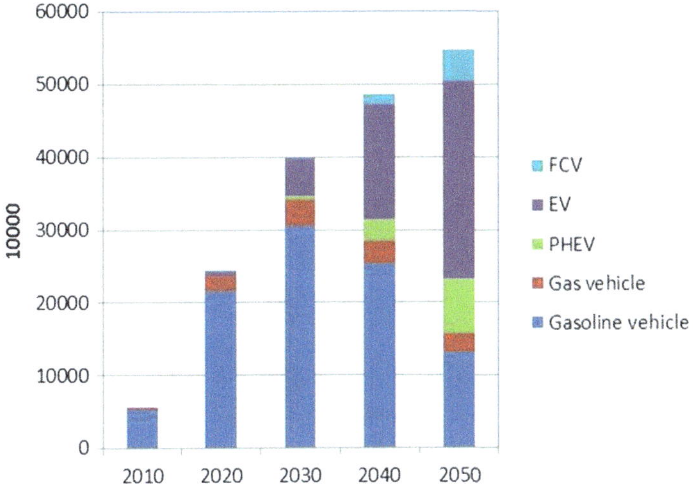

Fig. 12.57 Development of different driving technologies for passenger vehicles in the integrated policy scenario

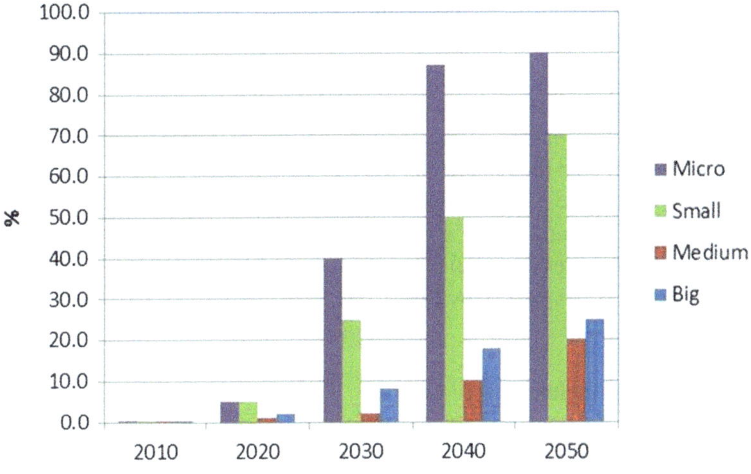

Fig. 12.58 Market penetration of different EVs (including PHEVs) among new vehicle sales in the integrated scenario

shown in Fig. 12.58. In 2050, the proportion of mini, small, medium-sized, and large EVs among new passenger vehicle sales respectively amounts to 90, 70, 20, and 25 %. It should be noted that PHEVs begin to play a more significant role after 2030; their proportion amounts to 14 % in 2050—up from 2 % in 2030.

The proportion of FCVs among the passenger vehicle population amounts to 0.5 and 7.8 %, respectively, in 2030 and 2050; those figures are 0.4 and 7.4 % higher

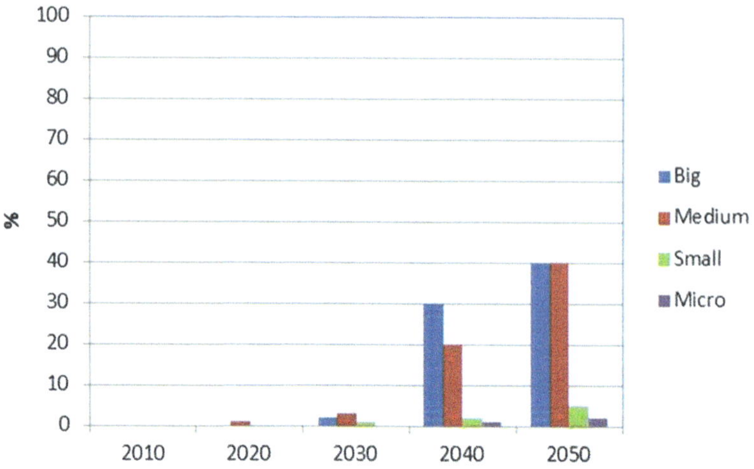

Fig. 12.59 Market penetration of different FCVs among new vehicle sales in the integrated scenario

than in the reference scenario. The market penetration of mini, small, medium-sized, and large fuel-cell passenger vehicles is presented in Fig. 12.59. In 2050, the proportion of mini, small, medium-sized, and large FCVs among new passenger vehicles sales amounts to, respectively, 2, 5, 40, and 40 %.

The proportion of passenger vehicles using different driving technologies under the integrated policy scenario is indicated in Fig. 12.60. The proportion of gasoline vehicle decreases, being reduced from 96 % in 2010 to 24 % in 2050. The proportion of gas passenger vehicles shows a relatively fast development up to 2030: it attains a peak of 9 % in 2030—up from 3 % in 2010—before being reduced to 4 % in 2050. EVs and FCVs play a more significant role after 2030, and their proportion rapidly rises: the increase is from 15 % in 2030 to 72 % in 2050.

12.6.2.2 Driving Technology for Buses

In the integrated policy scenario, the proportion of buses with different driving technologies is presented in Fig. 12.61. The proportion of diesel buses peaks in 2030 and thereafter declines. In 2050, the proportion of diesel vehicles among the bus population falls under 40 %. Gas buses show fast development up to 2030: their proportion increases from 0.4 % in 2010 to a peak of 6.2 % in 2030 before falling to 4.8 % in 2050.

The proportion of gasoline buses remains at about 15 % up to 2030 before dropping to 5.4 % in 2050.

Electric buses play a more significant role after 2030. The proportion of EVs among the bus population amounts to 5 and 20 %, respectively, in 2030 and 2050; those figures are 2.2 and 16 % higher than in the reference scenario. The market

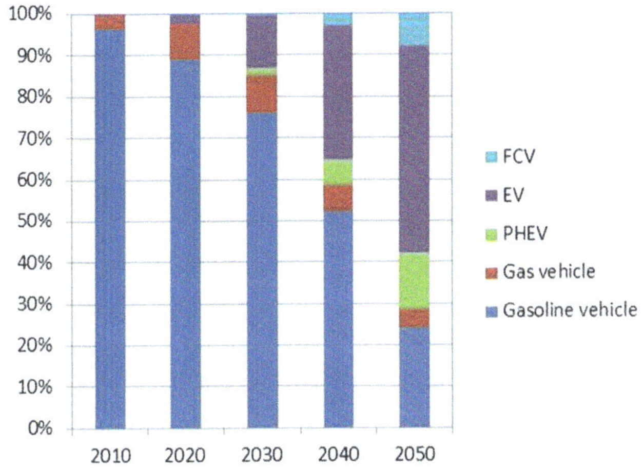

Fig. 12.60 Proportion of passenger vehicles with different driving technologies in the integrated scenario

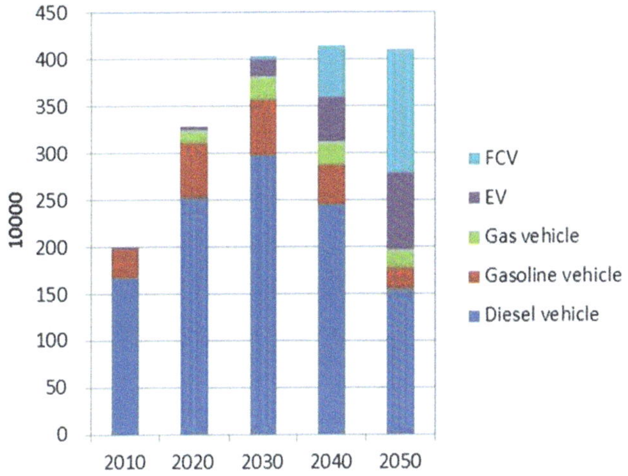

Fig. 12.61 Driving technologies among buses in the integrated policy scenario

penetration of electric and fuel-cell buses appears in Fig. 12.62. The proportion of electric buses among new bus sales is 25 % in 2050.

Fuel-cell buses play a more significant role after 2030. The proportion of FCVs in the bus population amounts to 1 and 32 %, respectively, in 2030 and 2050; those figures are 1 and 32 % higher than in the reference scenario. The proportion of fuel-cell buses among new bus sales amounts to 40 % in 2050.

The proportion of buses using different driving technologies in the integrated policy scenario is indicated as Fig. 12.63. The proportion of diesel and gasoline

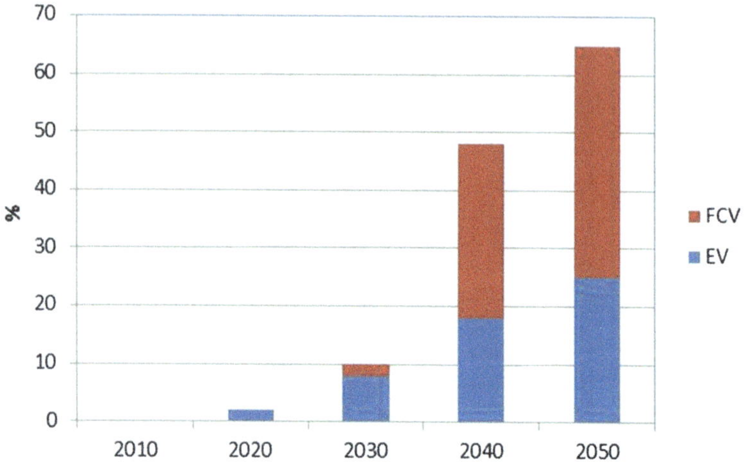

Fig. 12.62 Market penetration of electric and fuel-cell buses among new vehicle sales in the integrated scenario

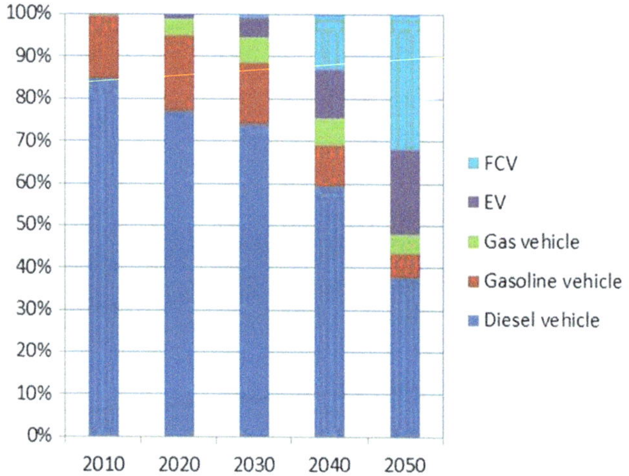

Fig. 12.63 Proportion of buses using different driving technologies in the integrated scenario

buses declines in the future: the respective reductions are from 85 and 15 % in 2010 to 40 and 5 % in 2050. The proportion of gas buses increases quickly up to 2030—from 0.4 % in 2010 to a peak of 6.2 % in 2030 before falling to 4.8 % in 2050. EVs and FCVs play a more significant role after 2030; their share rapidly increases from 15 % in 2030 to 52 % in 2050.

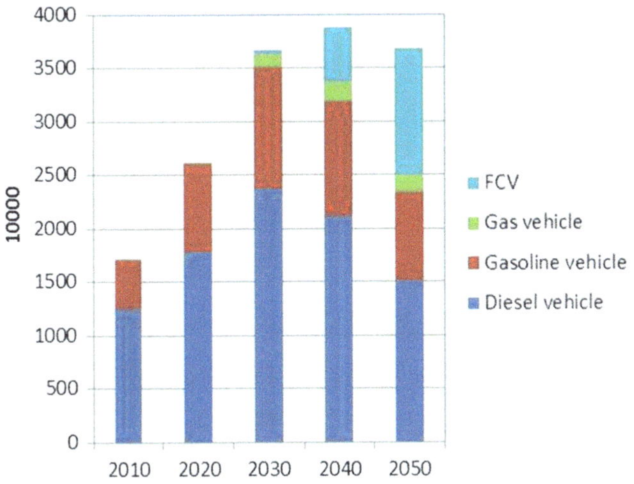

Fig. 12.64 Number of trucks with different driving technologies in the integrated policy scenario

12.6.2.3 Driving Technology for Trucks

In the integrated policy scenario, the proportion of trucks using different driving technologies appears in Fig. 12.64. The proportion of diesel trucks peaks in 2030 and subsequently declines. By 2050, the proportion of diesel vehicles among the truck population is under 42 %. The proportion of gasoline trucks increases up to 2030— from 27 % in 2010 to a peak of 31 % in 2030 before falling to 23 % in 2050.

The proportion of gas trucks remains at a low level up to 2030 and plays a more significant role thereafter, with the proportion being between 3.5 and 4.5 %.

Fuel-cell trucks play a more significant role after 2030. The proportion of FCVs among the truck population amounts to 1 and 32 %, respectively, in 2030 and 2050; that is 1 and 32 % higher than in the reference scenario. The market penetration of fuel-cell trucks is indicated in Fig. 12.65. The proportion of fuel-cell trucks among new truck sales amounts to 40 % in 2050.

The proportion of trucks using different driving technologies in the integrated policy scenario appears in Fig. 12.66. The proportion of diesel trucks decreases in the future, falling from 73 % in 2010 to 42 % in 2050. Gas trucks play a more significant role after 2030: the proportion is between 3.5 and 4.5 %. The proportion of gasoline trucks remains at about 30 % before being greatly reduced after 2030 and declining to 22 % in 2050. FCVs play a more significant role after 2030: the increase is rapid, rising from 1 % in 2030 to 32 % in 2050.

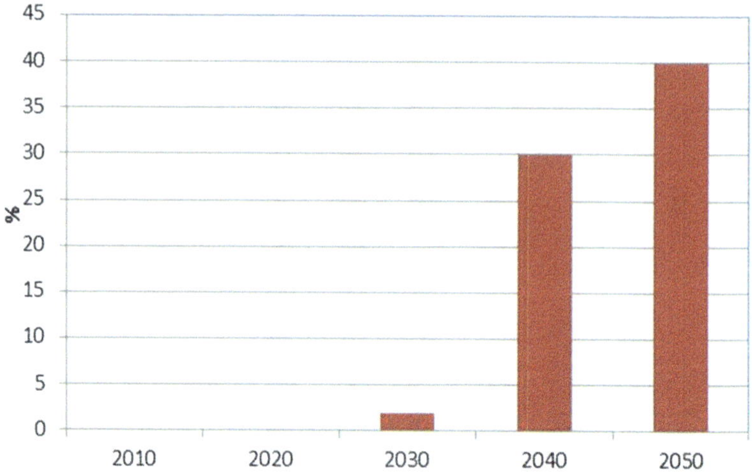

Fig. 12.65 Market penetration of fuel-cell trucks among new vehicle sales in the integrated policy scenario

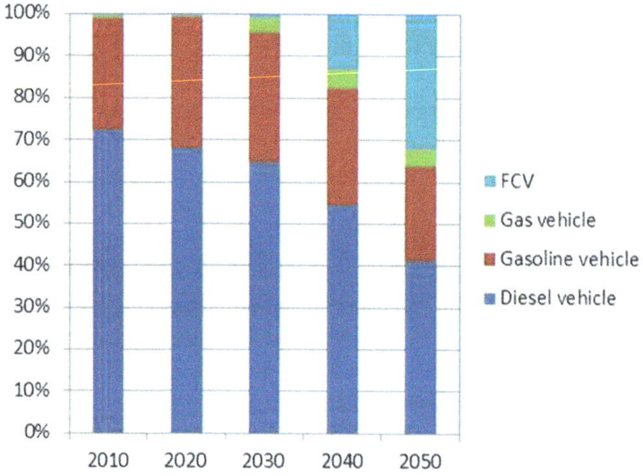

Fig. 12.66 Proportion of trucks using different driving technologies in the integrated policy scenario

12.6.2.4 Vehicle Energy

In the integrated policy scenario, China's vehicle energy consumption increases rapidly from 2010 to 2050, as shown in Fig. 12.67. Under this scenario, such energy consumption in 2020, 2030, and 2050 is, respectively, 0.282 billion, 0.358 billion, and 0.306 billion toe; those figures are 23.2, 27.6 and 45.8 % lower than in the reference scenario.

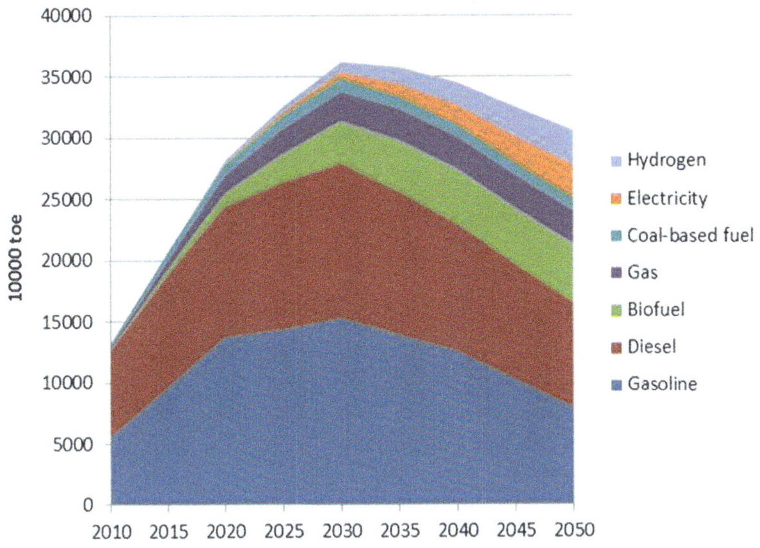

Fig. 12.67 China's vehicle energy consumption in the integrated policy scenario

China's total vehicle fuel consumption will continue its rapid growth up to 2030: the average annual increase is 5.4 % between 2010 and 2030. China's total vehicle fuel consumption peaks in 2030 and subsequently declines at an annual average rate of 1 %.

In the integrated policy scenario, diesel and gasoline show similar trends in terms of consumption. Diesel and gasoline consumption increases up to 2030, with the average increase being, respectively, 3.1 and 5.4 %. Total diesel and gasoline consumption peak in 2030 and subsequently decline, with the average decrease being, respectively, 2.0 and 3.4 %. The proportion of diesel and gasoline among total vehicle energy consumption decreases from 53 and 42 %, respectively, in 2010 to 28 and 26 % in 2050. The total proportion of diesel and gasoline declines by 41 %.

As a type of strategic fuel to replace petroleum-based fuels, biofuels play a more significant role after 2030. The supply of biofuels increases from 0.6 million toe in 2010 to 48 million toe in 2050, and their proportion grows from 0.45 % in 2010 to 15.8 % in 2050. It should be noted that the additional biofuel supply is mainly used to replace diesel (Fig. 12.68), thereby helping to moderate the imbalance in the supply and demand for diesel and gasoline in the vehicle fuel market.

The consumption of natural gas-based vehicle fuel maintains a continued increase up to 2030. The supply rises from 0.5 million toe in 2010 to 2.38 million toe in 2030, with the proportion increasing from 0.4 to 6.6 % over the same period. There is no obvious increase in the supply of natural gas-based vehicle fuel after 2030; this type of fuel thus shows the characteristics of a transitional vehicle fuel.

Coal-based vehicle fuel shows similar consumption trends to those of natural gas-based vehicle fuel; however, the proportion is only one-third that of natural

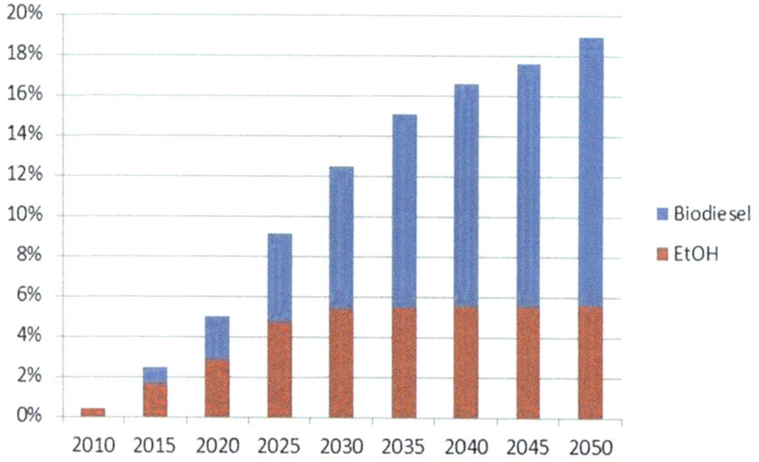

Fig. 12.68 Market penetration of biofuels in the vehicle fuel market in the integrated policy scenario

gas-based fuel. Moreover, the additional supply of coal-based fuel is mainly used to replace diesel. Coal-based vehicle fuel is thus a kind of accessory vehicle fuel.

The supply of electricity and hydrogen as vehicle fuels is rather low up to 2030. Their total amounts to 14 million toe in 2030, and they account for only 4 % of all vehicle fuels. However, the supply of electricity and hydrogen begins to rise after 2030: the supply amounts to 27.56 and 27.83 million toe, respectively, by 2050, thereby accounting for 9 and 9.1 % of total vehicle fuel consumption. In 2050, therefore, electricity and hydrogen account for over 18 % of total vehicle fuel consumption, which means that these two become the most important replacement for petroleum-based vehicle fuel.

WTW fossil energy consumption of China's vehicles from 2010 to 2050 under the integrated policy scenario is presented in Fig. 12.69. This energy consumption in 2020, 2030, and 2050 is reduced by 17.8, 24.1, and 39 %, respectively, compared with the reference scenario; oil consumption is reduced by 23.7, 34.0, and 63.4 %. In 2050, WTW fossil fuel consumption by passenger and freight transport is, respectively, 71 MJ/(100 km·person) and 51 MJ(100 km·tons). In the integrated policy scenario for 2050, the dependence on oil imports by the automotive transport sector is reduced to 48 from the 82 % in the reference scenario.

12.6.2.5 GHG Emissions

GHG emissions by China's vehicles from 2010 to 2050 under the integrated policy scenario are shown in Fig. 12.70. The total amount of such emissions by China's vehicles in 2020, 2030, and 2050 is, respectively, 0.852 billion, 1.062 billion, and 0.761 billion tons in the integrated policy scenario; this is a reduction by 20.1, 27.1,

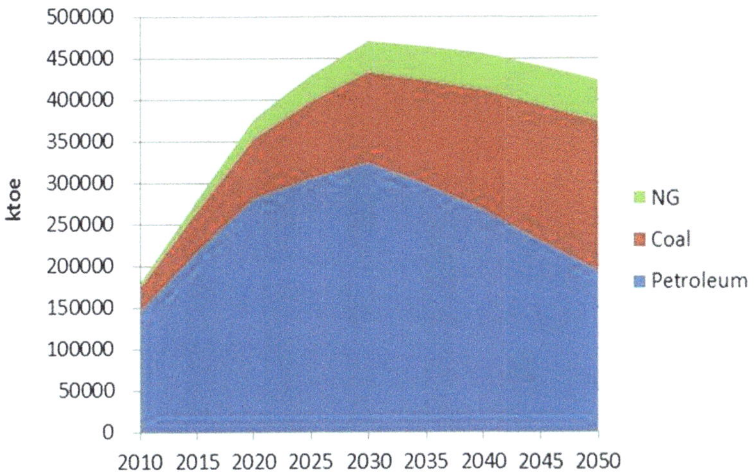

Fig. 12.69 WTW fossil energy consumption by China's vehicle in the integrated policy scenario

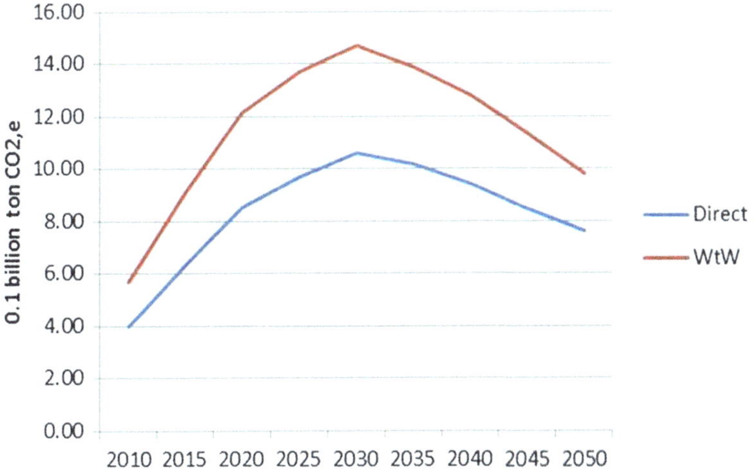

Fig. 12.70 GHG emissions by China's vehicle in the integrated policy scenario

and 53.5 % compared with the reference scenario. If China's population in 2050 is assumed to be 1.46 billion, per capita GHG emissions of automotive traffic amount to 0.52 tons.

WTW GHG emissions by China's vehicles from 2010 to 2050 in the integrated policy scenario also appear in Fig. 12.70. The total amount of WTW GHG emissions by China's vehicles in 2020, 2030, and 2050 is, respectively, 1.215 billion, 1.478 billion, and 0.998 billion tons in the integrated policy scenario; this amounts to a respective reduction by 20.9, 29.7, and 57.9 % in 2020, 2030, and 2050 compared

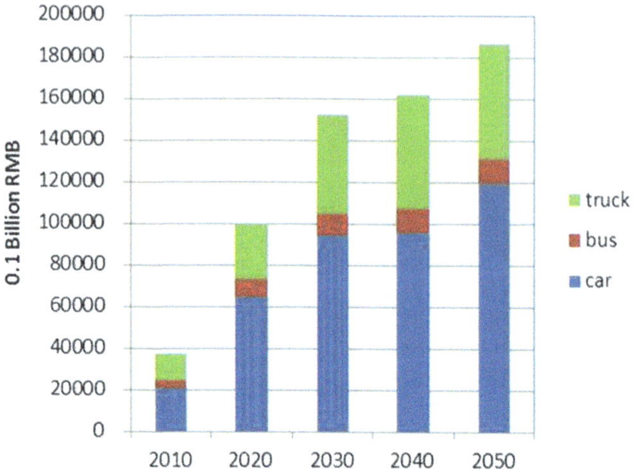

Fig. 12.71 Comprehensive cost of vehicles in the integrated policy scenario

with the reference scenario. Assuming, once again, a Chinese population of 1.46 billion in 2050, per capita WTW GHG emissions by automotive traffic are 0.68 tons.

12.6.2.6 Vehicle Transport Economy

The comprehensive costs of vehicles and their fuels under the integrated policy scenario appear in Figs. 12.71 and 12.72. The comprehensive cost of vehicles in 2050 is RMB 18.7 trillion, which accounts for 7 % of GDP. The cost of oil imports for the vehicle transport sector is RMB 0.26 trillion in 2050, accounting for 0.1 % of GDP. The cost of oil imports for this sector is reduced by 77.7 % compared with the reference scenario.

The comprehensive passenger service costs per person and comprehensive freight service costs per ton are presented in Figs. 12.73 and 12.74. Passenger transport and freight costs are reduced by 39.0 and 27.0 %, respectively, in 2050 compared with the reference scenario.

12.6.3 Scenario Summary

12.6.3.1 Energy

Vehicle WTW fossil energy, vehicle fuel, diesel, and gasoline consumption increase rapidly up to a peak in 2030 and thereafter rapidly decline. The proportion of diesel and gasoline among total vehicle fuel is reduced from 42 and 53 %, respectively, in 2010 to 26 and 28 % in 2050. Biofuels, electricity, and hydrogen play a significant

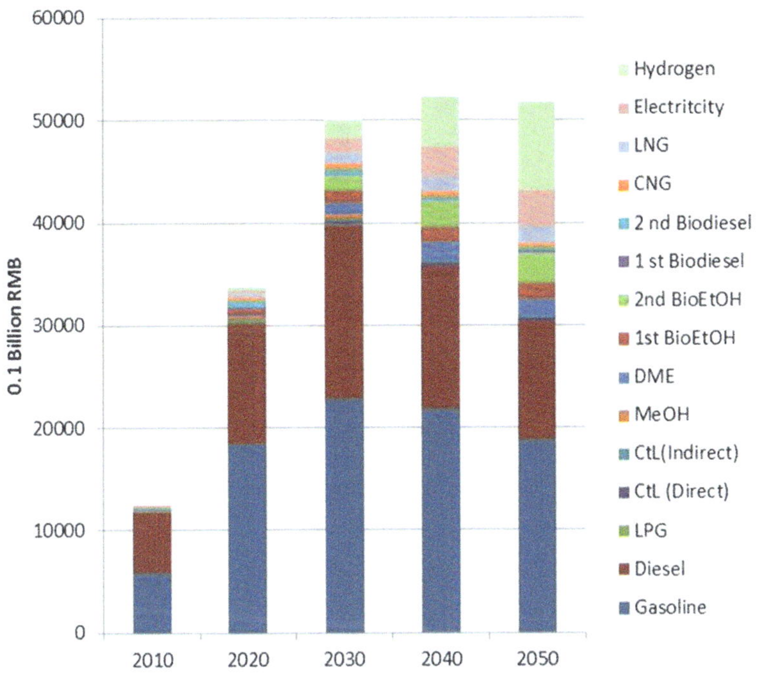

Fig. 12.72 Vehicle energy costs in the integrated policy scenario

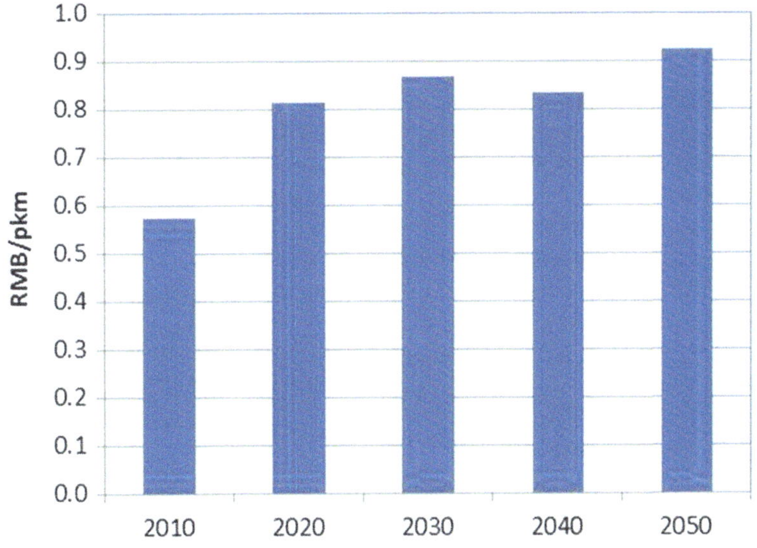

Fig. 12.73 Passenger transport costs per person in the integrated policy scenario

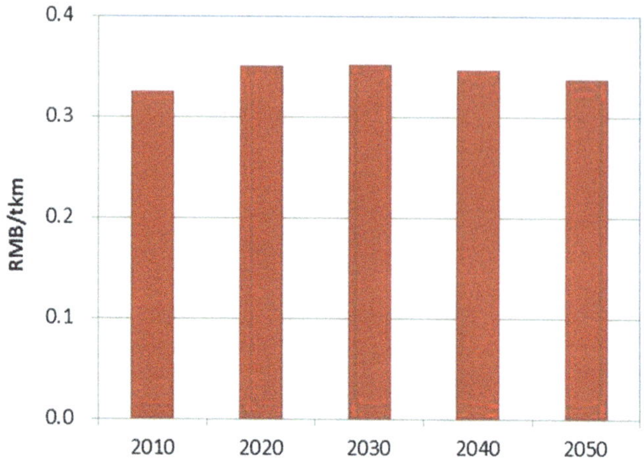

Fig. 12.74 Freight transport costs in the integrated policy scenario

role as replacements for petroleum-based vehicle fuel. The proportion of biofuels, electricity, and hydrogen amount to 10, 2, and 2 %, respectively, in 2030 and 16, 9, and 9 % in 2050. The proportion of natural gas-based and coal-based fuel to total vehicle fuel increases from 0.5 and 2.3 %, respectively, in 2010 to 8.9 and 3.3 % in 2050. In the same year, the dependence on oil imports by the automotive transport sector is 48 %. The demand ratio of vehicle diesel and gasoline is 1:1 in 2050.

12.6.3.2 Vehicles

EVs and FCVs play a more significant role after 2030. The proportion of gasoline passenger vehicles falls from 76 % in 2030 to 24 % in 2050, while that of EVs and FCVs increases from 14 and 1 %, respectively, in 2030 to 64 and 8 % in 2050. The proportion of diesel and gasoline vehicles among the bus population is reduced from 74 and 15 %, respectively, in 2030 to 38 and 5 % in 2050; the proportion of EVs and FCVs rises from 5 and 1 %, respectively, to 20 and 32 % over the same period. The proportion of diesel and gasoline vehicles among the truck population falls from 64 and 31 %, respectively, in 2030 to 41 and 23 % in 2050; the proportion of FCVs increases from 1 % in 2030 to 32 % in 2050.

12.6.3.3 Environment

WTW GHG emissions by vehicles reach a peak in 2030 and then decline rapidly. In the integrated policy scenario, WTW GHG emissions by vehicles are reduced by about 53.5 % compared with the reference scenario. China's per capita WTW GHG emissions by automotive traffic in 2050 are 0.68 tons.

12.6.3.4 Economy

The comprehensive costs of vehicles in 2050 amount to RMB 18.7 trillion, which accounts for 7 % of GDP. The cost of oil imports for the vehicle transport sector is RMB 0.26 trillion in 2050, accounting for 0.1 % of GDP. The cost of oil imports for the vehicle transport sector is reduced by 77.7 % compared with the reference scenario.

Chapter 13
Policy Recommendations Regarding Sustainable Development of China's Automotive Energy

Zhang Xiliang and Ou Xunmin

Abstract The ultimate goal in developing China's automotive energy is the establishment of a sustainable automotive energy system, which amounts to an ideal policy target scenario.

In this chapter, a method that can help China achieve sustainable automotive energy, involving a change from the reference scenario to the integrated policy scenario, will be summarized:

- Promote Long-Term Automotive Energy Saving and Improve Fuel Economy
- Promote Rapid Development of Electric Vehicles
- Promote Second-Generation Biodiesel
- Promote Fuel-Cell Vehicles
- Promote Natural Gas
- Optimize Transportation Modes
- Support Key Technologies and Automotive Energy R&D
- Give Fair, Effective Finance and Tax Policy for Automotive Energy

Keywords Automotive energy • Policy recommendations • China

The ultimate goal in developing China's automotive energy is the establishment of a sustainable automotive energy system, which amounts to an ideal policy target scenario. Six basic standards are used to estimate the sustainability of the automotive energy system in China: transportation economy, energy efficiency, greenhouse gas emissions, security of the energy supply, balance in the supply and demand for various fuel types, and competitiveness in the automotive industry. Analyses of automotive energy scenarios were presented in Chap. 12, among which were the

Zhang Xiliang (✉) • Ou Xunmin
China Automotive Energy Research Center, Tsinghua University, Beijing 100084, P.R. China
e-mail: zhang_xl@mail.tsinghua.edu.cn

CAERC, Tsinghua University, *Sustainable Automotive Energy System in China*,
DOI 10.1007/978-3-642-36847-9_13, © Springer-Verlag Berlin Heidelberg 2013

reference scenario and integrated policy scenario. In this chapter, a method that can help China achieve sustainable automotive energy, involving a change from the reference scenario to the integrated policy scenario, will be summarized.

13.1 Promote Long-Term Automotive Energy Saving and Improve Fuel Economy

Improving fuel economy is a very effective measure for the short-, medium-, and long-term control of petroleum-based vehicle fuel consumption and greenhouse gas emissions. It is in fact considered the second most effective measure (Figs. 13.1, 13.2, and 13.3). Improvement in the fuel economy of passenger vehicles, buses, and goods vehicles could be carried out on a continuous basis up to 2050. According to the scenarios presented in Chap. 12, improving the fuel economy of these vehicles will achieve a 31, 23, and 33 % reduction in automotive fuel in 2020, 2030, and 2050, respectively (Fig. 13.1). It will achieve a 25, 17, and 24 % decrease

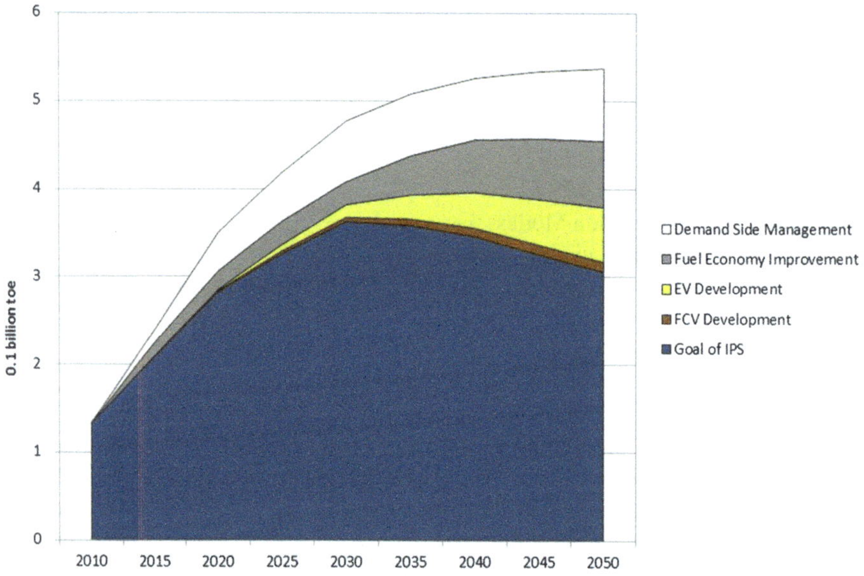

Fig. 13.1 Contribution by different measures to reduce automotive fuel consumption. *Note*: Demand Side Management includes demand management and optimizing the transport mode with respect to passenger and goods vehicles, such as using public transport instead of passenger vehicles and transporting road goods by other methods. Fuel Economy Improvement includes improving the fuel economy of passenger vehicles, using miniaturized and lightweight materials to manufacture automobiles, improving the fuel economy of commercial vehicles, and large-scale logistics optimization of goods vehicles. EV signifies electric vehicles, which include pure electric vehicles, plug-in hybrid vehicles, and range-extended electric vehicles

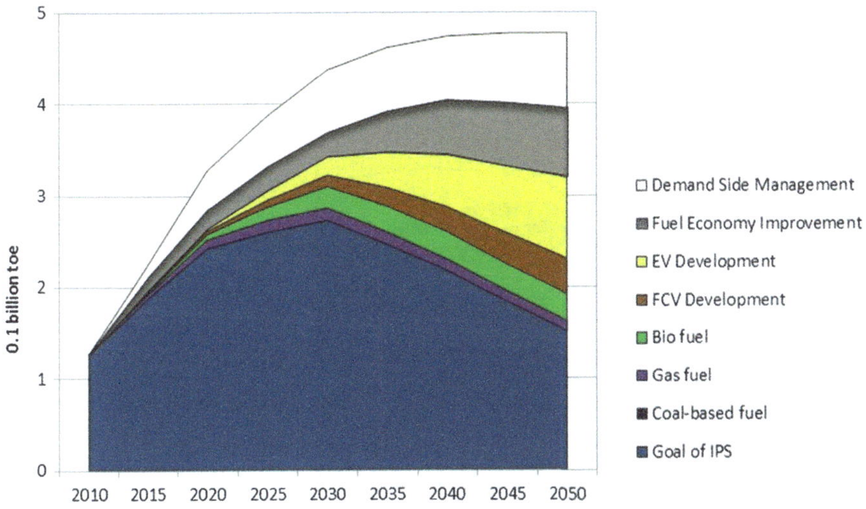

Fig. 13.2 Contribution by different measures in reducing fuel consumption by petroleum-based vehicles. *Note*: Demand Side Management includes demand management and optimizing the transport mode with respect to passenger and goods vehicles, such as using public transport instead of passenger vehicles and transporting road goods by other methods. Fuel Economy Improvement includes improving the fuel economy of passenger vehicles, using miniaturized and lightweight materials to manufacture automobiles, improving the fuel economy of commercial vehicles, and large-scale logistics optimization of goods vehicles. EV signifies electric vehicles, which include pure electric vehicles, plug-in hybrid vehicles, and range-extended electric vehicles

in the consumption of petroleum-based vehicle fuels in 2020, 2030, and 2050, respectively (Fig. 13.2). Improving the fuel economy will also bring about a 28, 18, and 23 % reduction in greenhouse gas emissions in 2020, 2030, and 2050, respectively (Fig. 13.3). Therefore, it is necessary to promote long-term energy saving by automotive technology and improvements in fuel economy. The benefits from reducing automotive fuel consumption and greenhouse gas emissions through improving the fuel economy of commercial vehicles (including passenger and goods vehicles) will contribute more than two-thirds.

In addition, to promote advanced technologies for vehicle fuel efficiency, it is necessary to enhance the quality of automotive fuel on an ongoing basis. Active upgrading of the petrochemical industry is thus required.

13.2 Promote Rapid Development of Electric Vehicles

Electric vehicles (including plug-in hybrid vehicles and range-extended electric vehicles) are one of the key areas for increasing automotive fuel efficiency and automotive fuel security in addition to reducing greenhouse gas emissions. These

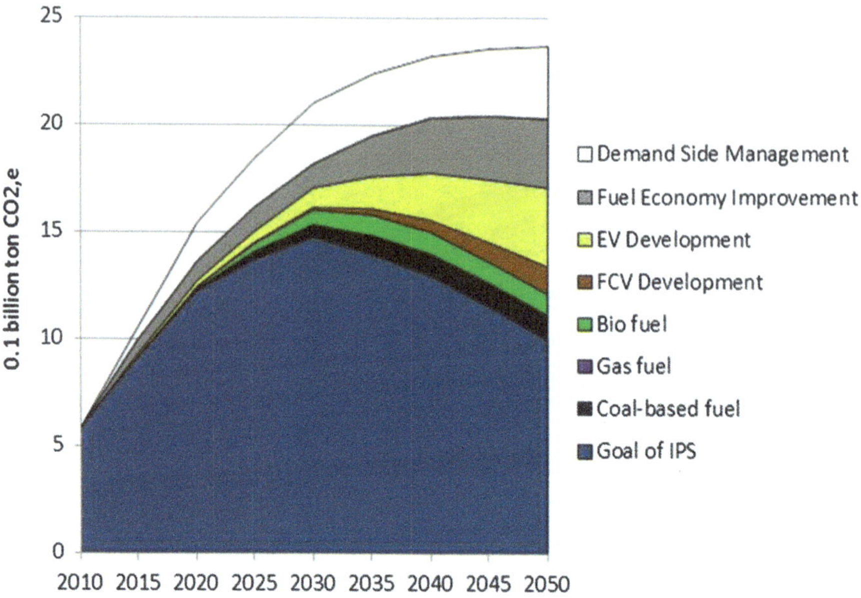

Fig. 13.3 Contribution by different measures in reducing greenhouse gas emissions during the life cycle of automobile traffic. *Note*: Demand Side Management includes demand management and optimizing the transport mode with respect to passenger and goods vehicles, such as using public transport instead of passenger vehicles and transporting road goods by other methods. Fuel Economy Improvement includes improving the fuel economy of passenger vehicles, using miniaturized and lightweight materials to manufacture automobiles, improving the fuel economy of commercial vehicles, and large-scale logistics optimization of goods vehicles. EV signifies electric vehicles, which include pure electric vehicles, plug-in hybrid vehicles, and range-extended electric vehicles

vehicles are also critical in terms of support technology toward achieving sustainable automotive energy. Electric vehicles, especially passenger vehicles and buses, need to be significantly promoted up to 2025. According to the scenarios presented in Chap. 12, EV will account for 64 % of the total car population and buses 20 %. Also according to those scenarios, electric vehicles in 2050 will produce a 27 % reduction in automotive fuel, 29 % reduction in petroleum-based vehicle fuel, and 27 % reduction in greenhouse gas emissions. An adverse result of large-scale development of electric vehicles may be a surplus in the gasoline supply and shortage in the diesel supply. This will be offset by developing petroleum-based diesel fuel, such as biodiesel.

Battery technology will be a major constraint in the large-scale development of electric vehicles. Toward a comprehensive and convenient charging infrastructure, which is a prerequisite for the large-scale development of electric vehicles, the government needs to coordinate between automotive and power industries, develop standards for charging facilities, and construct an effective charging infrastructure.

13.3 Promote Second-Generation Biodiesel

Developing second-generation biodiesel is essential in achieving sustainable automotive energy. This is because in addition to providing a direct substitute for petroleum-based diesel and reducing greenhouse gas emissions (Fig. 13.2), developing second-generation biodiesel can address the imbalance in diesel and gasoline consumption in the large-scale development of electric vehicles. With the new demand for biofuels after 2030, the development of vehicles powered by second-generation biodiesel will be in line with the development of electric vehicles. The development of second-generation biodiesel should enter a period of rapid growth up to 2025; R&D efforts and commercialization of vehicles using this fuel need to be promoted. In 2050, biodiesel will reduce petroleum-based diesel consumption by 28.5 %. However, major limiting factors for the large-scale development of second-generation biodiesel will be the supply of materials (cheap and abundant materials are a prerequisite) and overcoming technical difficulties. Such materials could be supplied through the full use of marginal lands, and advanced technology developed for planting crops and improving farming techniques while maintaining food security and agricultural land supply.

13.4 Promote Fuel-Cell Vehicles

Another key technology toward sustainable automotive energy involves developing fuel-cell vehicles. This is because in addition to achieving long-term security in the energy supply and reducing greenhouse gas emissions, these vehicles can help alleviate the imbalance in the amount of diesel and gasoline consumed. Fuel-cell vehicles require an efficient hydrogen supply system; toward this end, promoting R&D and commercialization of these vehicles will help accelerate their development and allow them to enter a period of rapid growth up to 2035. In 2050, fuel-cell vehicles will account for 16 % of passenger vehicles, 36 % of buses, and 16 % of goods vehicles. According to the scenarios presented in Chap. 12, fuel-cell vehicles in 2050 will produce a 12 % reduction in the consumption of petroleum-based vehicle fuels and 9 % reduction in greenhouse gas emissions. The large-scale development of fuel-cell vehicles is impeded by problems of energy consumption and CO_2 emissions during the process of hydrogen preparation and transportation.

13.5 Promote Natural Gas

Natural gas will play an important role in the diversification of automotive fuels and reduction of greenhouse gas emissions in the short and medium term (Figs. 13.2 and 13.3); it will also be helpful in the long term. Up to 2020, natural gas will be the third most important measure in the diversification of automotive fuels and reduction in greenhouse gas emissions (the first will be optimizing transportation modes, the second will be the energy saving in automotive technology). The natural gas supply in the international market and breakthroughs in biofuel technology will affect the long-term use of natural gas as an automotive fuel.

13.6 Optimize Transportation Modes

From the scenarios presented in Chap. 12, optimizing the mode of transport, such as the use of buses instead of private passenger vehicles and the use of rail freight instead of goods vehicles, will result in a 65, 60, and 36 % reduction in automotive fuel consumption in 2020, 2030, and 2050, respectively (Fig. 13.1). This optimization will produce a 52, 44, and 26 % reduction in the consumption of petroleum-based vehicle fuels in 2020, 2030, and 2050, respectively (Fig. 13.2). It will also result in a 56, 46, and 25 % reduction in greenhouse gas emissions in 2020, 2030, and 2050, respectively (Fig. 13.3). Optimizing transportation modes will be effective in controlling petroleum-based vehicle fuel consumption and greenhouse gas emissions in the short, middle, and long term; it will also be a significant measure up to 2030. Therefore, it is necessary to vigorously promote and develop the use of public transport and increase the proportion of rail freight in China.

13.7 Support Key Technologies and Automotive Energy R&D

Technological innovation will facilitate the sustainable development of automotive energy in China. The key technologies include the following: electric vehicle battery technology; fuel-cell technologies; second-generation biodiesel materials and preparation technologies; and efficient environment-friendly and advanced hydrogen technology, such as biomass-based polygeneration hydrogen production, coal-based polygeneration hydrogen production, and nuclear hydrogen production.

13.8 Specify Goals for Improving Efficiency of Automotive Energy

Vehicle fuel consumption standards will significantly affect automotive energy saving and technological innovation for vehicles. The goal of improving the system efficiency of automotive energy in China in the middle and long term needs to be clearly proposed. According to the integrated policy scenario presented in Chap. 12, the target of automotive energy consumption in the life cycle of passenger vehicles in 2050 will be 2.4 L/(100 km · person) gasoline equivalent and that the life cycle of goods vehicles will be 1.6 L(100 km · t) diesel equivalent. It is necessary therefore to improve existing fuel economy standards for passenger vehicles and commercial vehicles. China needs to establish such standards as soon as possible regarding the predicted transport situation in 2050.

13.9 Fair, Effective Finance and Tax Policy for Automotive Energy

Finance and tax policies are a market-based policy tool that can be used to encourage sustainable restructuring of the automotive energy system. Specific recommendations in this regard are as follows: investment subsidies from demonstration projects on electric vehicles, fuel-cell vehicles, and second-generation biofuel technologies; reform of the existing fuel tax system and close integration of fuel tax collection and revenue use for sustainable automotive energy transformation; exemption from all taxes for purchases of electric and fuel-cell vehicles; study tax-relief programs and action plans using electric vehicles and fuel-cell vehicles; and study proposals and action plans relating to a carbon tax in the transport sector.

The manufacturer's authorised representative in the EU is Springer
Nature Customer Service Centre GmbH, Europaplatz 3, 69115 Heidelberg,
Germany. If you have any concerns regarding our products, please
contact ProductSafety@springernature.com

Printed and bound by CPI Group (UK) Ltd, Croydon, CR0 4YY
30/09/2026
02225643-0014